Meltdown

Meltdown

The Earth Without Glaciers

JORGE DANIEL TAILLANT

Oxford University Press is a department of the University of Oxford. It furthers
the University's objective of excellence in research, scholarship, and education
by publishing worldwide. Oxford is a registered trade mark of Oxford University
Press in the UK and certain other countries.

Published in the United States of America by Oxford University Press
198 Madison Avenue, New York, NY 10016, United States of America.

Library of Congress Control Number: 2021941858
ISBN 978–0–19–008032–7

DOI: 10.1093/oso/9780190080327.001.0001

1 3 5 7 9 8 6 4 2

Printed by Sheridan Books, Inc., United States of America

To Ulises, my son, who constantly surprises me with youth's fresh and fearless, intrepid and baggage-free questioning of the status quo and with his envious unapologetic approach to life and everything.

Contents

Foreword

Between the time I submitted the first version of this manuscript in early 2020 and the time I submitted the last version in early 2021, the world changed.

Major events occurred in 2020 that were so great and so relevant to the topic of this book (climate change and melting glaciers) that it is hard to pinpoint a single change-event that is more significant than others. At the very moment I submitted this manuscript, Joe Biden and Kamala Harris were announced as the president elect and vice president elect of the United States of America—Joe Biden defeated Donald Trump with the largest amount of votes ever cast for a presidential candidate. Biden and Harris ran on a strong platform that included taking decisive action to stop climate change.[1] Since their taking office in January of 2021, they've ordered some of the most ambitious climate policies we've ever seen, anywhere.[2]

In 2020 we also saw the most devastating climate wildfires ever recorded in California, burning millions of acres of forest. California Governor Gavin Newsom announced, from the embers of a still-burning climate wildfire, that discussion was over, that climate change was real, that it was here and that we had to do *more*, that we had to do it *better*, and that we had to do it *faster*. He also announced that the targets we had set to reduce greenhouse gas emissions (and California has some of the highest and most aggressive targets on the planet) were insufficient. Soon afterward, California chose to eliminate the internal combustion engine for vehicles produced in the state by 2035 and for trucks by 2045.[3]

But clearly the change-event that will mark 2020 for all of history and for the entire planet was that the world was hit with a global health pandemic: COVID-19, the novel coronavirus, a new pandemic disease that radically altered our lives. I remember over the last several years speaking with my kids about climate change on many different occasions and expressing my frustration that the sorts of actions we needed globally and locally to really address climate change were not very likely to happen in our modern lives. They seemed too far-fetched and against the grain of our fast-moving society. In the mix of ideas we talked about, the world needed to stop for a few weeks each year, maybe even for a month, to help lower green house gas emissions, but this was nearly impossible to envision with our hustle and bustle lifestyles—industries, economies, and people simply could not just "stop."

Well, we did.

And guess what, in two weeks the air cleared all over the world. Suddenly people in cities hundreds of miles away from the tallest peaks of the Himalayas could see snow-capped mountains that they hadn't seen for decades, or ever. Large schools of fish returned to bays that they had abandoned due to the pollution and disturbance from commerce, industry, and shipping. Wild animals returned to empty urban environments, wandering around areas where people had abruptly vanished.

The sudden outbreak of a global health pandemic hit the planet like a global earthquake, revealing our extreme vulnerability to Mother Nature. It showed us (and still does as I submit the very final edits to this manuscript in April of 2021 with the COVID-19 infection rate suddenly rising again across the USA) that in the end, Nature and the state of the environment governs our lives and not the other way around, and unless we safely adapt to our environment and live in harmony with it, we are all highly vulnerable to its fluctuations, its reactions, and its abrupt changes, be they natural or human-induced.

COVID-19 taught us (and is still teaching us) several useful lessons. One is that we can stop pollution and improve our environment and that we can do it very quickly if we act collectively. We also saw that political will and trillions of dollars can suddenly materialize from political leaders to do what needs to be done to address a global emergency. In this case it was for a health pandemic, but why not for climate? We'll get into that question later.

I think it is also important to consider that these sudden changes in social, political, economic, and industrial behavior as a response to COVID-19 presented a critical opportunity and lesson for youth. I am of Generation X, born in 1968. At 53, I carry the baggage of the global failure to adequately address our environmental problems and the even more worrying failure of our incapacity to contain climate change. Over the years as an environmental policy advocate, I have learned how difficult it is to align political leaders, industry, business, and everyday people to do the right thing to address our environmental pollution and to work to halt and reverse our climate emergency.

Our world's youth, coming of age now during the COVID-19 pandemic, with respect to our climate challenge and to our climate emergency, are baggage-free. As young adults forging their social and political minds, my own two children, now aged 20 and 17, have no baggage as they walk into this climate emergency. They see examples and opportunities that I had never seen. In two weeks, at a global scale, we cleaned our air. Industry stopped. Pollution stopped (clearly not all of it, but much of it), and the planet healed for a short time. I've included a special chapter at the end of this book (Chapter 10) where I reprint an article that I published in April 2020 with a young women nearly one third my age, positing shared and different generational visions of what it takes to resolve climate change. I am happy to see that our youth have a more positive outlook on how things will play out. The planet needs it, and it needs leaders like Greta Thunberg, the Swedish teenage climate activist that has declared to our political leaders and to people like you and me, "how dare you!"[4] Everyone should listen to Greta's "how dare you" speech, take it to heart, and remember it always as we move forward. I encourage you to stop now for a moment and listen to Greta's speech before you read on. It's truly moving.

Our present leaders, but more importantly our future leaders, have learned through COVID-19 that we can turn things around, that we can make a difference, that we can stop polluting, and if we really want to do it, we can fix our climate problem. That gives me hope, and it should give our global society hope.

In early 2020 I had drafted a pre-COVID-19 foreword to this book, where of course, there was no mention of our global pandemic. As the first months of 2020 advanced, the manuscript was becoming quickly outdated on issues related strictly to the topic of this book, *glacier melt* and its relevance to our lives. Articles about rapid glacier melt were already appearing in early 2020 that showed things were changing quickly. I created a list of articles and studies that were popping up weekly, sometimes daily, of issues I should mention and reference in the various chapters of this book. The list kept growing, outpacing the time available to address each point. Glaciers were melting and they continue to melt, faster than previously thought and expected. Ice all around the world is disappearing, and it's disappearing fast.

During this calendar year as I went back and forth with the publisher with edits to the manuscript, troubling articles appeared, such as "Massive Ice Melt in Greenland Last Year Shattered Previous Records." This was a piece from August 2020 in the LA Times, indicating that 586 billion tons of ice (the equivalent of 140 trillion gallons of water) melted in Greenland during 2019. To get a sense of this volume, that's enough water to cover the entire state of California in four feet of water. Researchers are affirming over and over again that ice is not only melting, but that it's melting at a faster and faster pace.[5]

Another article that appeared while I was editing this manuscript indicated that the melting of the Greenland Ice Sheet is already past a point of no return and that efforts to stop global warming will not stop it from disintegrating. Basically it's too late. According to the scientists at Ohio State University who make this assertion, the Greenland Ice Sheet sheds about 280 billion metric tons of ice each year on average, making it the single biggest contributor to sea level rise. This meltdown of the Greenland Ice Sheet is so great that it actually changes the gravitational field over Greenland, and by the end of the century, global sea level will rise about 3 feet because of it alone.[6] This will change the landscape around the world, particularly in low-lying states such as Florida and in entire countries such as the Maldives, which will have to consider relocating most or all of its population.

Chris Mooney, on the Washington Post's Climate Change beat, published an article earlier in the year with the title: "Unprecedented Data Confirms that Antarctica's Most Dangerous Glacier is Melting from Below."[7] The article went on to describe the reasons for the melting of this faraway glacier called Thwaites Glacier (larger than the State of Pennsylvania or the entire island of Great Britain), and implications of the melting of Thwaites Glacier for the planet, which could cause global seas to rise 10 feet (3 meters) placing the homes of millions, if not billions, of people underwater. That's ten feet of sea level rise from only one melting glacier. Think about that.

A week or so after Mooney's article, yet another appeared in the New York Times. This one announced: "Antarctica Sets Record High Temperature: 64.9°F (18.3°C)." We've been hearing about rising temperatures in Antarctica for a while now, so this was not such a surprise. What I noticed immediately however, was the picture that accompanied the article, showing a series of building structures of the Argentine research station on Antarctica at the locality of Esperanza (*Esperanza*, ironically, means *hope* in Spanish). The buildings were on barren land. The snow and ice are gone.[8]

Articles like this kept popping up day after day, week after week, and month after month, delaying my manuscript for about a year. I was first aiming to publish *Meltdown* in 2019, when there were some, but not *so* many, articles in the press about melting glaciers. The topic was urgent and very appropriate at the time. My hope was to better educate society about this growing global concern, drawing attention to the issue of global glacier melt, but also educating about why it is important, both globally and locally. Worried that policy makers were falling behind on climate action, and very concerned that the United States in particular was pulling out of key global climate agreements (like the Paris Agreement) under the Trump Administration, I felt an urgency to get the word out before it was too late.

Then COVID-19 hit. All bets were off, anything could happen. Then climate wildfires hit. And then we came to the US elections, which opened up so many different political debates on the issue of climate. At some point I had to put a marker in the sand and publish. Glaciers are still melting and you've probably seen dozens more articles about their melt since I went to press and the mechanics of publication played out. So here it is. Outdated or not, the issue is still of extremely troubling relevance.

Meltdown is about glaciers, specifically about glaciers melting and how the melting of glaciers like Thwaites Glacier or Pine Island Glacier in Antarctica, or the Greenland Ice Sheet, or glaciers that are much closer to home such as in the Sierra Nevada of California, or in the Rocky Mountains of the United States and Canada, or in the European Alps, or the Andes Mountains of South America, or on Mount Fuji in Japan, or those in New Zealand, or in the Caucasus or in the Hindu Kush, in the Tian Shan, or the Himalayas or in the Karakorum Mountains, will actually change your life.

This book is not meant for scientists or academics, although if you are either you'll surely read *something* new or interesting in these pages that might inspire you to think differently about your work and your role *and responsibility* in fixing our climate— because we all have a role to play. It is meant for everyday people who are curious about the world around them, concerned about all of the new information that is surfacing about climate change, and likely have never thought about the relevance of glaciers to our global and local ecosystems, even though their demise may mean a radical change of lifestyle in the immediate years to come, possibly even in our lifetime.

I provide many references to the issues I present in this book, including from intergovernmental publications—such as the climate change reports coming out of the United Nations (UN) year after year—from academic publications that most everyday people don't read, and from streaming media sources, even from Wikipedia. I got some criticism from some academic experts after I wrote my previous book, *Glaciers: The Politics of Ice*, for utilizing non-academic sources in writing my book, but frankly, academic articles on glacier melt and academic information about climate change cannot keep up with the rapid advancement of reality and the pace of climate change and the impacts it is having on glaciers and communities that depend on them—and frankly, for most people, these publications are difficult to digest. Further, academic articles about glaciers and their melt don't necessarily capture how glacier melt affects real people, their lives, their stories, and their worries. Glacier melt is affecting our lives all around the world, even if academics have not yet done a study about it and about all the ways these impacts play out for all of us.

Sometimes local media news sources are the *only* information we have on the symptoms of climate impacts to glaciers and to other cryosphere environments, such as permafrost (permanently frozen grounds). If we expect society to take action as a whole in response to such a dire and urgent situation, we need to write *to* society in language and communicate *with* society in formats that people are familiar with, in ways that are understandable and accessible, and in ways that inspire attention and action. Our messages in social media may have to be reduced to 280 characters in a tweet, or to 60 seconds for a Tik Tok video, if that is what it takes to capture attention. We need to see glaciers on Instagram (please follow #glaciers) and on Tik Tok, on Snapchat or on Facebook. That's where people scroll to get their news, to see their friends, and to relate to the world around them. If it's not on Tik Tok, on Facebook, or feeding through Instagram, it's likely the world as a whole is not paying attention. If glaciers are only talked about in scientific publications, there is little chance people will be moved to do something about their vulnerability and to tackle climate change as we really should.

And so, I will double down on this approach of utilizing a variety of academic *and* non-academic sources for the information I convey in this new book. My intention is not to produce an academic publication, it is to produce a collection of information about melting glaciers that is digestible and that is useful for everyday people to understand how our climate is changing, and what this means for one of the planet's most important and most vulnerable natural resources: *glaciers*. And how the melting of glaciers will alter our habitat, our environment, our access to natural resources, and our lifestyles, forever. In fact, this is already happening.

Meltdown is meant to shed light on an issue that we hear about almost daily in the news: glaciers rapidly melting and their threat to the planet, to coastlines, and to society more generally. It is meant to bring to the fore this globally significant phenomenon—ignored by most people, although they may notice occasional stories in the news that they most likely do not read or relate to—and explain how it will change the way we all live. We can read academic materials about glaciers and glacier melt, but the reality is that most people (the target of this book) are not academics and won't read the UN's latest publication on the oceans and the cryosphere. Have you, the reader, read the Paris Agreement (I'm sure you've talked about it or at least mentioned it) or any of the IPCC reports on climate change? Do you even know what the acronym "IPCC" stands for or what it is?

But we do read our Instagram feeds, we scroll through Tik Tok posts, we read the news through our Twitter and Facebook accounts, and we occasionally (more often lately) read an article about the natural environment that mentions glacier vulnerability or a piece of ice bigger than a country breaking off of Antarctica. I am targeting *that* audience: everyday people who have a sense of the planet's vulnerability to climate change but don't necessarily spend their free time reading scientific journals. I certainly don't, and I'm dealing with these issues daily!

Most have heard about the melting Arctic, or have seen the famous image of a polar bear clinging on to an iceberg for lack of footing in the northern frozen oceans of our planet. Or they have heard that big pieces of ice the size of Texas, or France, or Europe, in the Antarctic region (the South Pole) are crumbling to the sea. But do we really know what this really means for our daily lives? Probably not. I'm hoping to take at least one step to change that with *Meltdown*.

For those curious people, *Meltdown* is for you, because I promise you that glacier melt *will* change your life in ways that you could not imagine! You may not even realize that glacier melt is already happening and that it is already affecting you, but it is. I predict that you will start to see and recognize those changes during your lifetime. You are actually already seeing them, you just need to open your eyes and connect the dots.

Glaciers are melting because of climate change. That's an indisputable fact that is free from political ideology. It is a fact that is independent of whether or not you believe that *people* are causing climate change or not. In the end, it really doesn't matter if you believe that people cause climate change, the fact is that climate change is happening, and either way, glaciers are melting and this will mean radical change for all of us, including you.

Glaciers melt away and reappear on the Earth cyclically, due to natural as well as external factors, such as a large meteor that might hit the Earth and darken the sky for long periods of time, or volcanic eruptions that can also abruptly change and destabilize our climate in ways that could impact glaciers.

Hundreds of thousands of years ago, Antarctica was largely ice free, while now it has layers of ice thousands of meters thick. Now Antarctica's ice is melting, because the climate of the Earth is changing. One day, maybe not too far off in the future, Antarctica will again be ice free. Glaciers come and go because of the Earth's constantly changing position in its oscillating orbit and the changing pattern of that orbit around the sun. These external and natural phenomena bring on and phase out our cyclical ice ages. We can argue as to whether or not unnatural phenomena like human activity can cause global climate changes that melt away glaciers, but again, the reality is that glaciers *are* melting and this is now more obvious than ever before. The question really is how this will change our lives and how we can address glacier melt and maybe slow it down to avoid the most severe impacts that it will bring. We must understand how it may make life difficult or even *unlivable* for certain segments of the population and whether or not we can stop it or at least slow the rate of melt before it's too late. As a very last resort, we should better understand the impacts of glacier melt so that we can better adapt to our climate as well as help others adapt who do not have the means or capacity to do so.

As far as natural Earth climate cycles are concerned, we are currently at the *end* of an ice age, which means much of the glacier presence on the Earth (compared to a full on ice age) has already dwindled from what it was 10,000 or 20,000 years ago. In the future, probably many millennia from now, vast areas of the Earth will once again be covered by ice, and the Earth will be much colder than it is today. For the moment, it is warming at an unprecedented rate in modern human history. That news bit from Antarctica about it reaching 65°F (18°C) is alarming. It is a record-setting temperature in a part of the world that should be cold, but suddenly, is not.

The issue of anthropogenic climate change (human-induced climate change) is certainly on the forefront of global political discussion and tension. It has even spilled over to dinner table discussions. This is so because we are seeing the instability of climate-related events growing in frequency. Unusually hot days in cold places—like that recent record high temperature in Antarctica—are becoming the norm. The recent fires we saw burn in places like Australia and California and the record-breaking temperatures of 130°F (54°C) in parts of the American West are indicators that something is out of sync. Scientific opinion has already converged on the idea that *humans* as opposed to natural phenomena are indeed impacting and changing the Earth's planetary climate. Some refer to this era we are living as the *Anthropocene*, an era where *humans*, as opposed to Nature, define our geological times.

Whether or not you, the reader, believe that people are causing climate change to accelerate, the fact is, we are losing our glaciers to a warming environment: glacier demise is not only real, it is also critically important to all of us. I accept the scientific consensus on anthropogenic climate change, but at the end of the day, for you, the reader, it doesn't really matter whether you do or do not. Nor am I writing this book to convince you that we are the cause of climate change. I am writing this book to draw attention to the risks of continued rapid glacier melt and to send out a warning and signal that we must act now, individually (*that's you*) and collectively (our governments, our companies, and our societies). If rapid glacier melt continues at the current pace, which seems likely, we're in trouble, as we will see drastic and catastrophic changes to our global ecosystems. In fact, we are already seeing these changes starting to happen.

We are in a position to do something about climate change, if we set ourselves to this task. Our power to change the global climate became evident during the COVID-19 pandemic. We cleaned the planet's air in mere weeks. Cleaning the air, for instance, is of enormous help in saving glaciers. We can do it again, and we can do more. We can make the changes we need to make. If we don't the consequences will be severe.

Meltdown is about the impacts of glacier melt to our planetary environment.

To properly understand the role glaciers play in our global ecosystem and realize the significance of their demise, we need to understand glaciers, which for the most part, few people do. Once we see and understand their majestic beauty and the critical role they play in keeping our climate stable, we can begin to explore ways to help conserve their delicate ecosystems, necessary for their healthy survival and sustainability. Knowing how melting glaciers are impacting our Earth, and how their demise will *further* impact our planet, is also important for us to prepare for change, not only for our living generation, but for generations to come.

We should also consider that the disappearance of our glaciers would impose severe limitations on our natural ecosystems, and so it is important to understand glaciers' functions and dynamics so that we not only protect them, but adjust our environments to their absence. In some places where glaciers have already dwindled or disappeared, or where water is becoming ever scarcer, some people have already begun to study how natural glaciers function and are taking action to generate artificial glacier dynamics by mimicking and recreating glacier ecosystems (I call them glaciosystems) where artificial glaciers can form and thrive, and in so doing, recreating their cooling and water conservation and production capacity. Yes, people are "making" artificial glaciers as natural glaciers melt!

Our Earth's climate can provide, for many more generations, and even for millennia, a comfortable human existence with livable ecosystems. However, if our glaciers melt away, those ecosystems will surely undergo major change, and become unable to sustain human life, or at least a version of life that we find appealing.

For this reason, *Meltdown: The Earth Without Glaciers* should be a book that is of interest to everyone, independent of their political ideology or their position on climate change.

I invite you to explore these pages, and hopefully, like me, you will be inspired to visit a glacier, and maybe even do something to help their survival. Maybe you already are!

JDT

Acknowledgments

Glaciers have changed my life. I didn't study glaciers at the universities I attended. I didn't even seek them out. But they were there, waiting. As I look back at my own glacial history, I think *they* came to me.

Back when I learned that a mining company was dynamiting glaciers to get at gold, I felt an urgency to learn everything that I could about glaciers, and so I did, and in order to do that I reached out to glaciologists all over the world. And they answered. Without their help, I wouldn't have been able to inventory thousands of glaciers that are at direct risk from industrial activity, or understand the mining impacts to glaciers, or come up with terms like "the glaciosystem" (something I defined to identify the specific glacier-ecosystem and the environmental characteristics necessary in order to protect them and make them thrive). It is to those glaciologists and geocryologists, who opened their minds and their hearts to this crazy environmentalist who didn't seem to care that saving glaciers sometimes seems like fighting windmills, that I am indebted.

In my last book, *Glaciers: The Politics of Ice*, I first thanked my Chilean friends who started a global movement to protect glaciers from mining impacts. We all owe you gratitude for waking us up to what was happening high up at thousands of feet of elevation in the Central Andes. Our first battle was against Barrick Gold and its ill-designed Pascua Lama gold mining project: that was the company that was dynamiting glaciers. A very long fight ensued to stop Pascua Lama. It wasn't only my fight, but that of hundreds and even thousands of people who understood that water was more valuable than gold and that it was profoundly wrong to dynamite glaciers. We took the side of the glaciers, arguing that placing a mega mining project in the middle of glacier terrain was crazy. But the miners called *us* crazy, they called us *alarmists*. In the end, and coincidently in the very year 2020 when this manuscript was completed, the Pascua Lama project was ordered permanently closed by Chile's Environmental Tribunal, precisely because of all of the "crazy" things we had been saying about glacier impacts and contamination—as it turned out, we were right. So thank you Leopoldo Sanchez Grunert, Rosana Bórquez, Sara Larrain, Rodrigo Polanco, Juan Carlos Urquidi, and Antonio Horvath in Chile who were the spearheads of the glacier protection movement, laying out the first proposal ever to come up with a glacier protection law. That law would die in Chile's Congress, but it lives on in so many other ways. Many would follow your lead, but you were the first! *Thank you!*

Special mention also goes to Marta Maffei, the mother of Argentina's national glacier protection law, along with her legal advisor Andrea Borucua of ECOSUR. Their contribution to glaciology and to the protection of glaciers around the world has been paramount and will survive for many generations to come. They hand-carried the failed Chilean glacier protection bill across the Andes and submitted it to Argentina's Congress, and it worked. I contacted Marta recently to congratulate her on the 10-year anniversary of the passage of the law. We've had to fight an ongoing battle against those working to weaken it or eliminate it, but the law is still holding strong. *Thank you!*

I am personally indebted to a handful of glaciologists who helped me learn about ice, about glaciers, about glaciosystems, and even about invisible glaciers. I may have not gone to a university to learn about glaciers, but grit, reading, walks on glaciers, many hours on Google Earth finding glaciers and rock glaciers that you helped me see for the first time, and long conversations with you guys made it happen. *Thank you!*

Cedomir Marangunic (Geoestudios, of Chile), Juan Carlos Leiva (IANIGLA, of Argentina), Benjamín Morales Arnao (*Patronato de las Montañas Andinas*, of Peru), and Bernard Francou (IRD, of France) were my personal and virtual instructors in the first course I ever took on glaciology organized by the UN's Environmental Program (UNEP) in Chile. I have kept in touch with them over the years and each time I have a question to answer, they oblige, sometimes with very lengthy emails.

Cedomir (Chedo to his friends) always has unique insight into the world of those enigmatic rock glaciers. Juan Carlos has inundated me with academic material, research papers, and studies. Benjamín has provided inspiration from Peru, where after decades of work to avoid glacier tsunamis in the Cordillera Blanca Mountains he is showing us that we can still do much to protect our glaciers and our communities. And then there's Bernard with whom I spent long coffee breaks and dinner hours during various glacier courses and on our visit to several glaciers in Ecuador and Chile. Bernard has shared his deep and first-hand knowledge of the receding glaciers of the Tropical Andes. Together we came up with the idea of a new term, *cryoactivism*, that is, activism to protect glaciers. *Thank you!*

I am especially in debt to three expert glaciologists and more specifically, *geocryologists* (those who study glaciers, ice, rocks, and their relationship) whose direct contribution to my work has been paramount: Juan Pablo Milana, Alexander Brenning, and Mateo Martini. They stuck their necks out and agreed to review my publications, signing their names to my glacier inventories as academic and professional reviewers of my work, when many others would not dare to have their name associated with me, one of *the crazy environmental activists*. Academics rarely walk across that bridge, and they did, for the betterment of society and of our planet. Juan Pablo is perhaps one of the most knowledgeable specialists of the periglacial environment, having spent the better portion of his career among the rock glaciers and frozen grounds of the Central Andes. Alexander Brenning deserves a fair share of gratitude (from everyone who cares about glacier protection) as he was one of the first (if not *the* first) brave souls in the academic glacier world to question and reveal mining impacts to glaciers. Glaciologists simply didn't do that. Those that studied mining impacts to glaciers were working for the mining companies, and if they did come out and publicly recognize the impact, they wouldn't get work. I remember when I did my first rock glacier inventory, I cold-called Alex in Canada, where he lived and worked at the time, and pleaded him to review my inventory and to teach me how to identify rock glaciers and other permafrost features on Google Earth. Thousands of rocks glaciers inventoried later, Alex is to thank. Mateo Martini, Doctor in Geological Studies of the University of Cordoba, Argentina, also contributed his time and his good name to undersigning my work. Without him, I would have never been able to legitimize much of the material to protect glaciers from mining impacts that I have produced over the years. *Thank you!*

I must recognize the folks at the IANIGLA, Argentina's Snow and Glacier Institute, particularly the contribution of Dario Trombotto, one of the world's most knowledgeable geo-cryologists and periglacial environment specialists. Dario also has offered his broad knowledge, freely, openly, and has patiently answered many questions about the dynamics of rock glaciers and permafrost and has shared much of his work, publications, pictures, and other materials that have been critical to my work and research. We are all indebted to Dario as one of the co-authors of the very first glacier bill drafted and eventually enacted into law and for having wisely included *rock glaciers* and the *periglacial environment* as protected resources in that bill. Ricardo Villalba, then Director of the IANIGLA, also contributed to ensuring the political support necessary to sustain the law and fight back against a presidential veto that undermined it. *Thank you!*

Thanks go also to Stephan Gruber, of the University of Zurich, who shared his permafrost zoning map tool with me and with whom I've had several communications over the years to discuss the particularities of his unparalleled tool for finding frozen grounds around the planet. *Thank you!*

I would also like to thank some of the key cryoactivists (you may not consider yourselves cryoactivists, *but you are*) who helped with the content of *Meltdown*, through interviews, through your own research and publications, and by answering many inquiries that I sent around. A shout out to John Englander, author of *High Tide on Main Street* (2012) and *Moving to Higher Ground* (2021) whom I met in Delray, Florida, not too long ago. We discussed approaches to climate change communication, something that seems to be simple to convey, but always falls short of achieving the action needed to revert our climate emergency. Thanks to Henry Fountain who I met to discuss his climate beat at the New York Times and his work on bringing global attention to glacier vulnerability. Your work is inspiring. *Thank you!*

I want to thank Jared Blumenfeld, currently head of California's Environmental Protection Agency. He's had pretty important job titles before, but what I take away from his friendship over the years is his ability to boldly lead, to take steps that really make a difference when others fear standing up for what may be hard but is also the right thing to do, like when he was San Francisco's Environment Director and made it the first city ever to ban plastic bags outright. That is merely one example of the many politically challenging actions he has taken and he keeps taking to fix our planet. Just as this book went to press, California's governor, Gavin Newsom, along with Jared and Mary Nichols of the California Air Resources Board, announced a ban on the production of external combustion engines in automobiles and trucks manufactured in California by 2035 and 2045, respectively. That's what I mean by bold! Not many will throw down the gauntlet in that way! I encourage you to listen to Jared's Podship Earth environmental podcasts, which are always unique and inspiring. One of these we did together and is about rock glaciers! *Thank you* Jared!

Thank you to Connie Millar, David Herbst, and Adam Riffle, who accompanied me and Jared on our trek to study the Sierra Nevada rock glaciers, those invisible subsurface glaciers that will survive when all of the other glaciers melt away. Connie is a fabulous human being who has devoted her life and heart to study the high mountain environments of the Sierra Nevada. Her work is inspirational. *Thank you!* David is a biologist who also has been a life-long activist, helping save Mono Lake and fighting for more sustainable ways to relate to our hydrology in the American West. *Thank you*, David! Adam helped the Center for Human Rights and Environment (CHRE) launch its Cryoactivism Program and now works in the forestry sector in Washington State. *Thank you!*

Special thanks goes to Durwood Zaelke, founder and CEO of the Institute for Governance and Sustainable Development (IGSD), a small but phenomenal and globally relevant environmental climate policy organization that has single-handedly led global leaders to act on climate. Durwood has provided incisive contributions that have helped steer the world toward some of the most critical and necessary policies and actions taken to date (and still to be taken) to contain and hopefully reverse climate change. Durwood's tireless energy and his unwavering commitment to stopping climate change is contagious and provides all of us a beacon of hope to help guide us and offer us the reassurance that through human ingenuity, passion and hard work, we will get the job done. To this day I receive emails from Durwood at late hours of the night or just before dawn as he works tirelessly to help mend our planet. I am indebted to Durwood for many reasons, but perhaps the most important one is that he believed in me and in us (at CHRE) when our nonprofit environmental and human rights organization was only an idea written on the back of an envelope in 1998. His unselfish support and his friendship along the way gave us the courage and the inspiration to do the impossible. *Thank you!*

To Gabrielle Dreyfus, or Gabby (the ice-cubologist as her dad called her), who showed up at the very end of this project and offered fascinating discussion about the possible onset of a new ice age, about Snowball Earth or Hothouse Earth theories that suggest we may have missed the onramp to the next ice age. Our latest discussion was about how the actual switching of the Earth's poles might radically change our climate. I don't know if we've gotten any closer to figuring out when the next ice age will come, but I truly enjoyed our nerdy ice age conversations about Milankovitch Cycles. Who else would enjoy that??? *Thank you!*

A young woman from Nantucket, Amelia Murphy, showed up one day online to offer her help to edit, correct, and research different sections of *Meltdown*. You know all of the work you've done to make this happen. More importantly however, in the middle of the COVID-19 pandemic, we wrote an article together, which is included here as Chapter 10, to look at how the world responded to COVID-19, and why we're not responding the same way to climate change. I am 53 and she's 21, not much older than my daughter. I was blown away by her wisdom, her freshness, and her uncanny ability to take new information and run with it. Whether she is researching the vulnerability of the Antarctic ice sheet or studying policy to reduce the urban heat island effect in Stockton, California, she comes through with marvelous content and analysis. She's still helping me edit this book (well now it's off to print, so she's done) but we are now working together to advance climate policy and environmental justice in California and around the world. Amelia, *thank you!*

Much appreciation to my editor, Jeremy Lewis, and assistant editor, Bronwyn Geyer, of Oxford University Press, who understood the value of tackling the issue of melting glaciers and have strongly supported my work. They have made this second book come together seamlessly. A shoutout also goes to Mark Carey of the University of Oregon, author of *In the Shadow of Melting Glaciers*. Mark introduced me to Oxford University Press before I wrote *Glaciers: The Politics of Ice*. This book would not exist had that introduction not occurred. I also cannot leave out the copyeditor who showed up for me (but not for you) in the "track changes" version of the manuscript: Bríd Nowlan. She read every word carefully, checked footnotes and acronyms, and made those many detailed grammatical corrections and stylistic changes that make everything for you, the reader, so much smoother. Another invisible but crucial person, James Fraleigh, the final cut proofreader, made the very last detailed checks, catching persistent typos and inconsistencies. Thank you both! Finally, special thanks to the editorial team led by Project Manager Ponneelan Moorthy and Production Editor Leslie Johnson. Ponneelan and Leslie kept everything on track and handled communication with me and between the various team members. *Thank you!*

Finally, my family. *Angelina, Ulises, and my wife Romina,* who are fed up with me always talking about glaciers. When we sit down to plan out family vacations, they always know what's coming. The last time we went to Alaska, and now, Colorado, of course, to visit glaciers. In Alaska, I dropped my wife down into a glacier moulin, and somehow she got out okay, *and frankly I'm glad she did.* On that family vacation, I had to mix things up a bit for them; the whole two-week trip was a surprise; they found out each day which glacier we were visiting. Ice-climbing, glacier traversing, moulin drops, canoeing by glaciers, or seeing them from the glacier train, it was an awesome trip! I am so glad I am part of my family. *Thank you!*

Jorge Daniel Taillant

Introduction

[If you have an Instagram account, follow #glacier and enjoy incredible glacier images while you read this book.]

The day I discovered that a mining company wanted to dynamite three glaciers to get at the gold beneath, I became a *cryoactivist*.

At the time, I didn't know what a cryoactivist was. In fact the words cryoactivist and cryoactivism[1] hadn't even been invented yet. They would be. I wasn't sure exactly *why* a mining company would want to put dynamite to ice, but it sounded profoundly wrong to me (and it was). That's where it all began and as time went on, my urge to protect glaciers and everything about them only became stronger.

As I jumped into all things related to glaciers, I discovered a world that was completely esoteric to me and realized that the same was true for most people going about their daily lives. I hadn't really ever thought about the importance of glaciers to our planet's environment, nor how their presence (or lack thereof) determines *how, where, and if* we can live on Earth.

Our Earth's surface comes in one of three varieties: water, land, or ice. The percentages are, respectively, 71%, 19%, and 10%. Now consider this. Of all of the water on Earth (including ocean water, rivers, lakes, and water contained in ice), only 2% is freshwater (that means that only 2% of the water on Earth is drinkable). The rest is in the oceans and is too salty to use, unless you spend lots of money and energy to desalinate it. Of that critical but minuscule amount (the 2%) of freshwater available for us to do things like drink, cook, bathe, flush toilets, wash dishes, water plants, irrigate farms, run industries, etc., an astounding 75% is in ice, *glacier ice* to be precise, most of it packed away in sheets of ice tens of thousands of feet (thousands of meters) thick at our polar extremes. Conclusion: most of our freshwater is in *glaciers*.

A significant portion of the water we use for drinking, however, is stored in glaciers found in high altitude mountain environments, like the Sierra Nevada of California or along the Rocky Mountains, in the Alps of Europe, in the Andes of South America, or in the Himalayas in Central Asia. The glaciers up in these, and in many other high mountain environments around the world, are *rivers of ice* flowing down the coldest mountaintops of the planet. Glaciers are where most of the Earth's stored freshwater is conveniently packaged away, ready for use when we need it most. The more I read and the more I delved into the world of glaciers, the more I discovered that glaciers were a critical part of our ecosystem, despite the fact that we knew (and that I knew) very little about them.

As I fell into the cryosphere (the Earth's frozen environment), I discovered unexpected things related to glaciers that completely startled me. This new information about this obscure ice located in remote places of the world took me to entirely new levels of awareness about my environment. Ironically, while glaciers account for about 10% of the planet's surface area,[2] most people have never seen a glacier. Most people have seen lakes, rivers or streams, mountains, the sea, the plains, and forests, but have you seen a glacier? Probably not. Most people around the world drink glacier melt-water, but are likely unaware that the water they drink is at least partially derived from a glacier. They also ignore the fact that if glaciers disappear (and they *are* disappearing), many communities may no longer have access to freshwater in their local environments.

As I read more and more about glaciers—collecting books, pamphlets, and academic papers—watched documentaries about the cryosphere, and visited glaciers and their surroundings, I learned that you can travel in time through glacier ice and breath air that existed over a hundred thousand years ago, even close to a million years ago. Glaciers are a remarkable natural safe-deposit box of our planetary history.

I also learned that some glaciers, because of their massive size and instability, can cause ferocious tsunamis that can abruptly and without warning kill thousands of people and flatten entire towns. I learned that glaciers are like black holes that devour things coming too close to their gravitational energy, trapping and preserving prehistoric creatures, swallowing up entire airplanes, and burying mysteries for generations and generations, even for millennia. I learned for instance that some ice in our cryosphere, incredibly, *is flammable*. Yes, ice can catch on fire! In fact, there is more methane gas in permafrost (ice in frozen grounds) than in *all* of the rest of globally available fossil fuel reserves combined! Who would have thought that ice contained fossil fuels! I surely didn't.

I read about armies that decided to *bomb* glaciers with attack planes fearing that they might take over communities as they advanced over land—not surprisingly, the bombs made no visible impact on the impervious glacier that simply went about its normal flow of affairs. I even discovered that some glaciers are *invisible*—that's right, you can't see them, until someone shows you *where* and *how* to look at the Earth, and then, by the art of magic, colossally large glaciers appear right before your eyes where before there was nothing but earth and stone! It never fails, each time I find one of these magnificent invisible glaciers on a mountain I am blown away. You will learn to find these invisible glaciers by reading this book, and I assure you, you will have the same exciting feeling of discovery each time you locate one!

Curiously, when I started trying to educate myself about glaciers, I found it both remarkable and frustrating that there is no college degree in *glaciology* per se. You may have heard of *glaciologists*, but actually, they too, like glaciers, are an intangible enigmatic bunch. Formally, glaciologists don't really exist, that is, they are instead *geologists*, *geographers*, or *hydrologists*, or like me, a new variety of glaciologist, they can also be *political scientists*, drawn by the majestic beauty of the cryosphere and the conviction that through public policy, we should be aware of and protect this delicate natural resource. Anyone, from nearly any field, can be, if they so desire, a glaciologist, and study the relationship of our icy eco-friends to their field. There are even anthropologists that develop expertise in the relationship between people and glaciers.

As a career environmental policy expert (not necessarily a glacier specialist), I thought I knew about our global hydrological resources. In fact, two decades ago, I was part of a fairly small number of environmental and human rights activists trying to convince global governance institutions like the UN that people had a "right to water." What I didn't realize then was that glaciers hold most of our freshwater!

And they play a critical role in how our global freshwater supply gets fed into the environment.

Here I was, fighting for the *right to water*, globally, and before some of the world's most important human rights agencies, but ultimately, without glaciers the *right to water* would be practically unattainable! There was a gap there that I needed to fill. For that reason, and urged on by my good friend John Bonine, an environmental law professor at the University of Oregon, and founder of the Environmental Law Alliance Worldwide (ELAW), I eventually wrote a paper that was published in an environmental law journal, positing the importance of establishing "the human right ... *to glaciers?*"[3] It was at a symposium at the University of Oregon, to honor John's late wife Svitlana Kravchenko, a bold human rights activist, that I met Mark Carey, author of *In the Shadow of Melting Glaciers*, who introduced me to my publisher.

I would also discover a special breed of glaciers, special because they are in fact *invisible*. I'd like to take a moment to explain this incredible phenomenon which most people, even many glaciologists, know little or nothing about. As I have said, most people are fairly impervious to glaciers, and figuratively we might say that glaciers are invisible to the greater population, but there are glaciers that in fact actually are *invisible*.

These "invisible" glaciers exist *beneath* the surface of the Earth, in places we would never imagine they might. These are glaciers that you can't see until someone tells you where to look for them, and then, suddenly, as if a curtain is pulled away from Nature, we begin to find ice where before the only thing we saw were rocks.

In California for example, the Sierra Nevada has nearly lost all of its many hundreds of visible surface glaciers, and yet, there are over a thousand *invisible* subsurface *rock glaciers*[4] (follow #rockglacier on Instagram to see some amazing photos of these frozen beasts) that few Californians know anything about. Some of these subsurface glaciers are hundreds of feet thick and up to a mile or more wide and long. They are rivers of thick solid ice, beneath the surface of the Earth, displaying many of the same properties of visible white surface glaciers, such as the fact that they slither down mountainsides.[5]

Accompanied by rock glacier specialist Connie Millar, a paleo-ecologist, and Jared Blumenfeld, head of California's Environmental Protection Agency, I organized a scoping exercise of some of these magnificent cryospheric reserves, which, unbeknown to most Californians, are critical to the state's water supply. You can listen to a short podcast about this wonderful trip we did to the top of the Sierra Nevada at Jared's Podship Earth podcast series.

In the world of glaciology, these subsurface rock glaciers form part of what is called the "periglacial environment," which includes permafrost (or permanently frozen grounds) rich in ice and water. These subterranean ice reserves in a drought-stricken area like California or the northern Central Andes, for instance, provide critical water supply year round to the communities and ecosystems below them, and yet, residents in these regions generally are oblivious to the existence and much less to the hydrological dynamics and importance of this water resource. Even the most seasoned environmentalists rarely know of the existence of this critical water supply found in rock glaciers and other elements of the periglacial (permafrost) environment.

Californians, for instance, are obsessed with the yearly snow pack—which they rely on to provide water to their drought-stricken state, yet they are woefully unaware of the hundreds, even thousands, of invisible rock glaciers and extensive ice-rich frozen grounds in the Sierra Nevada just above their noses. These are a truly important piece of their hydrological lifeline after the seasonal snow has melted, during droughts and in future climate scenarios for California when all of the visible surface glaciers finally wither completely away. Rock glaciers and frozen grounds of the periglacial environment (known more commonly as permafrost) provide water to all of the downstream communities beneath them, and yet, try to find information about this critical natural resource and your search will not be easy.

The more I discovered about glaciers *and rock glaciers*, the more I realized they existed in nearly all parts of the world, and the more I found that they really were an important and critical part of the delicate balance of our global and local ecosystems. I was even more startled to learn that despite this monumental importance of glaciers to our global ecosystems, and to our daily hydrological needs, there were *no* laws *in any* country on the planet specifically designed to protect them. Let me repeat that because it is truly startling. *There were no laws anywhere on the planet to specifically protect glaciers.*[6] That's a big statement for such an important natural resource, given that we have policies and laws to protect so many other natural resources, including all sorts of species small and large. That was one of the most sobering discoveries I made about glaciers, along with the fact that they are for the most part withering away due to climate change. All of these things were new to me, as they may be to you reading this book now.

Until the day that I learned that a mining company was purposefully destroying glaciers to get at gold, I had never given glaciers a second thought as such an important natural resource. I might venture to guess that you, the reader, probably have not either, simply because most people do not.

That time for me (in the mid 2000s) coincided approximately with the world tour of former US Vice President Al Gore, who was going around the planet showing his now very well known "Inconvenient Truth" PowerPoint presentation, warning people about climate change and its implications. That was a climate awakening time for the world, and for me, that included an awakening to the world of glaciers.

It was a time when society began to understand that there was something profoundly wrong with our planet's environmental and climate health. We had already been deep in discussion about environmental sustainability and pollution, but the worry over climate was only just beginning. The scientific community latched on to glaciers as an indicator and thermometer to show the problem that we were facing and that we continue to face today.

Lately, it seems we hear about glacier melt almost daily, as a sign of a collapsing climate ecosystem, but rarely do we delve into the fundamental reasons why glaciers melt and how it really *does* matter to our daily lives that they are melting and that as time goes by, they are melting faster and faster. When I awoke to these issues, I personally began to *see the ice* and felt a chilling urgency to do something about it.

We know less about glaciers because we've probably never been to a glacier. Glaciers are hard to get to. They are either at very high latitudes north or south of the equator (near the Arctic or in the Antarctic regions) or they're up there, at high elevations, above 4 or 5 thousand meters (13,000–16,000 feet) where it's cold and hard to breathe, in some of the most remote and inhospitable areas of the planet. Some people simply cannot visit glaciers because they quickly get ill at high elevations. Anyone that has ventured into glacier terrain in high mountain environments for the first time will attest that your head starts to hurt, your legs get shaky, your heart starts to pound, and your breath gets heavy. Mountaineers in the Himalaya refer to the very highest regions of the mountains as "the death zone," where the body starts to shut down and literally, die. This remoteness of glaciers, and the harsh environment they are located in, is one important reason for our global ignorance of such a significant resource.

In the Central Andes for example, you need to drive six or seven hours on dirt roads, sometimes traveling on horseback or on foot for days, to see a glacier. The weather at these sites is often inclement, and if you get caught in a snowstorm, or in really cold weather, which can happen any day of the year, even in the middle of summer, the environment can be deadly. At Pakistan's highest peak, K2 (8,611 meters or 29,029 feet), the second-highest mountain in the world after Everest, one in four mountain climbers attempting to get to the top of this remote glacier-topped mountain dies during the ascent. K-2 is so remote and difficult to get to that it is believed few people even knew it existed, and hence it didn't even have a name when modern society discovered it. That would make sense because of the difficult terrain that must be crossed to get to the mountain, and the fact that it is not visible from the closest community (Askole) to the peak. Some call K2 the *Savage Mountain*. The name *Chogori*, meaning simply "big mountain" in the local language, has been suggested. An Italian mountaineer named Fosco Maraini didn't like the name K2, so impersonal and unbecoming to such an impressive peak. This is what he had to say about "the big mountain . . . just the bare bones of a name, all rock and ice and storm and abyss. It makes no attempt to sound human. It is atoms and stars. It has the nakedness of the world before the first man. Or of the cindered planet after the last." [7]

It's no wonder we never see glaciers—unless it's from a comfortable cruise ship hired in Alaska or Patagonia to watch glacier calvings (when large pieces of ice fall off of the face of a glacier) or viewing an environmental documentary narrated by Sir David Attenborough,[8] who has done so much to raise the global awareness of our fragile environment. It's hard to empathize with, or to truly understand, something that we have never seen or felt before in person. What is certain is that those who *have* been to glaciers walk away with unforgettable breathtaking memories of Mother Nature's awe-inspiring beauty and power. The sheer size of glaciers, when we first see them, shocks and humbles, as Mother Nature reminds us that comparatively, we are truly insignificant.

My first time at a glacier was in Patagonia, at the Perito Moreno Glacier in Argentina. I was 18 years old at the time and have that moment etched in my memory as I sat for hours drinking maté (an Argentine tea), watching colossal pieces of ice the size of buildings fall in what seemed like slow motion onto the lake water below, creating a thundering sound that echoed throughout the valley seconds later. It was truly a majestic, surreal, and humbling experience that I will never forget.

Where we *do* come into contact with glaciers, as lately, is through news media. Already for the past several years, we've been hearing recurring news briefs (more recently it's been almost daily or weekly) about melting polar icecaps and massive cracks forming in Antarctic ice sheets. We *hear* and can actually *see* now that the ice covering the South Pole as well as glaciers and sea ice in the Arctic are rapidly dwindling. We hear about giant pieces of ice breaking off of large ice sheets the size of countries or even continents. Enormous schisms are appearing in once solid country-sized ice blocks of Antarctica at the South Pole near the Amundsen Sea.

We hear about glacier pieces the size of Manhattan or others much larger suddenly tumbling off of glaciers, rolling and collapsing into the ocean, and becoming floating icebergs.[9] Once these monstrous ice blocks completely melt away, the largest ones have the potential of raising the level of the sea by several feet, and this will occur *during* our lifetimes. We also hear that climate change is happening faster at the world's extreme polar environments, at rates double or triple of what is occurring around the rest of the planet. Melting glaciers are revealing landforms beneath the ice all over the Earth. Alaskan glaciers for example, according to new studies, are actually melting up to 100 times faster than previously thought.[10]

The emerging news about glacier melt is spilling over into new areas, and into our daily lives. For most people this is all quite esoteric. It sounds alarming, but we are for the most part still detached from the news. We might think that glacier retreat simply means the ice disappears from the mountain and that our only loss is not being able to see these fantastic and massive ice bodies before they're gone. We don't really fathom how these catastrophic environmental events at the extremes of the Earth will affect us. What happens at one end of the Earth where we seldom or never go is rarely linked in our minds to what happens in our local environment. But this is slowly changing.

With sea level rise flooding well-known areas and some of the world's most famous cities, with collapsing forest ecosystems, with climate wildfires destroying millions of acres of land, with recurring and more severe storms and extreme inclement and unstable weather becoming the norm, with many bizarre natural phenomena occupying daily news feeds and exploding in social media and other news outlets, these phenomenon are in part attributable to glacier melt. More and more, the consequences of glacier melt are becoming visible and tangible to most of us, even if we are not making the connection between glaciers and climate change. The link is slowly becoming clear, and little by little people are learning about these trends and their origins. We are beginning to connect the dots, and perhaps this is why in the last year or so, there has been far more political and policy attention, and media attention, to climate change and to the urgent need to address it.

In the middle of this growing ebb and flow of news about our changing climate, natural phenomena involving glaciers still seem alien because we are still far from the sources of the events. For most, glacier melt is still intangible because we can't *really* imagine these enormous natural occurrences happening so far away from us. We can't put them into perspective. We don't really understand how *big* these glaciers really are and *how significant* their breakup, destruction, and melting really is. Imagine flying from Los Angeles to New York, or from Paris to New Delhi, from Cairo to Johannesburg, or from Sydney to Beijing, and all you see beneath you is ice, for hours and hours on end! That's how big the biggest glaciers (the ice sheets) are. Now imagine these entire regions suddenly breaking apart into colossal chucks of ice, slowly converted into meltwater eventually filling the oceans and overflowing onto lands.

The consequences, whether we see them or not, whether we can fathom them or not, are huge.

Glaciers from pole to pole, from mountain top to mountain top in places like the Andes of South America, the Alps of Western Europe, the Sierra Nevada of California, the Rocky Mountains of the United States and Canada, the Caucasus of Western Asia, the Zangezur Mountains of Armenia, the Karakorum of Pakistan and China, the Tian Shan of Kazakhstan and Kyrgyzstan, the Himalayas of Nepal and China, or the ice-covered peaks of Kilimanjaro, are quickly retreating, melting away into their ecosystems below; they're melting faster than they can be replenished by snowfall. That's how you measure a glacier's health. If a glacier can replenish the melted ice from its mass each year by trapping newly fallen snow during the winter and converting it to ice, it is in ecological balance. If it cannot, then it progressively melts into oblivion. Where glaciers now exist and are melting, sooner or later, there will be no more glaciers.

All around the world, glaciers are retreating. Despite recurring snowfall in the winter season, and even despite occasional intense snowstorms, average year-round ambient air temperature is rising faster and faster, making it more and more difficult for ice to survive the spring, summer, and fall months. If more snow and ice melts off than is replenished during the following winter, glacier sizes are reduced, and each year, the glacier's mass declines further. As this pattern repeats itself year after year and decade after decade, we can see glacier retreat very clearly in most glacierized systems of the world. Some of these glaciers may never return to their former splendor and provide those critical functions they do to downstream communities and ecosystems. Some will have entirely vanished within our lifetime. Some are already gone.

The European Alps have lost half of their total glacier ice in just over a century, and since 1980 melting is accelerating, says Dahr Jamail, who has recently published moving book about glacier retreat called *The End of Ice*. The same is occurring in the Himalayas, while massive areas of Antarctica are "hemorrhaging ice as glaciers there retreat faster than anywhere on Earth." Jamail quotes Donald Blankenship of the University of Texas, who studies Antarctic glaciers: "The fuse is lit. We're just running around mapping where all the bombs are."[11]

Glacier National Park in the United States, for example, is quickly becoming a *glacierless* national park. When first created by President Taft and opened to the public in 1906, Glacier National Park had about 150 glaciers (not all glaciers were registered at the time). A 2017 study by the United States Geological Survey (USGS) revealed that the famous glaciers of the only glacier park in the United States have deteriorated on average by one third since 1966. Today there are fewer than 30 left, and ten of those have lost more than 50% of their volume over the past 50 years. Among these, Boulder Glacier has lost a whopping 85% of its size during that period, while Blackfoot Glacier, once the park's largest, has lost 25% of its size. At the current pace of climate change, Blackfoot is likely to disappear entirely by 2030.[12] Daniel Fagre, one of the most well-known glaciologists who studies the glaciers of Glacier National Park (and whom I'd like to meet one day) said recently: "Things that normally happen in geological time are happening during the span of a human lifetime. It's like watching the Statue of Liberty melt."[13] I checked up on Fagre's work at Glacier National Park, and since the study, just two years later, four more glaciers vanished (or are no longer considered glaciers because they have lost key characteristics, such as movement). There are now twenty-six left, and counting.[14]

The glacier reserves in other parts of the United States also show alarming rates of deterioration. According to Andrew Fountain, a professor of geology at Portland State University, the approximate 1,700 glaciers and snowfields in California's Sierra Nevada have dwindled by about 55%. In the Colorado Rockies, the deterioration of glaciers is 42%, in the Cascades Range of the Pacific Northwest, it is 48%.[15] Canada's British Colombia and Alberta are projected to experience 70% glacier loss by 2100.[16]

According to data provided by the National Aeronautics and Space Administration (NASA), glacier retreat around the world is alarming. In the Himalayas, which, after the polar icecaps, have the greatest number of glaciers, 95% of glaciers are retreating at alarming rates. Some 30,000 square kilometers of glacier ice, providing 8.6 million cubic meters of water per year are rapidly disappearing. Glaciers near Mt. Everest and in the Indian Himalayas are retreating between 20 and 30 meters each year (65–98 feet). These glaciers feed critical rivers such as the Ganges, as well as rivers in Nepal and Tibet.[17] Kilimanjaro's glaciers in Africa (yes, there are glaciers in Africa) have melted by more than 80% since 1912, while arctic sea ice has dwindled 10% in just three decades.[18]

The UN's Intergovernmental Panel on Climate Change (more commonly referred to as the IPCC), the global multilateral scientific body most responsible for addressing the science and politics of climate change, observed in its 2019 report, *The Oceans and Cryosphere in a Changing Climate*,

> Snow-cover duration has declined in nearly all regions ... Low elevation snow depth and extent have declined, ... Mass changes of glaciers in all mountain regions (excluding Canadian and Russian Arctic, Svalbard, Greenland and Antarctica) was very likely [negative] ... Sparse and unevenly distributed measurements show an increase in permafrost temperature [that means permafrost is melting] ... Other observations reveal decreasing permafrost thickness and loss of ice in the ground[19].

The IPCC report goes on to stress glacier melt processes also underway in Africa, Asia, and Europe.

> Other factors, such as changes in meteorological variables other than air temperature or internal glacier dynamics, have modified the temperature-induced glacier response in some regions. For example, glacier mass loss over the last several decades on a glacier in the European Alps was intensified by higher air moisture ... changes in air moisture have also been found to play a significant role in past glacier mass changes in Eastern Africa, while an increase in short wave radiation due to reduced cloud cover contributed to an acceleration in glacier recession in the Caucasus. In the Tian Shan mountains (in Asia) changes in atmospheric circulation in the North Atlantic and North Pacific ... resulted in an abrupt reduction in precipitation and thus snow accumulation, amplifying temperature-induced glacier mass loss.[20]

Global warming is withering away glacier ice, melting glaciers faster than winter snowfall replenishes them, progressively diminishing our planet's permanent storage of freshwater. The larger glaciers of Antarctica and the Arctic Regions are melting fast, faster in fact than glaciers in other parts of the world.

If you have never seen glacier melt in action, or personally witnessed the startling evidence of receding glaciers, or even if you have, I suggest you watch the award-winning glacier documentary *Chasing Ice* by James Balog.[21] This documentary captured glacier retreat in the Arctic in one of the most astounding live videos of a collapsing glacier terminus the size of Manhattan Island. We are not used to seeing things this big actually move so suddenly. The images of the film are beyond astonishing. The magnitude of the events filmed by Balog in the Arctic are simply beyond our imagination. This documentary should be watched by everyone. I was fortunate to see it on the big screen for one of its premier showings during a snowstorm in New York City, where I had a speaking engagement at Columbia University, coincidentally on actions to preserve *the right to water*.

So you might be thinking, OK, so glaciers are melting, why is that so important?

To begin with, when glaciers melt, the water they release goes downstream. That in itself doesn't seem to be a problem, since we *need* and *want* to consume glacier meltwater for our various uses. Some might think that the melting of glaciers is an opportunity to tap into some of that freshwater so that we can increase our consumption, and that might be the case for some communities, like Californians who so often have droughts that leave them without freshwater, but the benefits of suddenly increasing glacier melt are generally short lived and very limited, and often outweighed by the costs.

Our ecosystems actually need glacier meltwater to remain *constant* and *healthy*. The problem is that glaciers hold a lot of water, and for the most part, for many thousands of years, the meltwater coming from glaciers into our ecosystems was balanced and in equilibrium with our environments. We had just enough meltwater to keep all the plants and other species happy and thriving. Now, with accelerated and imbalanced glacier melt, this water is flowing too much, too cold, and too quickly. This greatly alters downstream environments, setting all downstream flora and fauna completely off-kilter.

Meanwhile the glaciers we need most for freshwater consumption are also dwindling away at alarming rates. According to the IPCC these critical glaciers will shrink by between 15% and 55% by the year 2100 and could even shrink to 85% of their present size if we are not able to contain global warming trends. That means that the 2% of the planet's water that is available to us as freshwater is rapidly decreasing and literally draining into the oceans (causing even more instability in our ocean ecosystems— we'll talk more about this in Chapter 7). Our planet's ice is vanishing and with it, our global freshwater supply.

Glacier melt will affect our planetary ecosystem in ways that were unimaginable to us until recently. We'll get into some of those ways later in subsequent chapters. But here is the real question.

Does glacier melt really matter that much? And will it really change your life?

This is the question that you may be asking at this point, as will many others. *Does it really matter that much*? So what if the ice melts? Should I really care? How will glaciers melting in the Arctic or in mountains that I will likely never see, far from my home, really change *my* life?

As a cryoactivist, worried about glacier melt, the question and the responses I sometimes hear scare me, but I get it. The question and the responses scare me because they are full of apathy—apathy that derives from disconnection. But it is a very legitimate question for most people to ask. Just as we don't really know much about glaciers, many of us can't quite fathom what glacier melt really means and implies for us and for our daily lives. Does it really matter that ice that is so far away from where I live melts away into the sea or that our planet's natural solid freshwater storage facility dwindles away?

It will keep raining, so doesn't rain provide us with all of the water we need? We'll get water from the snowmelt and it will fill our dams each year. And if it doesn't snow much we'll build new dams to capture more snow. Won't that be enough? When I was visiting the Root Glacier in Kennecott, Alaska, I overheard someone at dinner at the Kennecott Lodge (a person who, like me, was visiting the glacier) say to a friend, "I actually think it's a good thing that glaciers are melting, since before, the world was covered by ice and the melting of glaciers has made the planet more habitable."

He was technically correct. The planet *is* actually more habitable than it was during peak ice ages. Just think, during the peak of the last ice age, just a few tens of thousands of years ago, most of the northern United States and all of Canada were covered with uninhabitable ice. Where this person was wrong, however, is that the question is not really about peak ice ages (say 20,000 years ago, for example), but rather, the temperate climate that we live in today, and the role that glaciers play in *that* climate. In other words, we shouldn't be thinking about glacier *maximums* tens of thousands of years ago, when lots of the planet was covered in ice and was uninhabitable, but rather our place today in the glacier *minimums*, when we are (or have been for thousands of years) in rather pleasant inter-glacial times.

We're pretty lucky to be living in this era, when for the most part, the planetary climate is comparatively benevolent. In fact, our place on Earth is largely thanks to the fact that we're living in a glacier minimum, between ice ages. Our modern civilization probably wouldn't exist as it does today if it were not for this very comfortable inter-glacial period we were born into. It's incredible to imagine what global living conditions for humanity would be if half of North America were covered by massive ice sheets, and fortunately for us, we don't live in such an environment. Except for Hollywood depictions of what life would be like if we were frozen over, few can really comprehend what an ice age would look like, and how it would change our lives if one suddenly took place (something that in nature actually takes tens of thousands of years to play out). If you haven't seen the movie *The Day After Tomorrow*, I suggest you do to get a sense of what it would be like if the Earth were to suddenly freeze over!

The Earth's surface goes through glaciated and non-glaciated phases, and our existence as a human race today is taking place during a very small portion of these phases, when glacier presence, surface area ice coverage, and our climate are just right, so that our existence on the planet is mild, plentiful, and sustainable. For the most part, as a human population we live in areas *not* covered by glaciers, and the glacier-cover that *is* on the Earth's surface helps cool our planet and make our lives livable. The current melting of our planet's glaciers that is occurring however, and its relevance for our way of life, suggests that *not-so-good-times* are coming.

To answer the question at the top of this section, surely *it does matter* that glaciers are melting.

Most people however, don't really understand *why* it matters, but we do realize that if one of the most significant natural resources on our planet, none other than freshwater, disappears, something is wrong or will be wrong, and that it will have terrible consequences for our way of life. But let's be more specific.

Sea Level Rise

As the first draft of the manuscript of this book was finalized in early 2020, a news bit about flooding in the Florida Keys was circulating among what had been for the last several years a mostly climate change skeptic Florida population. The former governor of the state at one point even barred his employees from mentioning the term "climate change" at public events.[22] That would change! The article about flooding in Stillwright Point, on one of the Florida Keys, that appeared in the Miami Herald opened with:

> Saltwater has been flooding the low-lying streets of a Key Largo neighborhood for more than 40 days, leaving many residents there trapped unless they can walk or are willing to sacrifice their cars to the nearly foot-high corrosive seawater.[23]

I checked back on the Key Largo neighborhood a few weeks after that, and it was still flooded. It turned out that the floodwaters would remain for about 90 days before they subsided. A newspaper article celebrated the water's retreat, with one local resident, Bill Marlow, exclaiming in exasperation, "Freedom"![24] Residents are now pressing local and state government officials to help build up their roads to stop the water, which they expect will return. The Florida government, meanwhile, is suggesting to Keys residents in low-lying areas that maybe it's time to leave.

Does this have to do with glaciers melting? The simple answer is yes!

The oceans of the world rise and fall as glaciers form and melt and as we move into and out of ice ages, or glacial cycles. Glaciers are formed when ocean water is converted to snow and packed in glacier ice. As the climate goes into warmer phases, glaciers around the world melt and return the water they hold to the oceans. These phases between cold and warm global climate oscillate every 100,000 years or so, influenced by variations of the Earth's elliptical orbits, the Earth's tilt, and its positioning relative to the sun. These oscillations cause a very significant change in global temperature, which results in an effective cyclical *freezing* and *thawing* of significant portions of the planet.

This in turn influences sea level oscillations that vary by more than 600 feet (about 200 meters) between the highest and lowest points. Stop and think about that for a moment, that's the height of a 60-story building or two football fields placed lengthwise on top of each other. Two hundred meters is almost as tall as the Golden Gate Bridge. Imagine the sea rising and falling by that height. We are presently *at the end* of an ice age, where much of our glacier ice that once covered the Earth (more than 60%) has already melted into the ocean. We have about 30–40% of the world's ice cover left, and when (and if) all of that has melted, the oceans will rise further by some 200 additional feet (more than 60 meters). That would take ocean water level to the roadway of the Golden Gate Bridge. That may occur well into the future, or it may happen very quickly. A significant portion of this remaining glacier melt could even occur within our lifetimes. A not insignificant rise is already occurring. Ask the people of Key Largo. Imagine that a big storm or other weather event floods your neighborhood, which may already be at or close to sea level, and several weeks go by, but the water doesn't recede. In fact, it never recedes. In the case of Key Largo, it stuck around for 90 days, but that is probably not the last time such flooding will occur.

In Florida, while the government and real estate agents don't like to talk about this much, this is happening. Government agencies are telling people in the Keys in particularly low-lying areas that recurrently flood, and that have lost property, that they should consider leaving.[25] It's only a matter of time before those messages will begin to go out to people living on the mainland in South Florida, in places like Miami-Dade County or Broward County and as far up as Palm Beach County (where I live). All three of those counties, under a full glacier melt scenario (which has happened many times during Earth's geological history) have been at the bottom of the ocean. And they will be again. I describe past and current flooding in South Florida in Chapter 2.

So while you may live in the comfort of a mountainous environment many feet above sea level, for anyone living *at* sea level, even the rise of a few inches of the ocean could have catastrophic consequences.

Water Supply

Glaciers are important for many reasons, not just because they determine sea level. Glaciers store water in ice form. Glaciers contain over 33 million cubic kilometers (km^3) of freshwater.[26] It's hard to grasp just how much ice that really is. The volume is simply far too large for our minds to understand. Let's try.

First imagine a smaller (albeit still pretty large), more imaginable volume of ice that we can get our minds around. Let's say a big cube of ice the size of a city block. I've never seen a man-made ice cube that large, and you probably haven't either but let's try to imagine it anyhow. In fact, why don't you actually walk outside now and look down the block to get a sense of the volume we're talking about. Look down the block, from corner to corner, and imagine the entire city block formed by all four corners. Now look up, and project that block upward for the same distance as the entire block. That's the cube, completely filled with solid white ice. That's a pretty big cube of ice. That already is a structure larger than most structures you've ever seen. Now multiply that cube by 10 city blocks, to the left, to the right, and up into the sky. Now that's an *enormous* cube of ice. It's a stretch of the imagination. That's about one kilometer cubed of ice. It's already becoming unimaginable because most of us have never seen anything made of ice or of any other material that is that big. Not even the largest buildings are that big. And now comes the real three-dimensional stretch of your imagination.

Explode that volume to 33 million of these huge cubes! That's equivalent to taking that big cube of ice 10 city blocks long and placing it together with another cube of the same size, and adding cubes one by one until you've created an ice cube path that wraps around the Earth 825 times! Impossible to image, right? That's why we can't really fathom just how big the ice on Earth really is. And it is that big.

That's a lot of water, too much water to truly imagine. And yet, that's how much ice is on our planet, and that's how much freshwater is stored in ice.

In fact, 75% of the world's freshwater is in glacier ice. That statistic alone explains one of the key reasons why glaciers are so important. They hold most of our freshwater. They are natural water reservoirs larger than any dam or storage container society could ever build. Even the totality of all human-made dams all around the world could only hold a minuscule amount of the freshwater held by glaciers.

Now here is an oddity about melting glaciers that for most people is completely counterintuitive. If glaciers are melting faster than usual, then shouldn't we have *more* freshwater water becoming available to us? It makes sense? No? And yet, here is what most of the Earth's climate scientists agree on: The UN's IPCC report on the Oceans and Cryosphere notes in 2019 that,

> Changes in snow have changed the amount and seasonality of runoff in snow-dominated and glacier-fed river basins with impacts on agriculture.... In some glacier-fed rivers, summer and annual runoff have increased due to intensified glacier melt, but decreased where glacier meltwater has lessened as glacier area shrinks. Decreases were observed especially in regions dominated by small glaciers, such as the European Alps. In some areas, where glacier and snow meltwater has decreased, especially where other climatic drivers or socio-economic stressors are also present, agricultural productivity has declined, e.g., in the Western USA, High Mountain Asia and the tropical Andes."[27]

So, while one would think that melting glaciers would provide *more* water to agriculture, in fact, because glaciers are melting fast (and subsequently shrinking), in some key regions where small glaciers are very significant contributors to local water supply, shrinking glaciers actually have progressively *less* surface area and volume, so there is less to melt each season. Bigger glaciers, particularly in the Polar Regions, which oftentimes melt into the ocean instead of into water-dependent communities, melt off more water than smaller glaciers ... and so we can thus have a net *reduction* in glacier meltwater available even though glaciers melt more and more each year!

In sum, the melt from larger polar glaciers has a huge impact on sea level rise, while the melt from smaller glaciers in the mountains in populated regions poses risks to communities and delicate ecosystems downstream from high mountain environments that depend on ecologically stable mountain glacier meltoff.

Water Basin Regulation

For communities fed by mountain rivers that are born at glacier terminuses, glacier melt affecting water basin regulation is a key issue. This has to do with one of the key "functions" of a glacier for the ecosystem.

Glaciers are considered "water basin regulators." That's a sophisticated way of saying that glaciers control the *flow speed* and *amount* of freshwater feeding into the environment below. At the top of the water chain in some of the world's highest and coldest peaks, stored glacier ice retains water after springtime and summer snowmelt. Yes, we get lots of water when the winter's snowfall melts away in the spring, but large ice bodies (glaciers) in mountains keep water in a solid state after that seasonal snow melts, and a glacier's extremely cold temperature is able to hold back water in the form of ice for an extended period (melting very slowly over the post-snowmelt months, until the next winter snowfall).

In this way, glaciers provide steady and slow ice melt into the waterways downstream. So after the rush of springtime seasonal snowmelt, glaciers keep the water flowing (albeit at much slower rates) for the entire year. Think of glaciers acting like a big water faucet on top of the mountain, opening slightly to let water trickle out little by little to meet the ecosystem's ongoing needs. Without glaciers, our mountain rivers and streams would dry up relatively fast after the springtime snowmelt. In this way, glaciers feed downstream ecosystems, agricultural lands, and communities that depend on yearlong water flow.

Similarly, during an especially dry year, or a drought that may last several years, mountain glacier ice will also keep a minimal amount of water flowing into the ecosystem until nature brings back the snow and the rain. Glaciers hence provide critical water supply for drought years to carry us over until the next wet season. Again, without glaciers, we're in trouble.

Reflectivity/Albedo

Glaciers cover 10% of the surface of the Earth. That means that 10% of the surface of the Earth is white (or off-white, since most glaciers aren't actually perfectly white), and because lighter colors reflect more sunlight than darker colors, 10% of the Earth's surface is adequately adapted to *reflect* sunlight instead of absorbing it. This role of the Earth's glacier surface is critical to our global climate. Wear a *black* shirt on a very hot day, and you feel very hot (your black shirt absorbs heat); wear a *white* shirt, and you'll feel a little cooler. Glaciers are white, so they make the planet cooler by reflecting solar heat. Remove the glaciers and Arctic ice (as is occurring rapidly through climate change and resulting glacier melt), and the surface below (land or water) is revealed (which makes the Earth, as seen from space, darker). The brown, black, blue, green, and other hues of the Earth all absorb much more light than the white of ice, which repels sunlight. That causes the Earth's surface and surrounding atmosphere to warm. Conclusion: melting glaciers (and the darker surfaces of the Earth that are revealed) cause accelerated global warming and climate change through decreased reflectivity and surface heat absorption. The more glaciers melt, the hotter our planet will get.

Ocean Temperature

Glacier melt also affects ocean temperature. When the glacier and sea ice covering our planet melts away, the ice-covered areas of the oceans go from white to blue, and the blue of the sea reflects less than the white of the ice, augmenting ocean temperatures because the ocean acts as a heat-trapping body instead of having solar heat reflected off of ice cover in its coldest regions. This is already happening. The year 2018, according to a study published in *Advances in Atmospheric Sciences*, was the warmest on record for our oceans. And the trend? The 2018 record beat the previous record set, you guessed it, in 2017. By the time I sent this manuscript to press, in 2021, the National Center for Atmospheric Research (NCAR) had reported that 2020 broke the ocean temperature record. Ocean temperature increases also result in the expansion of the ocean, increasing sea level rise, not only because it has more water, but because as water warms it expands, causing the entire volume of the ocean to expand. Warmer oceans note NCAR can cause a number of impacts including supercharged hurricanes and more intense storms.[28] The oceans are also great energy conservers so that the warming of the ocean will continue long after the warming effects take place. Coastal dwellers beware! Changes in ocean temperature not only affect the species that live in the ocean, they also affect the cooling properties of global ocean streams and can affect the surface air temperature, which subsequently affects the air streams. In this way, melting glaciers can have effects on ocean temperature around the world.

River Temperatures and Impacts to Flora and Fauna

Rapidly accelerating glacier melt can also alter river waters and related lake environments. If you consider that many species of flora and fauna live in and along glacier-fed rivers and streams, it can be expected that changes in river flows and temperature and other glacier-derived modifications to the natural environment will affect the natural habitat of glacier-connected and glacier-dependent ecosystems, including the flora and fauna that live in these ecosystems.

Spawning grounds, for example of salmon, trout, and other fish species that migrate through glacier- and permafrost-fed river systems (such as those in Alaska, Canada, or the mainland of the northwestern United States) or in other high mountain environments, are directly impacted by alterations of local ecosystems and particularly of hydrological systems that are in constant change due to glacier melt. A New York Times Special Report by Henry Fountain looked at this delicate balance in Washington State, "When the Glaciers Disappear, Those Species Will Go Extinct." Fountain reflects eerily on the plight of river streams in the Pacific Northwest:

> As surely as they are melting elsewhere around the world, glaciers are disappearing in North America, too. This great melting will affect ecosystems and the creatures within them, like the salmon that spawn in meltwater streams. This is on top of the effects on the water that billions of people drink, the crops they grow and the energy they need. Glacier-fed ecosystems are delicately balanced, populated by species that have adapted to the unique conditions of the streams. As glaciers shrink and meltwater eventually declines, changes in water temperature, nutrient content and other characteristics will disrupt those natural communities.[29]

Toxic Emissions/Methane Release

Glaciers and other permanently frozen environments of the Earth also hold danger in the form of organic material buried in *and under* ice. Entire forests and other plants and wildlife that existed tens of thousands or even hundreds of thousands of years ago, lay in frozen state. Now they are thawing as glaciers and frozen earth saturated in ice melts and these prehistoric organic wastelands are being unearthed. Revealing this organic matter causes gases to be released into the atmosphere, including CO_2 as well as ancient methane gas, which is contributing to accelerated climate change. In certain ice, particularly in permafrost regions of the periglacial environment (extensive swaths of frozen terrain which occupy 25% of the Earth), we find compacted methane crystals, which are flammable. These are called methane hydrates or *gas hydrates*. This is the "ice that burns." Light a match to it and the ice actually catches fire.

Methane is a potent greenhouse gas that causes accelerated climate change at much higher rates than CO_2. A statistic that will probably surprise most people is that there are greater fossil fuel reserves in ice crystals buried in frozen grounds than in all other sources of fossil fuels combined. If our glaciers melt, in addition to the other impacts already mentioned, CO_2 and methane released from thawing frozen grounds will only further accelerate global warming and climate change.

Glacier Tsunamis/Flooding

If you've ever witnessed or even heard of a *glacier tsunami* you might begin to realize the enormous power of glacier impacts. Scientists refer to glacier tsunamis as glacier lake outburst floods (GLOFs). [30] I think glacier tsunamis sounds more like what they really are.

GLOFs are sudden collapses of large pieces of ice that fall off of unstable glaciers (they are unstable because of the warming climate), pound into glacier lakes, and cause deadly waves. The lakes located immediately below glaciers are formed by receding melting glaciers. These receding melting glaciers leave large holes in the earth where they once stood, and then through accelerated climate change and resulting ambient warming, they melt to form lakes in those holes. When big pieces of ice falling off of unstable melting glaciers (very big pieces the size of large buildings), collapse onto the glacier lake water below, massive waves ensue.

These deadly waves come raging down ravines, and mountain valleys, taking out anything in their path. The IPCC's 2019 report on the oceans and cryosphere states on this point:

"Glacier retreat and permafrost thaw have decreased the stability of mountain slopes and the integrity of infrastructure. The number and area of glacier lakes has increased in most regions in recent decades." [31]

This tells us is that in the future we are likely to see an increase in glacier tsunamis. How serious are these events? One glacier tsunami that occurred in the mid-20th century in Peru rumbled down a mountainside killing thousands of people. One of the last cries heard in the village of Huaraz before the glacier flood buried the town in ice, mud, rocks, and trees, killing thousands, was: "RUN! The mountain is coming!" Moments later, an enormous flood of mud, rock, tree, water, and ice ran over the town, in minutes, decimating everything.

Prehistoric Diseases

Dangerous elements can be buried underneath millenary ice. Disease strains contained in frozen burial grounds, for example, or in sick animals that have died from disease and have been conserved in ice, or in plant matter, that have wiped out species thousands of years ago are being unearthed, potentially bringing these human, animal, and plant diseases that were eradicated long ago, back to life.

Electricity Generation

As noted earlier, glaciers in high mountain environments create water flow through their melting cycles. Societies throughout the world have utilized this glacier-melt flow by harnessing the force of water to generate electricity. Either by placing generators in river streams or by damming glacier meltwater and creating large high-mountain lakes, waterpower can be generated from glaciers and converted into an energy source. Increased glacier melt can provide a temporary windfall of available water for power generation, as some engineers are anticipating for places like the European Alps. For other areas, however, as climate change modifies glacier-melt flows, and in some cases, *reduces* glacier-meltwater availability (which will eventually happen even in places like the European Alps), the continued use of glacier water to produce electricity is at risk.

Tourism and Natural Beauty

Glaciers are truly majestic in their colossal size and breadth. They have attracted nature lovers as far back as history has been recorded. Anyone who has been to a glacier can attest to the awe-inspiring beauty of glaciers. In places like Alaska, Patagonia, Canada, Greenland, Iceland, or the European Alps, glaciers generate significant tourism revenues and jobs associated with the commerce that comes with people wanting to see glaciers. As glaciers retreat, become more unstable and dangerous, or completely disappear, so will the allure and majestic beauty, and the tourism revenues they generate. We will invariably lose this attraction and with it significant parts of local economic prosperity. But we will also lose the aesthetic value they contribute to our society. The deterioration of glaciers in Glacier National Park, that will soon become *glacierless*, is a perfect example.

These are but some of the consequences of the melting of glaciers that we are already beginning to understand. Glacier ice melt is revealing a story to us that we are piecing together slowly, but which is becoming ever clearer as we learn more about how the cryosphere interacts with our global ecosystems. Something is evidently profoundly wrong with the natural balance of things. Glacier melt, while we don't necessarily see it in our local environments, is a symptom that is calling out to us, and its consequences, which we may not fathom entirely, are enormous.

What Can We Do?

Shouldn't we be doing something about glacier melt? *Can* we do something? Is it really that important? Is it even possible to *save* glaciers?

As an activist working for a nonprofit environmental organization in South America for nearly two decades (we have since moved to Florida), I *thought* I knew quite a bit about environmental laws and policies. I *thought I knew* what the most critical environmental problems of our times were. And yet I didn't realize that even though glaciers held over three quarters of the natural resource most precious to us (besides air), there wasn't a single law on the books anywhere on the planet to specifically protect glaciers. As mentioned above, few if any laws even mentioned the word "glacier," this obscure frozen state of water high up on the water chain.

How could that be in a world so preoccupied with environmental degradation? I thought surely countries like the United States, Canada, Sweden, Austria, Nepal, France, Switzerland, China, or many of the world's other countries characterized by the presence of ice, *big ice*, had realized the importance of glaciers to our planet and had drafted policies and laws to protect these magnificent bodies of ice, if only for the purpose of sustaining our livelihoods.

No. Not so. With all of the environmental movements and causes out there around the world, and with all of the awareness of the importance of the environment and sustainable development that has occurred globally over the past decades, glaciers were *and continue to be* mostly unprotected.

My quest to become a *cryoactivist*, a term that is simply defined as *someone who works to protect our frozen world* (which includes glaciers), led me to many discoveries about the particularities of glaciers, which I had no idea about before embarking on the path of cryoactivism.

I knew very little about glaciers before that momentous day someone showed me how mining companies were dynamiting glaciers to extract gold. I immediately felt a burning need to do something about glacier vulnerability. I had to find a way to protect glaciers.

Little did I know that by opening the Pandora's Box of glaciology I would become a cryoactivist, falling uncontrollably into a frozen crevasse of natural beauty and significance beyond anything I could ever imagine.

It was a glaciologist named Bernard Francou who once said to me as we were on our way up to visit the Antisana Glacier in Ecuador, "your work to protect glaciers is like that of a *cryological* activist." We looked at each other and almost mouthed in unison, coining that day the term "cryoactivist."

There are many cryoactivists in the world, some with us today, some long gone, that have marked a clear path to valuing and respecting our planet's vulnerable glaciers. Louis Agassiz, a Swiss-born naturalist who first proposed the existence of the ice ages that shaped our present-day planet, was perhaps one of the world's first cryoactivists.

John Muir postulated, to much skepticism of the time, that Yosemite was formed by flowing glaciers. Most didn't believe him. Muir helped establish one of the world's most impressive natural laboratories to study the influence of glaciers on our natural environment, Yosemite National Park.

Then there are modern-day cryoactivists who have understood the increasing vulnerability of glaciers and are today contributing to the creation, conservation, protection, and even fabrication of glaciers critical to human and natural existence. Here, I'd like to name a few that I have met or that have influenced my work.

They include: Chewang Norphel, the Ice Man of India who decided one day to build his own glaciers to help provide more water supply to his environment; Dario Trombotto, a geo-cryologist (someone who studies ice and its relationship to rocks) of Argentina who was the inspiration to draft the first piece of legislation *ever* to protect the periglacial environment; Alexander Brenning who wrote one of the first and most influential academic articles on the substantial impacts of mining on rock glaciers in Chile, sparking a social movement to protect the hitherto unknown periglacial environment; Daniel Fagre of USGS who has devoted his life to understanding and communicating information about deteriorating glaciers at Glacier National Park; Connie Millar who has devoted her life to studying the enigmatic contribution of rock glaciers to the hydrological systems of the Sierra Nevada; Juan Pablo Milana, a geo-cryologist of San Juan, Argentina, who was one of the first to warn of mining impacts to glaciers in the high Central Andes on the border region between Chile and Argentina; Cedomir Marangunic, a glaciologist (also a geo-cryologist) working in Chile and devoting his career to better understanding glacier and rock glacier dynamics; Benjamin Morales Arnao of Peru, who has worked to understand glacier tsunamis and their risks to protect downstream communities; Bernard Francou of the Institute for Research and Development in France, who has delved into glacier dynamics and steered his culminating glaciological studies to look at the relationship between indigenous cultures and glaciers; James Balog[32] who produced the phenomenal documentary *Chasing Ice*—that everyone must see—capturing time-lapse images of collapsing glaciers in Greenland and around the world that was never before seen; and Kalia Moldogazieva of Kyrgyzstan who has fought tirelessly to try to get a glacier protection law passed in her country, where mining impacts continue to harm sensitive glacier environments. These are but a few of the cryoactivists I know that are out there, fighting to save glaciers. There are many others I have not mentioned.

Meltdown shares the inspirations and concerns that I have come across in my quest to protect glaciers. It is an attempt to convey in simple and understandable terms, the urgent need to address climate change, not only to protect these magnificent natural water towers of insurmountable beauty, but also to draw attention to how climate change is affecting our world. It sets out to teach the general public, people who, perhaps like me not too long ago, know little or nothing at all about glaciers or the cryosphere. It is also an effort to bring the attention of our society to the need for a new path if we are to avoid losing glaciers all together.

For the scientists out there who often focus solely on the academic dimensions of glaciology, you may not find material in this book that is relevant to your scientific knowledge about glaciers, but I challenge you to reflect on its social, political, and community relevance, to explore the ways *your* scientific work can contribute to our political and social understanding and to the policies to protect and conserve the frozen domains that keep glaciers vividly healthy. What can *you* do through your work and your profession to alter our global path to ensure that glaciers and glacier environments and our climate are protected?[33] That is a challenge that we should all mutually strive for.

Meltdown is a book written to answer the questions: Why are glaciers important and why is their demise (glacier melt) important to you and to all of us in our ever-changing environment? You are a part of this changing environment and climate, and you are also a cause of the impact our climate is suffering, and I hope that this book helps you take a step, however big or small, to lead the way to recuperate our climate, *and our glaciers.*

1
And Then There Was Ice

[follow #glacier on Instagram to receive daily images of some of the world's most fascinating glaciers]

[and have your smart phone ready for this chapter, you'll need it!]

Have you ever been to a glacier? Probably not.

So let's go to one right now: pull out your smart phone and open a map app, such as Google Maps. Put the app into satellite mode and in the search box type the following GIS address, exactly as it appears here, including all commas, periods, and spaces:

45 52 28.85 N, 6 55 35.26 E

You may have to zoom in or out to place the area in context and to see the full breadth of the image.

This is the Mer de Glace (Sea of Ice) Glacier in the French Alps near the town of Chamonix (fig. 1.1). This textbook *valley glacier* is nestled in cirques (the ridges that form a sort of mountain armchair) of very high mountains where it snows a lot and ice forms through prolonged snow accumulation.

The Mer de Glace Glacier is about 10 kilometers long (that's about 6 miles) as the crow flies, and about 2 kilometers wide (1.6 miles) at its origin (the top of the mountain) and depending on how you measure the length, it winds about 12 kilometers (7.5 miles) down the mountain—the real length if you had to ski down the glacier. It is about 200 meters deep or about 650 feet (that is the equivalent of a 65-story building).

Figure 1.1 The Mer de Glace Glacier in Chamonix, France.
Source: Creative Commons Zero. GIS: 45 52 28.85 N, 6 55 35.26 E

It flows downward approximately 100 meters per year. It is the longest and largest glacier in France.

And yes, people do ski on the Mer de Glace: one of France's most notorious ski resorts is at Chamonix.

If you're like most people, you've never seen a glacier in person. Why is that so? Well, first of all glaciers are usually in very *remote* places. They are also in places where it's very *cold*, since if it were warmer, glaciers could not survive. The snow that falls in the area would melt away in the spring and summer. Glaciers form because some of the snow that falls in the area survives season after season, and as it accumulates from year to year, a glacier is born.

For a glacier to form *and survive*, daily temperature must average *below* freezing for the year. That makes sense. Remember Frosty the Snowman's tale, he's made of ice, and melts away when the temperature gets too warm. Glaciers are exactly the same. They need cold environments to form, to thrive, and to persist. If the air temperature gets too warm on average, they melt.

Most people on our planet choose warmer climates to live in and to spend their leisure time in. They don't generally like to be where it's always cold, and less so where it's freezing for most of the year. This is why we have generally *not* ventured to the places where glaciers thrive. It's also part of the reason humanity as a whole knows so little about glaciers.

Glaciers are thus in very remote locations, either at the planetary extremes close to the polar ice caps (also very remote and hard to get to) or they are at very *high altitudes*, generally above 4,000 meters (above 13,000 feet) because as you go up mountains, the temperature drops quickly. The tops of very high mountains are generally freezing for most of the year, ergo, that's where glaciers live and prosper.

Humans don't fare very well at such high altitudes. The air is thin, the body is weak, we immediately get headaches when we climb high, and our energy levels are low. We also don't like the cold very much and will get frostbite if we stay for too long in these environments, unless we're clothed appropriately. In sum, we don't generally visit glaciers because they exist in places that are *too remote*, *too cold*, or *too high* for our liking.

If you are one of the relatively few lucky people to have seen a glacier, particularly the very large glaciers in places like Alaska, Patagonia, the Himalayas, or Antarctica, it's safe to say it was probably one of the most exhilarating and unforgettable experiences of your life. If you are not, hopefully by the end of this book, you'll be planning a trip to see a glacier. It is an extremely memorable and for some, a life-changing experience. It was for me!

Simply, the awe that comes from seeing such a massive natural wonder up close, feeling its presence and energy, hearing its thundering cracking noises, and standing before this enormous wonder of nature or walking on its surface is humbling to anyone. And while hopefully you witnessed the glacier environment on a bright sunny day (see the Perito Moreno Glacier in Patagonia, fig. 1.2) , to fully grasp its majestic beauty, you may have also experienced the unforgiving fury of the cold environment glaciers thrive in, sensing your puny mortal vulnerability before the powers of this beast of nature.

There are approximately 200,000 glaciers around the world containing some 726,000 km^2 (280,000 mi^2) of ice and covering about 10% of the Earth's surface. In past eras

Figure 1.2 The Perito Moreno Glacier in Patagonia, Argentina.
Source: Luca Galuzzi. Wikipedia Commons. GIS: 50 28 2.65 S, 73 2 49.06 W

(millions of years ago), glaciers covered up to one third of the Earth's surface (or more). This represents nearly 16 million km² or 6.2 million mi² of the surface of the Earth.

Most glacier ice (about 97%) is in the polar regions (Antarctica and Greenland), very far away from where we carry out our lives day to day.[1] About 3% of glacier ice (about 500,000 km²) is in the form of ice caps and *smaller* glaciers located in high mountainous areas (such as the Rocky Mountains, the European Alps, the Central Andes, the Tian Shan, or the Himalayas, just to name a few).

Glaciers are very important because they basically do three critical things for us and for our environments:

1. They hold our freshwater and keep it from flowing into the sea (avoiding sea level rise).
2. Their extensive white surface area helps reflect solar sunlight, containing global warming.
3. Their cold temperature helps keep their environment, the oceans, and the planet cool.

As such, these polar glaciers are very important to our planetary health, and ultimately they are critical to our everyday lives. But there are also other glaciers that are extremely important to our daily existence, though they may account for a lower volume of ice. These are *mountain* glaciers. Almost every continent has mountain glaciers. We can find glaciers nearly at every latitude of the planet, including at the equator if you go high up on the mountains of these warmer middle latitudes of the Earth (such as in Africa, South America, Indonesia, etc.).

Definition

So what exactly is a *glacier*? Only by understanding what a glacier is and why it is important can we begin to understand its role in our planetary environment and why glacier melt really does matter.

Let's go first to my teenage kids' reference source, Google:[2]

Glacier: A slowly moving mass or river of ice formed by the accumulation and compaction of snow on mountains or near the poles.

In layman's terms,

A glacier is a body of ice that survives year to year, that moves because it is heavy, that accumulates new snow each winter converting it to ice over many months, and that melts off much of that snow and ice in the summer providing our ecosystems with valuable freshwater. If the glacier accumulates more snow than it melts off each year, the glacier grows. If however it melts off more than it acquires, it dwindles in size. And if melt and accumulation are even, the glacier maintains its size in ecological balance (a rare occurrence).

Let's look at a few more technical definitions:

Wikipedia:[3]

A **glacier** (US: /ˈɡleɪʃər/ or UK: /ˈɡlæsiər/) is a persistent body of dense ice that is constantly moving under its own weight; it forms where the accumulation of snow exceeds its ablation (melting and sublimation) over many years, often centuries. Glaciers slowly deform and flow due to stresses induced by their weight, creating crevasses, seracs, and other distinguishing features. They also abrade rock and debris from their substrate to create landforms such as cirques and moraines. Glaciers form only on land and are distinct from the much thinner sea ice and lake ice that form on the surface of bodies of water.

Another more technical definition is offered by the USGS, one of the world's foremost authorities on glaciers:[4]

A glacier is a large, perennial accumulation of crystalline ice, snow, rock, sediment, and often liquid water that originates on land and moves down slope under the influence of its own weight and gravity. Typically, glaciers exist and may even form in areas where:

1. mean annual temperatures are close to the freezing point
2. winter precipitation produces significant accumulation of snow
3. temperatures throughout the rest of the year do not result in the complete loss of the previous winter's snow accumulation.

Over multiple decades this continuing accumulation of snow results in the presence of a large enough mass of snow for the metamorphism from snow to glacier ice to begin. Glaciers are classified by their size (i.e., ice sheet, ice cap, valley glacier, cirque glacier), location, and thermal regime (i.e., polar vs. temperate). Glaciers are sensitive indicators of changing climate.

Over the course of the last decade, we've also seen *legislative* definitions of glaciers, which vary somewhat from the scientific definitions, including in some important ways. These are new, because in the last decade the world's very first glacier law was passed. You can read more about that in my book *Glaciers: The Politics of Ice*. The nuances may seem trivial, and while some scientists do not agree with or find problems with the definitions that legislative representatives have produced to define glaciers, there are some very well-founded reasons for the divergence between science and public policy.

Scientists generally aim at defining and registering their observations from a precise technical standpoint. The size or very specific location of a glacier may be a key element that scientists determine helps them categorize glaciers. So for example, the arbitrary decision to define a glacier as a body of ice that is 1 hectare (2.5 acres) or larger offers a useful limitation for scientists that are inventorying glaciers, in a world where smaller bodies of ice may simply be too small and too numerous to be able to realistically locate and count, even though the sum of many small glaciers may actually be equivalent to, or even larger in volume than, some of the largest glaciers in a given region.

A scientist may also consider that the size of a glacier will affect its deformation, and therefore the arbitrary cut-off at a hectare in the definition makes sense because at that size deformation may occur more readily than say at half a hectare. A scientist may say (and I've heard it many times) that a tiny perennial ice mass the size of a football field is not actually a glacier because it is not big enough to display movement and deformation, as most glaciers do. From the legislator's or public official's perspective, however, the functional role of that tiny glacier, to provide water or regulate waterflow for example, may be much more important than its size.

A public official working for an environmental protection agency, for example, may be focusing on different glacier dynamics than an academic, such as the relevance of a glacier to water storage value or how a glacier provides water to a given ecosystem or the glacier's ability to ration water flow to a downstream basin and ecosystem. With such a view in mind, the totality of very small glaciers (which scientists may prefer to call glacierets or ice patches) may have a very relevant input in the hydrology of an ecosystem even though some scientists may resist technically calling them glaciers.

While a scientist may not want to call a glacieret a glacier, a legislative representative may insist on that because a glacieret or the sum of all glacierets in a given mountain region is just as important, or more important, in hydrological value. It is for reasons such as these that the world's very first Glacier Protection Act, adopted in Argentina in 2010, defines a glacier as:

All perennial stable or slowly-flowing ice mass, with our without interstitial water, formed by the recrystallization of snow, located in different ecosystems, whatever its form, dimension and state of conservation.[5]

As is clear from these various definitions, scientific or political, there are some key elements that define a glacier, that recur from definition to definition, while others vary somewhat or considerably. I summarize here some of the key defining characteristics of a glacier and the environments in which they exist.

1. *Glaciers are formed by snow that has accumulated in a specific place*, and over time, turned into ice. It takes several weeks and even months to turn snow to ice through what's called the "diagenetic process." Over many years, that ice is compacted into dense glacier ice. The accumulated ice must survive year to year *and for at least two years* to be considered glacier ice. We refer to this as "perennial ice," much like we refer to the perennial leaves of a tree as "evergreen." While some large glaciers may form over hundreds or thousands of years, some smaller glaciers may form in just a few years or decades, survive for similarly short periods such as years or decades, or even a few centuries, and then vanish. For those periods, their ecosystems functions may be very important.

2. *Glaciers move.* Either they creep over land or they have internal structural movements. A glacier on a mountain slope may move downhill, sometimes in daily advances of many feet, or with little movement of just inches over an entire year. They move because they are on an incline, because they are heavy, and because their base is wet from melting snow, lubricating the floor, making it slippery, and assisting glacier flow. They are also moving because as more snow falls on their higher areas, the excess weight deposited on the glacier surface deforms the ice mass and pushes the glacier downhill. The Aletsche Glacier in Switzerland is a typical glacier that flows downhill. You can see it on your smart phone by typing the following address in the search box of your map app (placed in satellite mode). Remember to zoom in or out once you go to the site, and type all commas, spaces, and periods as follows:

    46 29 19.33 N, 8 3 11.08 E

    Certain landscapes may also *contain* glaciers and restrict their movement. Consider for example a large ice body formed in the crater of a dormant volcano. This is also considered a glacier. Such a glacier (which may survive for many years, even centuries, as long as there is no volcanic eruption) may still have internal structural movement caused by melt and reaccumulation of snow and ice. You can see such a glacier at the Teardrop Pool on South Sister Mountain in Oregon, right on your smart phone. Type the following GIS address exactly at it is here (don't forget to zoom in or out as needed!):

    44 06 05.73 N, 121 46 16.25 W

3. *Glaciers melt* during warmer months or years providing water to our environment. When they melt they get smaller (scientists call that *ablation*) and then *glaciers grow* again during colder months or years, as they accumulate snow and ice. You probably didn't realize that even if you've never been to a glacier, you can easily distinguish old ice from new ice on a glacier body, and quickly identify the accumulation and ablation areas on the glacier using commonly available internet software like Google Earth or right on your phone. Type the following GIS address into your maps app (remember to be in satellite mode):

    16 06 38.91 S, 68 17 36.86 W

Zoom in and you'll see exactly what I am talking about at this glacier-covered mountainside in the Bolivian Andes. The areas closer to the top, or the *higher* points of the visible glaciers—*the accumulation zone*—are often whiter (new snow and ice) than the *lower* zones—*the ablation zone*—where older, dirtier ice is located (it is dirtier from the accumulation of dust and other atmospheric contamination that is deposited on the surface of the glacier over the years). Over time, glaciers may grow because accumulation of snow is greater than snowmelt, maintain their size (this is called glacier equilibrium), or they may decrease their size because snowmelt is greater than snow accumulation. Climate change is causing widespread reductions in glacier size due to warmer climates resulting in greater melt than accumulation. As glacier melt increases beyond sustainable levels, you will notice that the whiter accumulation zones begin to dwindle, leaving larger grayer areas below in the ablation zone (see fig. 1.3).

4. Although glaciers can be of any size or shape, scientists generally prefer to call a body of ice a glacier if it is at least 10,000 meters^2 (just over 100,000 feet^2) in size (or about the size of a football field, i.e., 2.5 acres). Nonetheless, technically, glaciers *can* be smaller. Smaller glaciers are sometimes called glacierets or ice patches. It is important to note that even small glaciers can contain lots of ice, and collectively on a mountain, or throughout a mountain range, the total ice

Figure 1.3 The accumulation zone (white new snow/ice areas) and ablation or melt zones (dark old snow/ice areas) are easily distinguishable on high mountain glaciers in the Bolivian Andes.

Source: Google Earth. GIS: 16 06 38.91 S, 68 17 36.86 W

in the totality of all glacierets may be larger than the largest glaciers in a given area. Some studies have shown in fact that in terms of water provision, smaller glaciers collectively can provide more meltwater to the environment than larger glaciers. Keep in mind that a large glacier has a lot of ice deep in its core, while a smaller glacier has relatively more ice on its surface that is exposed to the warmer air, making its melt contribution potentially greater.

Take a look at the high mountain environment on the border between Argentina and Chile on your maps app at (zoom in and fly around the area to see many small glaciers, *or glacierets*):

27 7 20.58 S, 68 32 45.42 W

As you zoom in to look at the small glaciers in the image, you'll notice numerous small lakes in the vicinity. These lakes are being filled constantly by glacier water that derives from the melt of small glacierets.

5. Glaciers form where the environment is conducive to their sustainability. Since glaciers must constantly receive snowfall, and because they depend on cold temperatures, their local environment and its regularity over time is key to their survival. Unlike a tree that might survive alternating temperature environments or snowfall patterns, or even forest fires, glaciers critically depend on the sustainability and regularity of their ecosystems to survive. Because the scientific world doesn't have specific terminology for this ecosystem specific to a glacier, I invented the term "glaciosystem." (For more information about the glaciosystem, see the link in the footnote[6] and the definition in the text box—we'll talk more about this later.) For instance, snow may accumulate more on one side of a mountain than another. Wind patterns may cause snow to drift through the air to a specific spot. If you live in an area where it snows in winter, you've probably noticed that when it snows overnight, snow may accumulate in a particular part of your yard more so than in another area. This may be because of prevalent wind patterns that take snow to a specific spot. Also, areas that receive less sunlight are colder than areas that receive direct sunlight. So with an equal amount of snow falling throughout a given area, the snow that falls in the places with more shade during the day is likely to outlast snow that melts quickly in sunny spots. If new snow falls again on these shaded areas, over time it will accumulate. During the day, sunshine will melt the snow in less shady areas, meaning that over several days or weeks, you may see ice forming in shady areas, while sunny areas are cleared daily. If that environment were to persist over long periods of time, a glacier could form. The shape of a mountain's surface terrain may also help accumulate and trap snow in specific spots. A mountain slope may be too inclined to allow for snow or heavier accumulated ice to stay on the incline, while in other spots, the incline may be just right to allow for snow accumulation and for the diagenetic process to begin (the conversion of snow to ice). These are all factors that contribute to the ideal glaciosystem for a glacier to thrive and survive. If you change any one element of the glaciosystem, you may affect the glacier's health. Warm the air, cut out a piece of the terrain, diminish snowfall, or change wind patterns and you could choke the glacier's lifeline. I've met glacier makers—yes, there are people that actually *make* glaciers—and they understand the idea of the glaciosystem. In fact, that is what they create to

make glaciers, a system that is conducive to glacier formation and sustainability. They might want to make glaciers so that they can have extra water reserves in the spring or summer and they can do this by modifying natural structures in the environment to encourage snow accumulation at a given spot, and voilà, they have made artificial glaciers where none formed before! When we get into policy discussions about protecting glaciers, a term like the glaciosystem becomes a key concept to consider and utilize to lay out the rules for glacier protection! Take a look at a beautiful glaciosystem in the southern region of New Zealand, at:

44 27 18.7 S, 168 36 22.2 E (you'll probably have to zoom out of this image to get a full view!)

You can distinguish the mountain features that help contain snow that is eventually converted to ice to help glacier formation over the many cyclical years of snowfall and ice accumulation in winter. You can follow the glacier flow and see the clear relationship of the glacier to the ecosystem below, forming rivers and lakes all fed by glacier meltwater.

6. Glaciers, and more generally, frozen mountain areas that are usually a little below visibly glaciated areas, may have a lot of rock debris. Combined with cyclically freezing and thawing temperatures, this may cause active interaction between ice, water, and rock, forming what are called "rock glaciers" or "debris-covered glaciers." We'll talk more about this in Chapter 8, but basically these are subsurface glaciers with high rock content that covers them to the point of making them invisible. You can see a typical rock glacier in the Tian Shan Mountains of China at:

43 30 14.57 N, 86 5 25.07 E

Adjust your phone map app to get a perspective of the mountainside. The rock glacier appears as flowing rock debris that looks like thick honey or molten lava oozing down the mountainside in a tongue-shaped mass of rock. This is not just rocks, it is actually an ice-rich rock glacier that is over 60 meters thick (180 feet) and over 2 kilometers (1.2 miles) long! There is lots of ice in there!

The Glaciosystem

The *glaciosystem* (or *glacier ecosystem*) is the glacier and its surrounding ecosystem that influences its constitution and composition, with respect to its water and ice accumulation and ablation, determining its biological process, its natural evolution during its periods of charge and discharge, and which, if affected, could impact or cause the alteration of the glacier and/or impact the ecosystem in which it exists.

The glaciosystem (or glacier ecosystem) includes elements such as:

Solids: Geological/rock formations surrounding the glacier, whose characteristics and orientation influence in the accumulation of snow, the valleys through which the glacier flows, walls, mountainsides, and the slope on which the glacier advances, rock debris and other natural materials in its vicinity or in its ice, the moraines formed and accumulated by its advancement, among others;

Biological: Flora and fauna and other biological organisms in its immediate sur-roundings, underneath, beneath, and inside of its ice;

Water, Snow, and Ice: Snow that accumulates in the glacier through precipita-tion, water that flows on the surface and inside and underneath the glacier, ice with varying densities and in different stages of compacting, other glaciers that unite with the glacier from higher water and ice basins, other glaciers to which the gla-cier unites, frozen grounds (permafrost) in the periglacial environment, natural or artificial lakes (dams) formed and nourished (even if only partially) by the glacier, natural or artificial meltwater at the foot of the glacier;

Air, Light, Shade and Atmosphere: The air surrounding the glacier, the amount of light and shade received, and the atmosphere in the zone of impact that can be affected by artificial changes in the topography that alter the natural wind patterns and shade on the glacier's surface, contributing to the natural accumulation of water and snow on the glacier, by contamination of the air with particulates that are de-posited on the glacier, and that contribute to the natural evolution of the glacier.

The glaciosystem (or the glacier ecosystem) can extend to zones including:

a)   in all directions surrounding the glacier;
b)   snow and ice on the glacier and above or to the side of the glacier as well as water immediately above, to the side of, and below the glacier;
c)   to the side of and on the valleys through which the glacier flows;
d)   in the immediate proximity to or at a significant distance from the glacier, depending on the specific case and on the relevance of an eventual impact in the ecosystem of the glacier;

Human populations (rural and urban), agriculture activity, and industries that are located in the vicinity of the glacier and that can be directly affected by the changes of the mass of the glacier and on the accumulation and ablation of snow, might depend directly or indirectly on the glacier and its glaciosystem.

The health of the glacier and its glaciosystem is evaluated by measuring and monitoring the evolution of the following variables related to the glacier:

Accumulation and Ablation of snow/ice
Line of Equilibrium
Mass Balance
Energy Balance
Temperature
Caloric Balance
Water Flow
Albedo
Impurities/Contamination
Air/Atmosphere in the Vicinity

Some of the newest glaciosystems we've discovered, thanks to the travels of NASA's New Horizons space probe, believe it or not are on Pluto, the furthest planet from the sun (fig. 1.4).[7] In 2015 the probe revealed spectacular images of mountain ranges, ice

Figure 1.4 Glaciers on Pluto.
Source: NASA

plains, glaciers, *and an atmosphere*. A giant ice plain the size of Texas, now named Sputnik Planitia, was revealed in the shape of a massive heart. Ninety-eight percent of Pluto's surface is comprised of nitrogen ice with an average temperature of −229°C (−380°F). Nitrogen ice this cold acts like water ice and that means it can flow, as glaciers do. Glaciers are visible and appear to flow around the heart-shaped plain, through valleys in the Plutonian mountains. There are even processes of sublimation (evaporation) visible on the ice plain, that form ice shapes called penitents, spike-like ice features that also form in arid glaciosystems on Earth.

In Chapter 8 we will examine a different type of glacier, an invisible subsurface rock glacier that has also been found in one of the celestial bodies, in this case on Mars. Later in this chapter, I will also mention that glacier ice has incredibly been found on Mercury of all places. Even there, on the planet closest to the sun, where the temperature is phenomenally high, a glaciosystem can exist and ice can accumulate to form a glacier. We will see shortly how Mercury's glaciers survive despite the hot temperature of the Mercurial surface, and like for many glaciosystems, *shade* is key.

This goes to show that if the essential elements of a *glaciosystem* are present— precipitation, ice formation, persistent cold weather, mountain niches, and gravity— the likelihood of finding glaciers is high, anywhere!

How a Glacier Forms

Glaciers are basically formed when snow accumulates in areas where it can survive the warmer days, weeks, and months of the year. Anyone that lives in an area that gets lots of snow in the winter (places like Chicago, Denver, New York, Montreal, parts of

Scandinavia or the French Alps, Patagonia, the Peruvian Andes, or at the base of the Himalayas) knows that most of the snow that falls in the wintertime melts away in the spring or early summer. However, in particularly high elevations (mostly in tall mountain ranges), if the temperature is low for much of the year, the accumulated snow may be able to survive the warmer months and persist on to the next winter. Glaciers can also form in much lower elevations but only at extreme polar latitudes, since at or near the poles, in places like Alaska or Patagonia, it is cold for most of the year.

Glacier formation in high mountain environments generally occurs in enclosed mountain ridges where snow might get trapped or where the snow may be located on shaded areas of mountainsides where it remains protected from the sun (particularly north-facing mountainsides in the northern hemisphere and south-facing mountainsides in the southern hemisphere). Remember that the sun appears to us facing the northern side of our skies when we are in the southern hemisphere and to the southern side of our skies when we are in the northern hemisphere. South-facing mountainsides in the southern hemisphere and north-facing mountainsides in the northern hemisphere, hence receive less sun during the day, creating ideal niches for glacier formation.

If lots of snow falls in these places, and the temperature stays low for much of the year, with average temperatures at or below 0°C (32°F), glacier ice may form.

Through a process called *diagenesis*, snow is compacted into ice over days, weeks, months, and years. As fluffy snowflakes fall and accumulate and start sticking to each other, the ice crystals compact onto themselves, and ice starts to form nodules that coalesce, eventually becoming more and more dense. Snow has turned to ice.

By the end of the winter season, the fluffy snow has become hardened ice. Keep in mind, fluffy snow is only about 6% as dense as water, whereas hardened ice is likely to be more than 90% as dense. If you look at the diagenesis process depicted in figure 1.5, the snowflake on the left is mostly air, while the small dense little ice ball on the right has barely any air in it at all, making it much denser than the snowflake. Keep in mind the small ball at the right has just as much ice as the snowflake on the left, it is simply much denser. You may have experienced the difference between dense ice and soft ice if you've been hit with a snowball. Being hit by the former is no fun!

Whether or not a glacier forms is determined by the warmth of the spring and the heat of the summer and whether or not they melt *all* of the snow (some of which has turned into ice) that fell in the winter. If they do not, then the ice survives and will be recharged with fresh, new snow the next winter and by the end of the cycle, the ice that survived will capture still more ice. Repeat that cycle over several years, several decades, or several centuries, and that's how you create a glacier!

Figure 1.5 Snow to ice diagenesis.
Source: Geoestudios

I most like Mariana Gosnell's description of the snow to ice process in her book *Ice: The Nature, the History and the Uses of an Astonishing Substance*:

> Snow changes slowly into ice, the colder the setting, the slower the change. First, new-fallen snow crystals settle within a fluffy snowpack, moving closer to each other and becoming rounded as molecules evaporate off their sharp points and condense in their hollows (where there are more molecules for them to join up with). These small, rounded crystals, which once they have lasted through a summer melt season, are called *névé* in French and *firn* in German, bond together, or sinter, under the weight of more snow above them as well as from the continued migration of vapor molecules, until eventually all air channels are blocked. When there remain in the pack only individual air bubbles, unconnected to each other, the permeable snow has become impermeable ice.[8]

Christopher White, author of *The Melting Point*, describes a mind-boggling assessment of the number of crystals that form ice to put the diagenesis process into a volumetric perspective:

> I pick up a dab of snow on my fingertip and inspect it. . . . the snow crystals are hardly bigger than a grain of salt. It's tempting to calculate how many snowflakes create this crevasse, that headwall, or the entire glacier. Untold zillions, I believe. The zeros are lost to the calculator and to the brain. But I try. A cubic foot of snow may contain up to ten million snowflakes, but once it (and more snow) compresses and mutates to a cubic foot of ice, the count is likely in the billions. A glacier may contain tens of millions of cubic feet of ice or more. All glaciers on Earth—nearly 400,000 of them—descend from snowflakes. The mind reels. Too many zeros. I give up.[9]

The *diagenesis process* combined with the sustainability of cyclical snowfall and sustained cold weather (yearly average of less than 0°C/32°F) as well as the natural geological features that are conducive to the accumulation of snow (a crevasse, a niche in the mountain, etc., and just the right incline so the snow doesn't slip down the mountain) is what creates glaciers. If more snow falls (accumulation) on the glacier than melts away each year (what glaciologists call *ablation*), the glacier will grow, if less snow falls than melts away the glacier will shrink, and if snow fall and snow melt are equal, the glacier will remain the same size (glacier equilibrium).

As this perennial ice accumulates, it gains weight and volume. And particularly if the ice is on a mountain slope (which is common in high mountain environments), it will reach a critical size where, because of its own weight, it may begin to slip down the mountainside.

This is aided by the fact that the underneath portion of the glacier where it touches the ground is lubricated by its own meltwater. So you can imagine this growing mass of ice permanently falling down the mountain toward warmer areas (where temperatures are higher than the freezing point). As the lower portion of the glacier has reached warmer environments further down the mountain (remember, as you go down the mountain, the temperature rapidly increases), the ice in that lower section

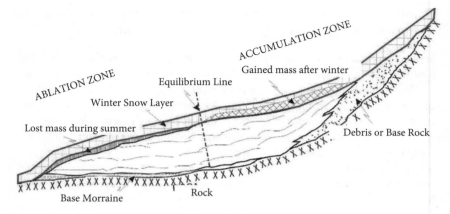

Figure 1.6 The composition of a glacier.
Source: Geoestudios

of the glacier begins to melt because the ambient temperature can no longer sustain the ice year to year.

All the while, at the higher elevations, the cold environment and recurring snow-fall continue to feed the glacier so that any lost portion in the lower section is replenished each year at the higher section. You can think of this glacier as a massive river of churning ice, flowing down the mountainside. It melts off once it reaches the part of the mountain where the average temperature surpasses the melting point (0°C/32°F) and recharges at the top of the mountain where on average it is very cold and substantially lower than the melting point. .

The lower sections of the glacier (that are more rapidly melting) are called the ablation zone (this is the melt zone) of the glacier and the higher sections of the glacier (where the glacier is colder and is recharging with new snow) are called the accumulation zone. In the middle is what we refer to as the equilibrium line. A cross-section of a typical glacier is shown in figure 1.6. The picture of the glacier in Bolivia (fig. 1.3) or the one in the Peruvian Sierra Blanca Mountains (fig. 1.7) show these distinct zones on the glacier (fig. 1.6) surface, which can be distinguished by lower and darker areas of the glaciers (the ablation zone) and higher and whiter areas of the glaciers (the accumulation zone).

As our climate warms, mountains with very cold ecosystems are also warming. And while colder temperatures may still occur at the higher parts of the mountain, the warmer *average* temperatures are slowly creeping up hill. Maybe 100 years ago, the average freezing point of the mountain was halfway up the slope, now with warmer average ambient air temperatures, it is three-quarters of the way up, which means ice cannot survive beneath that point. Since the surviving ice cannot go higher than the highest point on the mountain, climate change is effectively crowding the glaciers higher and higher up the mountainside. The perennial ice recedes further and further up the slope, only able to survive where average yearly temperatures remain below freezing. Once the ambient average yearly temperature moves above the freezing point in all areas where glaciers form, all glacier ice perishes.

Figure 1.7 A cirque glacier in the Peruvian Sierra Blanca.
Source: Google Earth. GIS: 8 57 54.00 S, 77 37 47.23 W

Types of Glaciers

We can see that there are many types of glaciers of all forms and sizes depending on the glaciosystems in which they were conceived.

The larger glaciated areas of the world, including the Antarctic region and Greenland, have enormous glaciosystems, which produce continental-sized ice sheets and ice fields. It's when we get to the mountain glaciers and smaller glaciosystems, that the forms, types, and sizes really begin to differ. Basically the form or *type* of glacier will depend on the geological features and environmental conditions of the glaciosystem that led to the glacier's creation. Some of the most typical forms of glaciers (or glacier ice) found are listed here (you can see all of them on your smartphone—remember to zoom in or out as necessary):

- Ice Sheets: Large unconstrained ice flow (Antarctica/Greenland)
                Greenland: 73 04 01.66 N, 42 51 10.98 W
                Antarctica: 73 10 29.75 S, 111 08 46.38 E
- Ice Shelves: Ice floating and protruding from land onto oceans
                Antarctica (Larsen): 65 35 17.65 S, 60 05 29.51 W
                (notice the massive cracks in the ice to the west and east of
                this point)

- Valley: Ice flows down the mountain into a valley
                     Aletsch Glacier (Switzerland): 46 28 42.84 N, 8 03 52.54 E
- Cirque: Ice trapped in an *armchair-form* geological formation on
                     top of a mountain
- Piedmont Lobes: Outflow of valley glaciers in lowland areas
                     Axel Heiberg Island, Canada: 78 27 53.48 N,
                     90 50 33.58 W
- Niche: Small ice masses trapped in crevasses on mountainsides
                     Glaciar du Mont Perdut, Pyrenees, Spain: 42 40 47.58 N,
                     0 02 12.35 E
- Glacierets: Small perennial ice masses formed by accumulation of
                     drift snow Mt. Lebanon, Lebanon: 34 17 55.63 N,
                     36 06 55.70 E

The Ice Ages and Interglacial Periods

Ice ages, eons, eras, periods, epochs, ages, and other subdivisions used to define moments of historical and geological times can be confusing. Let's put our ice ages in context so that we can understand the cycles that determine whether we go into or come out of an ice age.

The planet's geological history can be divided into very long periods that last tens of millions of years. The chart in figure 1.8 shows that that humans begin walking the Earth at the very end of this geological history, during the *Quaternary Period* (which began 2.58 million years ago). Dinosaurs, for example, lived between 100 and 300 million years ago, during various periods—Cretaceous, Jurassic, Triassic, and Permian.

The Quaternary Period (in which we live now) is divided into two epochs, the Pleistocene, which began 2.58 million years ago and ended 11,700 years ago, giving way to the Holocene (this is the epoch of our current time). Modern human history developed entirely within the Holocene (fig. 1.9).

I mentioned briefly in the introduction that the planet's glaciated surface varies over time. Between 10 and 30% *or more* (even much more) of the surface of the Earth may be glaciated at any given moment. There have been at least 5 *major* ice ages[10] (or *big* ice ages) over the Earth's multi-billion year existence that span these large periods (the Quaternary, the Cenozoic, the Phanerozoic, etc.), during which time glacier ice took over as the dominant paradigm.

The Five Major Glacial Ages (or Big Ice Ages)

Five major ice ages account for about 25% of the Earth's long history. Laura Geggel writes for Life Science about these colossal ice ages:

> The five major ice ages in the paleo record include the Huronian glaciation (2.4 billion to 2.1 billion years ago), the Cryogenian glaciation (720 million to 635 million

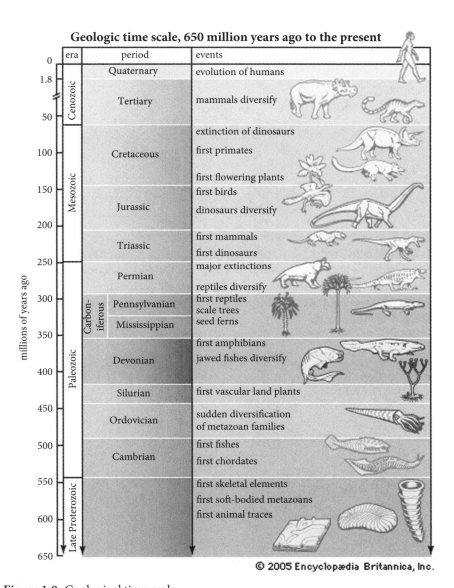

Geologic time scale, 650 million years ago to the present

era	period	events
Cenozoic	Quaternary	evolution of humans
	Tertiary	mammals diversify
Mesozoic	Cretaceous	extinction of dinosaurs first primates first flowering plants
	Jurassic	first birds dinosaurs diversify
	Triassic	first mammals first dinosaurs
Paleozoic / Carboniferous	Permian	major extinctions reptiles diversify
	Pennsylvanian	first reptiles scale trees
	Mississippian	seed ferns
	Devonian	first amphibians jawed fishes diversify
	Silurian	first vascular land plants
	Ordovician	sudden diversification of metazoan families
	Cambrian	first fishes first chordates
Late Proterozoic		first skeletal elements first soft-bodied metazoans first animal traces

millions of years ago: 0, 1.8, 50, 100, 150, 200, 250, 300, 350, 400, 450, 500, 550, 600, 650

© 2005 Encyclopædia Britannica, Inc.

Figure 1.8 Geological time scale.
Note the Quaternary Period at the top of the scale. We live in this period of the Earth's geological history, which began 2.58 million years ago.
Source: Encyclopedia Britannica

Quaternary Period with the Anthropocene Epoch

Eonothem/ Eon	Erathem/ Era	System/ Period	Series/ Epoch	Stage/ Age	millions of years ago
↑ Phanerozoic ↓	↑ Cenozoic ↓	↑ Quaternary	Anthropocene[1]		── 1950 CE
			Holocene		── 0.0117
			Pleistocene	Upper	── 0.126
				Middle	── 0.781
				Calabrian	── 1.806
				Gelasian	2.588

Figure 1.9 The Quaternary Period with the Anthropocene.
In August 2016 the Anthropocene Working Group (AWG), a special body created within the International Commission on Stratigraphy (ICS), recommended that the Anthropocene Epoch be made a formal interval within the International Chronostratigraphic Chart with the year 1950 as its starting point.
Source: Encyclopedia Britannica

years ago), the Andean-Saharan glaciation (450 million to 420 million years ago), the Late Paleozoic ice age (335 million to 260 million years ago) and the Quaternary glaciation (2.7 million years ago to present).

These *large* ice ages can have *smaller* ice ages (called glacials) and warmer periods called interglacials within them. During the beginning of the Quaternary glaciation, from about 2.7 million to 1 million years ago, these cold glacial periods occurred every 41,000 years. However, during the last 800,000 years, huge glacial sheets have appeared less frequently—about every 100,000 years.[11]

Within the Quaternary Period (our present period dating back about 2.5 million years), the amount of ice on the planet has fluctuated significantly. Sometimes we've had a lot of ice, and sometimes we've had a moderate climate with less ice. During this period's first million years or so (the early Pleistocene Epoch), cycles of "lots of ice" occurred approximately every 41,000 years, with about 50–60 minor ice ages within that period. For the last 800,000 years, such cycles occurred approximately every 100,000 years, and there were about 10 minor ice ages during this latest period until the present. The last time we had one of those "lots of ice" moments was about 11,700 years ago, which is when the Pleistocene Epoch came to an end; after that, things began to melt quickly.

The graph in figure 1.10 shows the high and low points of the various ice ages and interglacial periods. The high points denote high average global temperature so these would be "less ice moments" or interglacial periods. The low points on the graph denote low temperatures so these would be "lots of ice" moments. The last deglaciation of the planet started about 11,700 years ago (far right on the graph), at the beginning

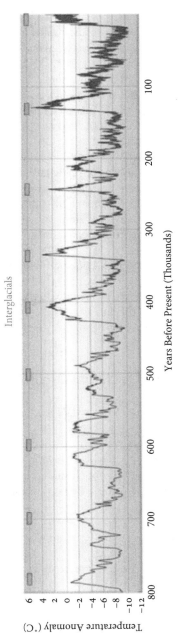

Figure 1.10 Interglacial periods occurring throughout the Pleistocene Epoch and up to the Holocene Epoch.

Source: NASA, Climate Science Investigations

of what we call the Holocene Epoch, which we are living in now. At some time in the future, the cycle will begin again, and we'll move into another "lots of ice" moment. Note that after low points on the graph ("lots of ice" moments), temperature increases very rapidly (the graph line becomes steep), this is because ice ages come to an end very quickly, with rapid ice melt. In other words, it takes a long time for an ice age to develop, and a relatively short time for it to come to an end. We're in one of those end-of-ice age moments now.

The Holocene Epoch (which we are now in), hence, is referred to as an *interglacial* period, which began at the end of the last ice age during the present Quaternary Period (at the end of the Pleistocene Epoch). The Holocene offers the planet a much more livable and comfortable climate, than the end of the Pleistocene, when ice covered a much larger portion of the Earth.

Recently, scientists have made three new subdivisions of the Holocene, but we won't get into those.[12] Scientists are also debating whether or not we should actually end the Holocene (as of 1950) and begin calling the present time the *Anthropocene*, an era defined by *human* ("anthropo") impact over our geological time.[13] That's because human-induced climate change is changing the natural geological and atmospheric evolution of Earth. No final decision has yet been reached on this categorization.

I utilize the term "ice age" throughout this book quite liberally, but what I am usually referring to (unless I specifically say "major ice age") are these moments over the past 800,000 years (during the Quaternary Period) when there was *lots of ice* as compared to the in-between years of interglacial periods that have occurred over the past 800,000 years where there was *less ice*.

The Earth's Glaciation Oscillation

Eccentricity (orbit shape), Obliquity (tilt), and Axial Precession (wobble)[14]

We understand the origins of ice ages through the theory of an astronomer named Milutin Milankovitch, known appropriately as "the Milankovitch Cycles."[15] The key determinant of ice ages has to do with how much sunlight the Earth, and particularly its northern hemisphere, receives during the summer months. That sunlight varies according to various factors including:

- the way the Earth orbits around the sun (on a circular vs. an elliptical path)
- the tilt of the Earth's axis
- the wobble of the axis (whether the axis faces to or away from the sun).

As the Earth changes its orbital path and its tilt, the amount of summer sunlight received in the northern hemisphere also varies, by plus or minus 15%. Just how much of this sunlight we're getting determines how cold it is throughout the Earth and whether we move into or out of an ice age. A little more or a little less tilt, either brings on the ice or withers it away.

The amount of sunlight received by the Earth is constantly changing. As the Earth orbits around the sun, because of the gravitational pull by large planets, such as Jupiter and Saturn, its orbit becomes more or less elliptical. The Earth's tilt also affects the amount of sunlight received: the tilt cycle occurs by about 23 degrees, and repeats every 41,000 years, while the Earth's orbital path changes cyclically from circular to elliptical approximately every 100,000 years. Both of these variables affect the amount of sunlight received. Finally the Earth also has a wobbling effect, which determines whether the northern hemisphere faces toward or away from the sun. This wobble has a cycle of 26,000 years.

Eccentricity (Orbit)

The shape of the Earth's orbit is called "eccentricity," the tilt of the Earth's axis is called "obliquity," and the direction of the axis (or wobble) is called "precession."

The orbit cycles, as mentioned, occur every 100,000 years, and it is when they are most elliptical that the Earth receives more solar radiation, about 23% more at the Earth's closest approach to the sun than at its further point (see fig. 1.11). We are currently at the closest to a circular path around the sun, when solar radiation is most evenly distributed throughout the orbit.

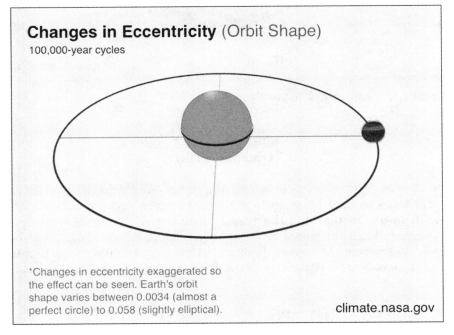

Figure 1.11 Changes in eccentricity (orbit shape).

Source: NASA

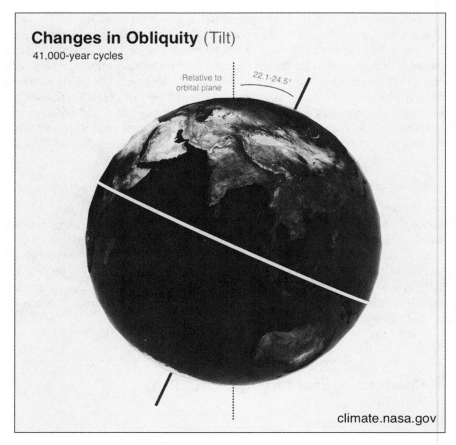

Figure 1.12 Changes in obliquity (tilt).
Source: NASA

Obliquity (Tilt)

The tilt of the Earth on its axis, or its "obliquity," causes seasonal changes. Over the past one million years the tilt has shifted from 22.1 to 24.5 degrees from perpendicular to the planet's orbital plane (fig. 1.12). That means the Earth is more tilted now than in the past. It's warmer when we tilt toward the sun and larger tilts favor deglaciation. We are currently tilted at about 23.4 degrees. The tilt cycles recur every 41,000 years. The last maximum tilt was about 10,700 years ago and we'll reach minimum tilt in about 9,800 years.

Precession (Wobble)

The gravitational force exerted by the moon and sun on our oceans affect the Earth's rotation, shifting the position of the Earth's access as it spins like a wobbling top

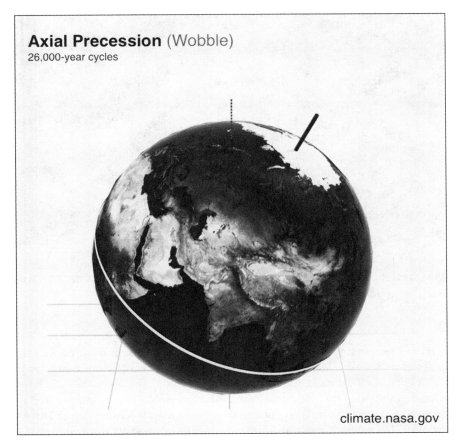

Figure 1.13 Changes in axial precession (wobble).
Source: NASA

relative to the fixed position of the stars (fig. 1.13) . This precession influence affects the intensity of winters and summers in the northern and southern hemispheres. Precession cycles occur approximately every 26,000 years.

All of these factors concerning orbit shape, Earth axis tilt, and wobble, combine in particular ways to determine sunlight positioning on the Earth, influencing ice formation, ice reflectivity, warming, ice melting, and glacier creation. This in turn affects the solar radiation reaching various parts of the Earth in different ways and intensity, which in turn affects temperature and snow patterns on the different continents.[16]

The closer the poles face to the sun, the warmer they get and vice versa. This has a real influence on whether or not ice accumulates at the poles. Even on Mercury, the planet closest to the sun where temperatures are incredibly hot, as high as 400°C (750°F), ice has been found in the polar regions, where because of a deep gorge on the surface, ice hides away in crevasses that receive no sunlight thanks to the tilt of

Figure 1.14 Glaciers on Mercury.
NASA has found glacier ice on Mercury, forming in unexposed niches in crevasses of craters on its pole.
Source: NASA

Mercury's axis relative to the sun (fig. 1.14). Ice can survive in these permanently shaded areas, veritable Mercurial glaciosystems![17] Yes, there is glacier ice on Mercury!

This oscillation causes large structural changes to our climate, greatly cooling or warming the polar caps, the hemispheres, and the Earth's overall climate. Ice ages come and go because these changing variables have a regular relationship and interrelationship that regularly repeat themselves, taking us into and out of ice ages.

Other Factors Influencing Glaciation and Deglaciation

Alterations to the Earth's atmosphere influenced by volcanic eruptions, large meteor impacts to the Earth (which lift impenetrable dust into the atmosphere), as well as the shifting of tectonic plates can also affect global temperatures and can cause the onset or the end of glacial eras.

We noted that in natural cycles, the *very glaciated* vs. *less glaciated* cycles range in periods of about 100,000 years. It's also important to note that it takes much more time to accumulate ice or *create an ice age* (about 80,000 years) and less time to melt the ice away or get out of an ice age (about 20,000 years). We see this represented in the interglacial graph in figure 1.10 with very steep moves from the low points of the graph to the high points over the oscillations between ice ages and warmer climates. This is so because melting rates are three to four times faster than snow accumulation rates.

Over the last 400,000 years these cycles have repeated pretty evenly, says Gabrielle Dreyfus, Senior Scientist at the Institute for Governance and Sustainable Development (IGSD). Gabby, as her friends and colleagues call her, got a PhD studying Antarctic ice, specifically stable isotope geochemistry, applying isotope analysis of oxygen, nitrogen, and argon to reconstruct paleo-indicators from Antarctica. That's a mouthful, even for her father, who preferred to put it in layman's terms and simply call her an *ice-cubologist*, since she spends her time cutting up ice cores taken from the world's oldest and deepest ice!

Gabby reminds us that we are in the dip *between* glaciations, in the dip between a time when we had lots of ice and the next time when it comes back. She shares the same challenge I do when trying to convey these alternating glacial periods to people that simply haven't given any thought to the fact that the world as we see it today is actually *not* like it has been for most of geological history, or for that matter, a mere 20,000 years ago. The issue really is that we're at the low portion of the dip, and we're either still dipping (expect still less ice), or we're about to shoot back up (slowly) into the next glaciation (expect more ice). Gabby is convinced we're not going to ice over any time soon.

Mark Maslin considered the issue of a likely ice age return in an article he wrote for Phys.org in December 2016. He suggests that during the last 2.5 million years we've gone through about 50 ice ages (he's referring to smaller ice ages). He points to the end of the last one, 21,000 years ago, when a two-mile-thick sheet of ice crossed North America from the Pacific to the Atlantic Ocean, at its deepest over Hudson Bay. It reached as far south as New York and Cincinnati. In Europe it reached as far south as Norfolk, England. There were also significant ice sheets covering major portions of South America, New Zealand, Africa, and Australia. There was so much ice in glaciers that the seas were about 125 meters (410 feet) lower than they are now!

Maslin points to a study by Jim Hays, Nick Shackleton, and John Imbrie that tackles the question of just what gets us into and out of an ice age. They note the significant impacts that small changes in insolation can have on climate feedback mechanisms, which are basically recurring loop systems in which for example, a rise in heat causes another reaction that in turn brings on even more heat. An example of such as feedback loop system could be melting ice that reveals darker heat absorbing surfaces. These feedbacks can start or delay an ice age. Ice reflectivity (which we will discuss in Chapter 4) has a big role to play in heating or cooling the Earth, as do the so-called greenhouse gases (GHGs), which in past ages occurred naturally but which are now intensified by anthropogenic activity. One idea suggests that the Earth's entry into the next ice age may have already been delayed. It suggests that long before the industrial revolution, the advent of human agriculture 8,000 years ago had already begun to generate methane emissions that are known to cause global warming. Add to that *current* rates of GHG emissions and we may have delayed the next ice age by up to half a million years, trapping us in *Quaternary Prison!*[18]

It takes a long time to construct an ice age and a short time to melt it away. This is one of the reasons that glacier melt is so dire today. Even if we stop glacier melt by stopping global warming, getting our glaciers back to a healthy state will be a lengthy endeavor, maybe beyond our abilities until the next ice age comes around and who knows how long that will be from now. In this time of human-induced global warming, we are in what some scientists have labeled the "Anthropocene," that is, an era in which humans are affecting *and defining* the planet's geological times (and in consequence, influencing glaciation).

This era is a rather benign time for human existence, during which temperatures on the whole are quite moderate and livable in many parts of the world. This era should, in the best of cases, survive for some considerable time, during which glaciers still have lots of melting to do, before we say goodbye to the non-glaciated era and turn back into an era of deep freeze.

The problem is that a global consensus among scientists now posits that in this *Anthropocene*, we are quickly altering this moderate and livable climate, and accelerating glacier melt speeds at an alarming rate, destabilizing our global climate and creating unlivable conditions that, for many millions of people are no longer tolerable. How much more ice will melt and how quickly will it melt, will depend on current climate trends, both natural and human-induced.

We don't know when the next ice age will arrive; it may not be for many thousands or hundreds of thousands of years. In the meantime, we are contributing to the deterioration of our moderate and very livable planetary climate. What we *do* know, with much certainty, is that the disappearance of glaciers in this process will radically change livability on Earth. Gabby, the ice-cubologist at IGSD, is concerned that we may be causing a blip in the glacial cycles, disturbing it by warming our climate so much, and injecting so much CO_2 that we won't only miss the onramp to the next ice age for a very long time, but that we won't be able to survive sustainably in the new and much hotter and unstable environment that we are creating.

Over the last million years, the Earth has more or less regularly aligned the right formula between its orbital path around the sun, its tilt, and its axis relative to the stars, with just the right atmospheric conditions to help us shift smoothly into and out of glacial phases. If we are not on the right path today because of anthropogenic climate change trends, we could shift into a Hothouse Earth model, one in which it just keeps getting warmer and warmer, says Gabby. She sites Will Steffan and his colleagues who argue that

> self-reinforcing feedbacks could push the Earth System toward a planetary threshold that, if crossed, could prevent stabilization of the climate at intermediate temperature rises and cause continued warming on a "Hothouse Earth" pathway even as human emissions are reduced. Crossing the threshold would lead to a much higher global average temperature than any interglacial in the past 1.2 million years and to sea levels significantly higher than at any time in the Holocene. If the threshold is crossed, the resulting trajectory would likely cause serious disruptions to ecosystems, society, and economies.[19]

That was Hothouse Earth, but then there is *Greenhouse Earth*.[20] Greenhouse Earth is a climate scenario that has actually been quite common when we look at the entire planetary history. During 70% of the past 2.5 billion years, the Earth has been in this state, with CO_2 levels 10–20 times higher than today and no ice anywhere on the planet. The dinosaurs lived during a Greenhouse Earth phase. Reptiles swam in the Arctic seas, and land animals covered the continent at the time. Greenhouse gas levels were high (including vapor and methane) and sea surface temperatures were also high. Several factors can cause Greenhouse Earth, including shifting tectonic plates moving the continents into planetary positions where they simply get too warm to collect and hold on to ice. Extensive volcanic activity can also create Greenhouse Earth conditions by increasing CO_2 and causing warming. Anthropogenic causes, human induced CO_2 emissions to the atmosphere, could be a new cause for the onset of Greenhouse Earth. It's probably something that we're not going to see in our lifetimes, but how soon could it happen? Who knows?

As already noted, glaciations come in cycles, and for about 1 million years those cycles have been pretty constant. For instance, at the end of one of those interglacial periods, about 800,000 years ago, the Earth's temperature was about the same as it is today, and CO_2 concentration was about 240 parts per million, which was similar to the concentration just before the industrial revolution when it was about 280 parts per million. We know what the CO_2 concentration was at the time because we can find traces of it, for example, in deep ice cores taken from the world's oldest glaciers.

To begin an ice age, the common belief is that we need less atmospheric CO_2 More CO_2, conversely, delays the onset. Since the industrial revolution, the level of CO_2 in the atmosphere has been increasing to just over 400 parts per million. That's double what we would expect for the onset of a typical ice age.

One theory suggested by climate scientists in Germany's Potsdam Institute for Climate Impact Research (PIK) is that humanity has become a geological force that is able to suppress the beginning of an ice age precisely because of the changes we've created in our atmosphere. Insolation and CO_2 concentrations in the atmosphere, say the scientists, have determined the last eight glaciation cycles of the Earth's history. A large carbon imbalance, they say, combined with other factors such as the eccentricity of the Earth's orbit, could delay the next glaciation by an unusually large period.[21]

Another theory suggests that the large ice melt of polar glaciers and sea ice now occurring due to accelerated global warming could actually bring about an ice age, by causing a disruption or slowing of the ocean currents (the *Thermohalian Circulation*; more on that in Chapter 7). This rapid ice melt could destabilize the delicate balance of saltwater vs. freshwater in the oceans. This could in turn create an even colder climate at the poles with an even warmer climate along the equator regions, throwing us into another Little Ice Age[22] (a time when glaciers suddenly grew for a few centuries) or even a bigger Medium-Sized Ice Age.[23]

Glacier Distribution Around the World

Every continent of the planet has glaciers, except for Australia, although technically, the Heard and McDonald Islands in the Southern Indian Ocean midway between Australia and Africa, but belonging to Australia, *do have* very impressive and active glaciers.[24]

See GIS: 53 5 41.52 S, 73 29 31.11 E

What's important to understand is that glaciers exist in every continental ecosystem (except continental Australia) and are in many cases key contributors to each region's hydrology.

We've already indicated that 97% of the world's glacier ice is either in Antarctica or in Greenland. That leaves 3% everywhere else. It might seem that the 97% is more important than the 3%, and on some levels this is true, particularly with regards to their relevance to sea level, to the Earth's reflective capacity, or to the influence polar ice has on global ocean temperature.

However, this 3% has enormous hydrological value, in terms of water provision for people and agriculture, much more so than the 97%, since practically no one consumes water from the Earth's larger glaciated surfaces in the polar regions. Meltwater from the large glaciers simply runs off into the oceans, losing their immediate hydrological consumption value. I should note, however, that some people are already contemplating towing pieces of this polar glacier ice, now floating in the oceans as icebergs, to their jurisdictions in times of drought![25]

What we need to realize is that this remaining 3% of glaciers in mountain ranges is likely to end up in your tap or watering the crops that you consume. That's why, from a consumption angle, mountain glaciers are critical to human survival and ecosystem balance.

So where are those glaciers of the 3% located? There are significant volumes of glaciers (including glacial ice in glacierets or in rock glaciers) in:

- North America: United States, Canada, and Mexico (although in Mexico they are almost completely melted)
- South America: Argentina, Bolivia, Chile, Colombia, Ecuador, Peru, Venezuela
- Western Europe: Austria, Iceland, France, Germany, Liechtenstein, Italy, Norway, Spain, Sweden, Switzerland
- Eastern Europe and Western Asia: Albania, Armenia, Azerbaijan, Bulgaria, Georgia, Lebanon, Montenegro, Romania, Serbia, Slovakia, Slovenia, Turkey
- Central Asia: India, Pakistan, Bhutan, Afghanistan, Iran, Tajikistan, Kyrgyzstan, Uzbekistan, Kazakhstan
- Southeast Asia: Myanmar (yes, Southeast Asia also has glaciers!)
- Asia: China, Mongolia, Nepal, Tibet, Japan, Russia
- Africa: Kenya, Tanzania, Uganda
- South Pacific: Papua New Guinea, New Zealand, Australia (Heard and McDonald Islands)

Basically, wherever you have tall mountains, generally above 2,000 or 3,000 meters (6,500–10,000 feet), the chances are you have glaciers. Even at the Earth's equator, where it is generally very hot, if the mountains are tall enough, above 4,000 or 5,000 meters (13,000–16,000 feet) you will find glacier ice or, at least, subsurface permafrost ice. And that's why in Papua New Guinea (at 4,700 meters/15,400 feet), Venezuela (at 4,900 meters/16,000 feet), Ecuador (at 5,700 meters/18,700 feet), and even in Kenya (at 5,000 meters/16,400 feet) you will find glaciers.

Let's turn now to one of the most important reasons why glacier melt and glacier vulnerability is critically important to nearly everyone on the planet.

Sea level rise.

2

The Rising Seas

A recent news item from the Florida Keys reported on a remarkable, *perhaps eerie*, flooding event. At the time I sat to draft this chapter, the floodwaters had not yet dissipated.[1]

> Saltwater has been flooding the low-lying streets of a Key Largo neighborhood for more than 40 days, leaving many residents there trapped unless they can walk or are willing to sacrifice their cars to the nearly foot-high corrosive seawater.
>
> Those who choose to drive through the brine have developed a way they hope will ward off a rotting undercarriage—parking their car over rotating sprinklers when they get home and turning on the hose. They also drive slowly to minimize saltwater damage to the rest of the car.
>
> "It takes me about as long to get out of here than it does to get to where I need to go," said C.J. Ferguson, who lives in the Stillwright Point neighborhood.[2]

Think about this for a moment. You live in your home (which is located in a coastal city), and one day, maybe because of heavy rains, maybe because of a storm, or perhaps just after a hurricane, your neighborhood floods. You deal with the flood, you struggle to keep your home dry, you try to save your furniture, and at least while the flood lasts, you're trapped in your home until the water recedes. And after a few hours or maybe even a day or two if it's really bad, you expect the water will go away soon and that you can get back to your normal life. Except well, maybe, what if the water doesn't leave? What then?

This is the future facing many coastal dwellers around the world as seas swell and floodwaters remain. It reminds me of a fishing trip for *reds* that I did with my son a few years ago to the Louisiana Bayou. We were out on the Gulf of Mexico, in the heart of the Bayou with Jason Catchings, a fishing guide out of Venice Marina. Jason took us out through the flats, through the low-lying watery grasslands and canals, and out into the Gulf. We spent a great day catching some of the biggest reds we've ever seen. At one point we started talking about the Deepwater Horizon oil spill and its impacts to the flora and wildlife in the area, including fish, and talk suddenly shifted to climate change. And then Jason said something startling. We were in about 3 feet of water. There weren't any trees, grass, or any other vegetation anywhere. We could still see the coastline, but we were definitely out on the Gulf. "This used to be land," said Jason. I thought for a moment he was joking. It seemed impossible. But surprisingly he wasn't. "Before Hurricane Katrina in 2005," he continued, "people would hunt here. You could walk around on dry land. The hurricane flooded this land, but the water never left."

Daniel Glick, for National Geographic, writes about Windell Curole of Louisiana, who is worried about rising seas. The coasts are literally sinking, about a meter (3 feet) per century, while the sea is rising. Curole is a seventh-generation Cajun and manager of the South Lafourche Levee District who lives with recurring floods, and says, "We live in a place of almost land, almost water." Rising sea level, sinking land, eroding coasts, and temperamental storms are a fact of life for Curole, as the Gulf of Mexico slowly creeps onto the lands he has grown up on, recounts Daniel Glick from his visit to the Louisiana Bayou.[3]

An article in the journal Nature Communications, which also appeared when I was drafting this chapter and was reported in Climate Central, painted a dire picture for global impacts of sea level rise.

> As a result of heat-trapping pollution from human activities, rising sea levels could within three decades push chronic floods higher than land currently home to 300 million people. By 2100, areas now home to 200 million people could fall permanently below the high tide line. The threat ... could have profound economic and political consequences within the lifetimes of people alive today.[4]

What Climate Central revealed, as an increasingly alarming future scenario in our climate-change defined era, is that we've grossly underestimated the near-term impact of sea level rise because our data on land elevations were wrong. Much of our coastal lands are actually *lower* than we previously thought and that means that *any* sea level rise for those lands on the coastline will have more impact than previously thought.

Over the course of the 21st century, according to Climate Central, global sea levels will rise between 2 and 7 feet, or more (0.6–2.1 meters). So if you currently live at sea level, that means that on high tides or in very inclement weather (with a strong storm surge), water may reach your thighs, or in a worse scenario, reach higher than your head. Based on sea level rise projections out to the year 2050 (that seems far off, but it's actually only 30 years out or about a single generation), the homes where 300 million people live today will be below the projected average flood level line.

By 2100 (that's in about 80 years), the average flood lines will be at the window and places where 200 million people live today will be permanently below the high tide sea level line.[5] As we saw in the Louisiana Bayou on our reds fishing trip, the water flooding the homes of those 200 million people will not recede.

Every once in a while we hear about the mystical and mythical underwater cities of Atlantis. Well, in a few centuries, some of our coastal communities around the world will provide the basis for those mythological stories and lost cities for future generations to find when they go scuba diving!

Glaciers have a lot to do, in fact, *everything* to do with this phenomenon, and understanding this relationship, while difficult to fathom for some, is critical to understanding the Earth's constantly changing hydrological environment (see fig. 2.1).

The IPCC published its most recent report on oceans and the cryosphere in 2019 stating that the average level of the sea is both rising and that this rise is accelerating, and the cause ... well, it's the sum of glacier and ice sheet melt.[6]

The other bit of information that followed this assertion suggests also that in past geological eras, global warming levels of about 0.5 to 1°C higher, and in some cases up to 2–4°C higher,[7] caused even higher sea levels than we have today. Remember,

Figure 2.1 Water gushes forth from the mouth of the Schlatenkees Glacier in Austria.
Source: SehLax, Wikipedia Commons. GIS 47 06 47.89 N, 12 23 48.81 E

the Earth cyclically enters and exits glacier eras. Ice melt flows into the oceans, raising sea level during non-glaciated times, and ocean levels fall as ice re-accumulates in the higher latitudes and the polar regions during an ice age (during glaciation).

The Earth moves naturally from glaciated to non-glaciated eras, that is, from periods when there is extensive glacier coverage to periods when there isn't (see Chapter 1). We don't fully grasp these cyclical global ice-cover changes because these eras are far longer than our common recognizable framework for gauging time. Most people tend to think of life on Earth in terms of years and decades. That's because these are *manageable* periods that we can personally relate to.

We recall what we did in the 1980s or 1990s, the 2000s or early 2010s, but nobody's memory really reaches further back than that in historical terms. Maybe we ponder what life was like 100 years ago, when our grandparents or *great*-grandparents were born and lived. We can do this because we might have some personal references such as antique pictures, or maybe a grandmother has told us of what life was like back then, or because we've seen movies about the period, or we might have read a book about events that took place in another century. However, we have no direct *personal* experience to gauge what life was like before recorded history, so we do not really relate to it.

Planetary eras, however, are a more complicated thing to imagine and understand because they're *far beyond* our comfortable time frame of reference. We don't have books, for example (not even fictional books), about human life in prehistoric times, say 10,000 or 15,000, or 100,000 years ago. This information is all in academic research papers that normal people simply don't read.

Isn't it at least curious that in modern times there are no romance novels about the daily lives of cavemen and cavewomen of the past?[8] What was Lucy's life like? These stories are not on the New York Times Best Seller's List (maybe they should be!). We have novels about life in the 19th century and even during medieval times. There are even novels about the Greek or Roman Empires before the common era (BCE), but that's about as far back as we go. The only references we might have to planetary eras long before that, before humans walked on Earth, and that common people may see, are a few animated depictions of the end of the ice age (Disney's *Ice Age* and its sequels reach back about 10,000 years), a few sci-fi novels featuring past geological eras, and select television documentaries focused on the Earth's formation or reaching back hundreds of millions of years to the times that dinosaurs walked the surface of the Earth. For most human reference, however, it's a time that is beyond the grasp of our imaginations.

The closest we've come to attempting to portray a large structural shift in a planetary era as we might live it today is the science fiction film *The Day After Tomorrow*, in which a massive cold storm hits the planet, buries a number of big cities under hundreds of feet of snow, and deathly freezing temperatures take over most of the Earth, turning it into something resembling the abrupt arrival of an ice age. But for the most part, in our collective imagination, we simply don't reach that far back to think about what our planet was like 100,000 years ago, and these large geological shifts in temperature and land, when the glaciers thrived or how we move from one geological era to another and the implications of those changes.

Glacier eras occur in periods that are 10,000 times longer than our usual time of reference, that is, in periods of about 100,000 years. Throughout these *geological* eras, the world goes through cycles of cold and warm climates fairly regularly every 100,000 years or so.

At peak cold moments, much of the Earth's ocean water is converted to and trapped in ice at the polar ice caps and in mountain glaciers that cover extensive swaths of land over most continents. These cold eras eventually give way to warming periods, or *interglacial periods*, when these large ice bodies melt and flow back into the oceans. Sea level rises accordingly. Once the peak warm climate is reached, and most *or all* glaciers have melted away and filled the oceans, the cold climate returns and through the conversion of ocean water to snow, the colder climate sucks the water out of the oceans and re-deposits it in the polar ice caps and in glaciers throughout the world. Much of the Earth is then, once again, covered in ice. The polar ice caps and mountain glaciers are thus regenerated, lowering ocean levels.

The Snowball Earth Theory

In a somewhat controversial theory, some scientists suggest that it is likely that in the past the Earth has become almost *completely* covered in ice. Paul Hoffman and

Daniel Schrag postulate in a 1999 scientific paper published at Harvard University that "many lines of evidence support a theory that the entire Earth was ice-covered for long periods 600–700 million years ago, during the Neoproterozoic eon (1,000–543 million years ago), when the first signs of life on Earth appeared." Others, like Joe Kirschvink and David Evans, point to even older periods of full Earth ice coverage, dating back over 2 billion years.[9]

Each of these extreme glacial periods lasted for millions of years and ended violently under extreme greenhouse conditions. Purportedly, these climate shocks triggered the evolution of multicellular animal life, and challenge long-held assumptions regarding the limits of global change.[10]

We refer to these periods of time when the Earth theoretically was *mostly* covered in ice as "Snowball Earth." During such periods, the majority, if not all, of the planet would be completely frozen.

According to the Dartmouth Undergraduate Journal of Science, there were "four ice ages between 750 and 580 million years ago, [that] may have been so severe that the Earth's entire surface, from pole to pole, including the oceans, completely froze over. Once the oceans began to freeze, more sunlight was reflected off the white ice surfaces and cooling was amplified," says Melissa Hage, an environmental scientist at Oxford College of Emory at the University of Georgia.[11] Earlier we spoke of feedback loops that compound warming; this is a feedback loop that compounds cooling.

Paul Hoffman and Daniel Schrag talk about "runaway albedo effect," that is, an accumulating process of sun reflectivity that would suddenly take over once certain thresholds were reached. According to a review of past scientific studies by Mikhail Budyko, at the Leningrad Geophysical Observatory, as the Earth's climate cooled from latitude to latitude, slowly reaching the middle latitudes, once ice formed at 30 degrees north or south of the Equator. That swath of land includes pretty much the entire United States, as far as the northern tip of Africa, and much of Asia in the northern hemisphere, as well as the southern third of Australia, all of New Zealand, most of Argentina and Chile in South America, and half of South Africa. This area is equivalent to half of the Earth's surface. The positive feedback would be so powerful that global temperatures would collapse and yield a completely frozen Earth.[12] Scientists estimate that the average temperature on Earth dropped to −50°C (−58°F) below zero at that time and that the water cycle of the Earth, shut down.[13]

Snowball Earth scenarios would not be fun cycles to go through. Most life on Earth cannot exist under such extremely cold global weather where everything, including the oceans, freezes over. According to one NASA publication, early models show that once ice cover reaches the middle latitudes of the Earth, a positive feedback loop would take hold, leading to even lower temperatures and more ice cover.

"The runaway effect," says Michael Schriber, writing for NASA, "would presumably continue until the entire planet froze over, with even the oceans covered with as much as a kilometer-thick layer of ice."[14] A Snowball Earth cycle could conceivably last about 10 million years. That's one long ice age! There is no widely accepted explanation of how the Earth comes *out* of these Snowball Earth phases, although some point to the possibility of extensive plate tectonics and volcanic activity that thaws the Earth anew, breaking us out of the Snowball Earth's grasp.

In this warming scenario, de-glaciation of Snowball Earth could occur quite rapidly. Raymond Pierrehumbert of the University of Chicago (quoted in Hoffman and Schrag) suggests that the change in albedo resulting from the thaw near the Equator would bring about rapid sea level temperatures of about 50°C at the equatorial regions. This could melt the massive sea ice sheets in a few hundred years.[15]

Could we see another Snowball Earth in the future? Maybe.

Hoffman and Schrag say the following in their 1999 evaluation of such a possibility:

For the last million years, the Earth has been in its coldest state since the Neoproterozoic period. We are now living in a relatively warm episode, some 80,000 years from the next glacial maximum, but some evidence suggests that each successive glaciation over the last several cycles has been getting stronger and stronger. During the most recent glacier event, 20,000 years ago, the deep ocean cooled to near its freezing point, and sea ice reached latitudes as low as 40 to 45 degrees north and south [including about 25% of the northern United States, much of Europe, most of Northern Asia, the southern tip of South America, and New Zealand], still far from the critical threshold needed to plunge the Earth into a snowball state. But could such a state be in our future?

Certainly over time scales of hundreds to thousands of years, we are more concerned with anthropogenic effects on climate, as the Earth heats up in response to emissions of carbon dioxide. But only time will tell where the Earth's climate will drift over millions of years. If the trend of the last million years of Earth's history is continued and if the polar continental "safety switch" were to fail, we may once again experience a global ice catastrophe which would inevitably jolt life in some new direction. Perhaps Robert Frost foresaw this in his poem, "Fire and Ice":[16]

> Some say the world will end in fire,
> Some say in ice.
> From what I've tasted of desire
> I hold with those who favor fire.
> But if it had to perish twice,
> I think I know enough of hate
> To say that for destruction ice
> Is also great
> And would suffice.

While thinking about these crazy frozen environments millions of years ago may seem irrelevant to what we are living through in our lifetimes, they actually can give us some insight into how our planet might survive its current state of carbon disequilibrium.

Linda Sohl of NASA, who also studies the Snowball Earth scenario, looks precisely at a period 3.2 million years ago, when carbon levels in the atmosphere were remarkably close to what they are today, at around 400 parts per million.[17] Seeing and understanding how we went into and came out of these scenarios can provide us with some insight as to how we might deal with climate-change-causing carbon concentrations on Earth today.

More Normal Glaciation Cycles

Going back to the *more normal* freeze and thaw cycles of the Earth in a more recent era, the Pleistocene, most of us probably don't realize that between the *high* (most ice) and *low* (least ice) points of glaciation cycles, sea level differs by more than 600 feet (nearly 200 meters)!

To be clear, at full glacier melt in a glaciation cycle, the ocean is 600 feet higher than at full glaciation, when the water has been sucked out of the ocean to make glaciers!

This is the beginning of one of the fundamental factors of our ignorance of glaciers. Because these cyclical periods are so long, we have no experience with the world at peak glacier melt, or at peak ice coverage. We don't see 600 feet of sea level rise or fall in a human lifetime, in fact, we don't even hear about it across human written history. As a human race, we simply weren't there to see it, or the people that *were* there to see bits of it, say 15,000 or 20,000 years ago, didn't have the means to register what they saw so that we could learn about it. Maybe a few hieroglyphic images in caves might have information about the end of the last ice age, but certainly not enough information to make a memorable impact on modern society.

And while Disney makes a great attempt to try to show us what living at the very end of a glacial era was like (in *Ice Age*), our only modern human reference is a blip in time, close to the end of a melt cycle that has left a few clues to what actually happened. While scientists have a pretty good sense of what went down during these times, as a society we really don't grasp or understand what the world was like before us, and we can't really imagine what's coming next!

The graph in figure 2.2 grossly over-simplifies how our planet moves from eras of *lots of ice* cover to eras of *very little ice* cover and how the climate temperature and the

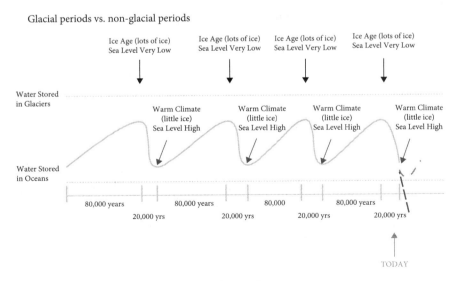

Figure 2.2 Glacial periods vs. nonglacial periods.
Source: JDTaillant

seas rise and fall with these 100,000-year cycles. Basically we're getting 80,000 years of cooling, and then periods of 20,000 years of warming.

Figure 2.3 depicts a graph that most climate experts are looking at closely. You may have seen it already. What's notable about the graph is that when global temperatures rise, CO_2 level rises, and sea level rises correspondingly. It's pretty much a direct relationship as can be seen in the graph. All the peaks and low points over a 400,000-year period correspond perfectly.

That means that as the climate warms further, and as CO_2 levels continue to rise, we can expect sea level to rise in response. It is a direct relationship. What's extremely troubling is the line in the middle, showing CO_2 levels. As is evident, CO_2 follows a natural fluctuating pattern throughout geological history, for hundreds of thousands of years. And then our period at the very far right hits and CO_2 goes through the roof into the graph above. You may recall former Vice President Al Gore climbing onto a mechanical lift while delivering his PowerPoint presentation "An Inconvenient Truth", reaching further and further up the screen to follow the CO_2 level increase in recent times.[18] It's a very tangible depiction of just what is happening with CO_2 levels. If anyone still doubts anthropogenic causes of CO_2 emissions, this graph fully debunks the idea. The question remains however, what happens to the other graphs (global temperatures and sea level) as the CO_2 line breaks out of the normal fluctuation?

We can learn from the past, say Chris Fogwill, Chris Turney, and Zoe Thomas, by looking at what happened about 129,000 years ago. During the last interglacial period,

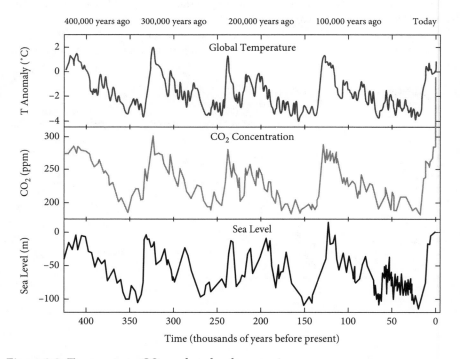

Figure 2.3 Temperature, CO_2, and sea level comparison.

Source: Englander from Hansen and Sato.

129,000–116,000 years ago, the melting of the West Antarctic Ice Sheet brought on sea level rise—more than three meters (10 feet) of average global sea-level rise. And, they say, it took less than 2°C of ocean warming to make it happen. They also point to the rapid melting of the Antarctic ice sheet as a possible culprit as well as the melting of methane trapped in the ice (more on that in Chapter 5).[19]

Today, glaciers occupy about 10% of the Earth's surface, and hold some 33 million cubic kilometers of fresh water. Millions of years ago, during previous ice ages, glaciers covered a more extensive area of the Earth—approximately one third of the Earth's surface was covered in ice.[20]

At that time, because the evaporation of ocean water transferred to the poles and to mountain glaciers around the world, our seas were much lower—as the water from the ocean was packed in ice. If you've ever been to Yosemite, for example, that wonderful canyon you drove through catching glimpses of Half Dome, was then *completely* filled to the brim, with glacier ice!

As glaciers retreated, the meltwater from that ice flowed down into California's rivers and out into the Pacific Ocean. This created natural environments that were more benevolent for animals, humans, and agriculture, and allowed for life to develop as it has in much of the modern world.

We are presently at the end of an ice age, in an interglacial period. Much of the ice cover from the last ice age has melted, and the ocean has already risen quite a bit since the time when glaciers covered much more of the Earth's surface. The oceans have risen during this melt about 400 feet (about 120 meters) already.

This means that as glaciers continue to melt, we have some way to go before oceans are fully filled with melted glacier water. Scientists estimate that with the current ice volumes in glacier ice, oceans can still rise 200 feet or more (about 60 meters or more) before they reach full levels. Think about that, if you have a one-story home at sea level, consider a 20-story building next to your home. Now look up. If all glaciers fully melt, the ocean will eventually cover that building you're imaging *entirely* with water. If you can survive until then (unlikely), you'd basically be swimming with the fish in the coral reef at the bottom of a 200-foot deep ocean! Your home and your entire city will be the source of one of those mythical Atlantis tales.

It's unlikely that all of the remaining glaciers on the planet will fully melt anytime soon, to create such a scenario, but even a few large ice masses that break off of glaciers at the Earth's poles can cause enough sea level rise to ruin your appliances and fully destroy your property! If you own a home along Florida's coastline, or in any of the world's low-lying coastal regions, I'd be watching the glaciers if I were you!

The problem with the current global climate is that we don't really know whether all or a significant portion of that remaining melt (or even a small portion that might still be very relevant to us) will happen in the next millennium, in the next century, or in mere decades, or why not, in merely a few years, as is already occurring in certain parts of the planet.

But the people in the flooded neighborhood of the Florida Keys, or the residents of the Bayou in Louisiana, have already reached this horizon! Ask the Venetians what they think of rising waters. The famous Italian city of Venice has been plagued by rising water levels, to the point that lately it seems human dwellings may soon be completely unsustainable.

What is certain is that if you live anywhere near a coast, you'd better follow forecasts and predictions closely, get out the measuring tape and figure out how soon your property will be underwater due to melting glaciers. The trouble is that you don't need much sea level rise to be forced to confront terrible predicaments, because it is not only average sea level rise that will get you, it's the high tide, especially when it combines with a particularly violent storm.

I recall staying at a beachfront property in Santa Barbara when my wife was teaching a climate change seminar at the UC Santa Barbara Bren School of Environmental Science and Management. The houses along the coast all have emergency barriers for when the tide is high and storms are brewing. When you know a storm is coming, you take out those barriers and put them at your doors and windows for protection. While we were staying there, some homeowners kept their barriers up, just in case. Some had waves crashing right up to their back steps or living room windows. A mere foot or two of sea level rise for those properties, combined with an especially high tide or storm, puts water right through the entire first floor of their homes. Simply google "erosion + pacifica" and you'll see frightful images of the type of coastal destruction that already happening along California's coast.

Try buying some beachfront property in Florida and see how insurance companies rate your mortgage if you live in a flood zone! Most Floridians living in non-flooding areas have a false sense of security, thinking that they will be spared from rising seas by building barriers to contain the ocean. This false sense of security is due to the failure of humanity to fully grasp the magnitude of geological eras and times. What Floridians living in South Florida don't realize is that it really doesn't matter if they live in a flood zone or not. During the peak of interglacial periods, half of Florida (most of which is now South Florida) is at the bottom of the sea. And the water will return one day. It's just a matter of time before it does.

The water contained in the frozen ice of the Greenland Ice Sheet, which belongs to some 200 glaciers, holds enough water to make the oceans rise 7 meters (23 feet). That's about the height of an average two-story home. Antarctica meanwhile is colossal by comparison. If Antarctica melts entirely (it has before and it's melting now), the oceans would rise 58 meters (190 feet).[21] That's about a 20-story building by the way, I'm just saying!

The images in figures 2.4 and 2.5 show just how extensive glacier melt can be for the world's land masses. In figure 2.4, we see the expanse of ice coverage in North America during the last glacial maximum about 20,000 years ago. This is a typical ice age. Notice *all* of Canada is covered with ice, with no land exposed, and this massive ice sheet extended well into what is now much of the northern United States.

In figure 2.5, we see how melting glaciers and resulting sea level rise affect our geography. Note the State of Florida in the southeastern United States. While Florida was never glaciated, melting glaciers once flooded what is the current geography of Florida, shown in bright white. Current Florida is outlined by a thick black line, while during maximum glaciation—when ocean water retreats into massive glacier ice sheets—the boundaries of the state are pushed into the sea, more than doubling its current land area (the dotted line).

All of the flooded areas during interglacial periods and full melt down were literally at the bottom of the sea at the time (see photo taken at the Longan Lakes Quarry in South Florida in figure 2.7) as all of the land mass seen in the picture was basically below sea level.

Figure 2.4 The extent of ice cover in North America during the last glacial maximum, 20,000 years ago.
Source: NOAA

When the planet returns to a global ice age, polar glaciers soak up the ocean, converting ocean water to ice, leaving much more extensive areas exposed. As we can see, whether we are in an ice age, or in a hot period, the difference in our geography and climate, and for our survivability, is enormous.

The Florida Keys at the end of the last ice age, about 100,000 years ago, would have been part of the continental land mass. Eventually as the glaciation and interglacial periods come and go, the Keys also come and go, becoming at times the bottom of the sea and at others *terra firma*. Today they are islands peeking up above sea level ... for now (see fig. 2.6).

Jerry Wilkinson has studied the Keys geology and points to examples of exposed coral visible in the Keys today that are proof that the islands were once underwater.

Coral rock is a generalized word to define the aggregate of the "corals" and all of the other calcium carbonate-producing organisms. Splendid examples can be seen

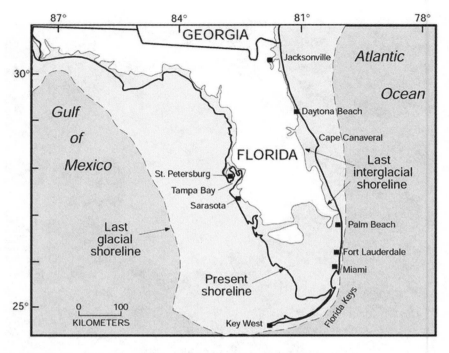

Figure 2.5 Florida sea level fully glaciated vs. interglacial periods.
Source: Muhs, Daniel R., et al. 2011

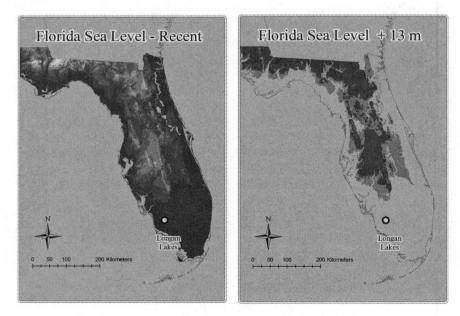

Figure 2.6 Florida at current sea level vs. a sea level 13 meters higher.
Source: Paul Octavius Knorr

Figure 2.7 A former sea bed surface about 8 meters high can be seen at Longan Lakes Quarry in Florida.
Source: Paul Octavius Knorr

at the walls of the Marvin D. Adams Waterway, or the Windley Key Quarry. As time passed, well-defined islands evolved and mature botanical growth flourished. The water levels continued to rise and formed the reformed channels. Wind, birds and water carried in botanical life that slowly took root and grew. ... Geologists are certain that as the polar ice melted, Hawk Channel changed from dry or damp land to swamp land and finally as it is today. Florida Bay was the Everglades of yesteryear and the Everglades area was a savanna.[22]

James Aber describes the rising and falling sea levels for South Florida, and the differences are eyebrow-raising. High sea level peaks occurred around 130,000 years ago, 180,000 years ago, 236,000 years ago, and 324,000 years ago. Through remains of coral reef found today in Key Largo, which would have been underwater at the time of their forming, exposed limestone shows that the sea 130,000 years ago would have been at least 5 to 8 meters higher than it is today (16 to 26 feet higher). That's about a two-story house worth of sea level rise.[23]

According to Aber, the highest sea level occurred in what is called the Sangamon Sea, some 130,000 years ago. At that time, the Florida peninsula's most southern extreme was about 150 miles (or 250 kilometers) farther north than it is today. In the

Bahamas, sea level fluctuated rapidly, reaching some 2 meters higher (6.5 feet) than it is today, but all of a sudden rose another 4 meters (nearly 13 feet) for a short period of a few centuries. These rapid changes, especially in sea level rise can be explained, says Aber, by sudden surges in glacier entry in the oceans, such as the breaking off and melting of a large glacier ice sheet (as is occurring today in parts of Antarctica).[24]

A dissertation by Paul Octavius Knorr written in 2006 argues that "the case for high-order, Pleistocene sea-level fluctuations in Southwest Florida is rather ominous indicating that based on analysis of sea bed deposits the high strata studied in Florida, Pleistocene sea-level was between 11.2 and 14.4 meters above current sea level."

Bruce Douglas of Florida International University estimates that for every inch (2.5 centimeters) of sea-level rise you have a resulting 8 feet (2.4 meters) of horizontal encroachment of the sea onto the beach—erosion basically. So it is not only a vertical problem we face, but a horizontal one that actually may be more problematic than the vertical rise. Impacts of sea level rise also include the intrusion of saltwater into freshwater supply and ecosystems, which could be devastating for the Florida Everglades. We tend to think of water level rising, but it is also coming *at us* much faster than it is going up vertically.[25]

The minimum sea level, conversely, and as measured in data collected in New Guinea, occurred about 18,000 years ago during what is known as the Wisconsin glaciation, when the sea was about 130 meters (426 feet) below current levels.

In the Florida map shown in figure 2.5, I live near that little black dot that marks the city of Palm Beach, so I am basically underwater in such a scenario. But I decided to go see for myself. My daughter is studying for a Bachelor's of Fine Arts at the New York Film Academy in South Beach, Miami, right near downtown. I figured there should be some signs in the area of past marine life right in the city, and maybe I could get a sense of just how high the ocean rose in the past. So I drove around for a few hours looking for signs of past sea-level rise. First I drove out to Key Biscayne, which was not too far, but not finding anything useful except for some fossilized coral under the sea, I decided to head back into town, since that's where much more of the population lives. Plus, it's not the same to tell Floridians that the Keys will flood (again, that's removed for most Floridians). If they knew Miami would flood and displace millions of people, that would be a different story. It didn't take long to find just what I had feared.

Right downtown, exactly five blocks from the heart of Miami's most expensive, financial, district and the famed Brickell Ave (which happens to be one of the oldest streets of Miami), on the corner of SW 3rd Ave and SW 7th St, (on Google Earth the street corner is at 2 meters above sea level), I found the first exposed coral site (fig. 2.8). Not too far from that site, about 7 blocks or so northwest, I found a second site (fig. 2.9) , and a little over a mile (2 kilometers) due south, at Alice Wainwright Park, a third (fig. 2.10).

Even more alarming to me, was that the first two locations seemed to be at the highest points of the neighborhoods in which they were located. Indeed, if you take a look at the altitude readings on Google Earth for the first two sites, it indicates 2 meters (6.5 feet), whereas in most areas surrounding these sites, it is at 1 meter (3.3 feet). At both locations my measurements indicated about 2.3 meters (just above 90 inches) of coral above the ground level. That would make the highest points of the exposed coral at both locations somewhere between 4 and 5 meters above the

Figure 2.8 Exposed coral (former sea bed) in downtown Miami, at the heart of the financial district. The top of this coral is nearly 5 meters (16.5 feet) above present sea level. The distance from the ground to the measuring tape was 2.3 meters (more than 90 inches). Google Earth indicates the ground is 2 meters (6.5 feet) above sea level. (Location 1 in Google Earth image in fig. 2.12.)
Source: JDTaillant. GIS: 25 46 00.94 N, 80 11 57.83 W

present sea level (13–16.5 feet). That means that at some point in the past, the ocean covered this relatively high piece of ground by at least a meter of water or more (see fig. 2.11).

I reached out to Daniel Muhs of the USGS, who has studied coral fossils and sea level in South Florida, to see if he could confirm this. According to a paper he published in 2011, during the last interglacial period, sea level got significantly higher

Figure 2.9 A driveway carved out of former seabed coral in the heart of downtown Miami. The highest point of the coral base found at this site is between 4 and 5 meters above present sea level (13–16.5 feet). (Location 2 in Google Earth image in fig. 2.12.)
Source: JDTaillant. GIS: 25 46 12.83 N, 80 12 22.53 W

in South Florida, and this likely lasted about 9,000 years. His calculations suggest sea level during the last interglacial was at least 6.6 meters, and as high as 8.3 meters higher than it is at present.[26] I was already calculating prior sea level at 4–5 meters above where it is now, but Muhs notes in his research that ooids present on the surface of the coral suggest that there would likely have been an additional meter of water over the top of the corals. That gets my back of the envelope calculations to 5–6 meters higher than Miami area sea level at present, right on target with his estimations.

Clearly, the sea has been here before, and the rise occurred precisely at the time of the last interglacial period. Now we're in the *next* interglacial period, so it seems reasonable to believe, unfortunately for the people living and the infrastructure located today in Miami, it will come again. The local population is knowingly (or unknowingly) already taking action to address this problem. The picture in figure 2.13 of flood barriers that have started to appear in Miami each time a storm brews shows that the adaptation agenda, preparing for sea level rise moments, is already upon us. People may attribute the need to put up barriers to recurring and more intense heavy rains, or to drainage systems incapacity to deal with more intense inclement weather, but the reality is that sea level rise is pushing forward and will likely continue to show its wrath.

Figure 2.10 Exposed coral (former sea bed) at Alice Wainwright Park. The top of the coral is about 3 meters (10 feet) from the ground. Considering there is about 1 meter to the water level and probably an additional meter or so of former sea level above the top of the coral, that places former sea level at about 5 meters (16.5 feet) above present sea level. (Location 3 in Google Earth image in fig. 2.12.)
Source: JDTaillant. GIS: 25 44 50.38 N, 80 12 20.29 W

Does It Matter?

It is normal that our planet goes through ice ages and warmer interglacial periods from era to era. So should we just accept that this is the natural change of our global environment and that sooner or later, it will happen anyway—why worry or do anything about it, such as fight to hold back climate change?

Well, one thing to consider is that our natural environment has been pretty livable for many thousands of years, giving rise to our modern society as we know it, and we should likely have thousands more years ahead of us offering a favorable living climate. We certainly shouldn't knowingly and intentionally destroy that livability, should we?

Geological eras such as the one that we are living in now can last for thousands of years, or longer. We are presently coming out of an ice age that peaked some 20,000 years ago.

Natural glacier melt out of that ice age gave us the pleasant climate that our planet has experienced until recently. We were supposed to have many hundreds or even thousands of years more of this pleasant living environment with a gradual shift into the next ice age over thousands of years. Rapid climate change, however, is now accelerating our passage through this warmer climate phase, rapidly destabilizing the atmosphere and precipitating glacier melt at unnatural rates. We are effectively

Figure 2.11 Miami coral up close. You can see the remains of shells in the limestone bedrock. This is located in the heart of the financial district, evidence that in the past the sea level has been at least 5 meters (16.5 feet) above the present sea level. (Location 1 in Google Earth image in fig. 2.12.)
Source: JDTaillant. GIS: 25 46 00.94 N, 80 11 57.83 W

destroying our benevolent climate and pushing ourselves toward one that may simply no longer sustain life in such a pleasant condition.

The change in coastlines has profound consequences for nations around the world, beginning simply with the fact that as coastlines move inland, they alter in shape and land shrinks accordingly. The map in figure 2.6 showing the state of Florida in times of glaciation vs. times of non-glaciation is remarkable because of the breadth and extent of change in something we know and can touch. And we are now beginning to

Figure 2.12 Downtown Miami coral locations.
Source: Google Earth

see substantial variations in and impacts of sea level rise in areas like the Florida Keys. Christian Maldonado, a good friend of mine who runs a fast food delivery service in Miami tells me that when it rains heavily in South Beach and certain parts of Miami he knows already that certain neighborhoods become impossible to enter. How long will it be in those neighborhoods, as in parts of the Keys, before the incoming water simply no longer recedes?

These big changes, like the shape of Florida (fig. 2.6) , occur over many hundreds or even thousands of years, but *other* changes (some small and some large) can occur within our own lifetimes. Just as Florida transforms in shape over millennia, for very low lying countries, slowly rising seas could quickly invade inland areas and conceivably suddenly change the shape of entire countries. This could alter borders, or erase limits entirely, swallowing up lands that once were exposed but are suddenly under the sea. All it takes is one large storm or hurricane, coupled with rising seas to radically change our living environments. A massive flood of these proportions occurring in low-lying areas doesn't need to fully and permanently flood lands. One big flood could destroy property and ecosystems to the point that reconstruction is impossible, or economically unwise, forcing thousands or even millions to relocate.

Rising and shifting seas could also shift maritime territories inland, and for some countries and economies, the fishing areas they depend on or coral islands they live on could be lost due to changes in ocean levels.

I regularly take strolls with my wife along the beaches in South Central Florida, and am always surprised to see just how much the sand on the beach has shifted around

Figure 2.13 Flood barriers like this one go up in Miami each time a major storm approaches.
Source: Atlantic Shutters, Inc.

after a big storm. Sometimes it is piled up high in the middle of the beach. But a single storm can flatten out the entire beach, with everything smoothened out overnight. A big storm, with a storm surge of 10 or 12 feet, coinciding with extremely high king tides, can cause massive damage. Recently, Palm Beach County public works has been dredging the canal waters off the coast and has dumped ocean bottom sand on our beaches to try to recuperate lost beach sand to the changing climate. This may help us gain a few feet of beach sand, but it causes havoc to the local ecosystem (the fish are gone now probably due to changing nutrients of the beach sand). It's a temporary patch to a problem that will surely only get worse. Some towns on the east coast regularly lose beachfront houses and even entire beaches to storm surges and heavy surf. The increasingly unpredictable and changing weather patterns that greatly modify coastal landscapes, compounded by increasing and prolonged storm surges, could permanently affect coastal houses and developments, and not in very good ways. Increasingly unpredictable, changing weather patterns greatly modify the landscape for coastal Floridians and if storm surges keep increasing or become prolonged, some of the houses we see along the coast could be permanently affected (see fig. 2.14).

Sea level rise, as the recent IPCC Report on Oceans and the Cryosphere informs us, can also impact coastal ecosystems, which are already highly impacted by human activity. Sea level rise can affect saltmarshes, mangroves, vegetated dunes, and sandy beaches, which can be pushed and pulled in response to sea level rise.[27]

Figure 2.14 A house on Sheep Pond Road in Madaket loses its battle with the ocean, Nantucket, Massachusetts, 2015.
Source: Peter B. Brace, Nantucket Chronicle

Figure 2.15 An aerial view of the icebergs near Kulusuk Island off the southeastern coastline of Greenland, a region that is exhibiting an accelerated rate of ice loss. As the region warms, several times quicker than the rest of the planet, glaciers lose pieces of ice that collapse as icebergs and melt into the sea. *Source*: NASA, Goddard Space Flight Center

A warming climate and subsequent glacier melt can also cause the reverse, that is, land masses previously under ice are exposed to the atmosphere, creating new land territories. Such is the case in the Arctic region where melting glacier cover is changing geopolitics, with oil reserves now becoming accessible and trade routes opening up through seas that were previously unnavigable due to the presence of ocean ice.

In Russia recently, the melting Vylki Glacier off the coast of the Novaya Zemlya archipelago in the Arctic Ocean ice has revealed five new, previously unknown, islands. The country is adding this territory to its national territory and will now proceed to name the islands. These newly revealed lands have likely been buried under ice for tens of thousands of years.[28]

Scientists tell us that the Arctic region is heating up twice as fast as the rest of the planet, placing enormous stress on the local environment and ecosystems. The Greenland Ice Sheet, also in this part of the world, lost 11 billion tons of ice in 2019 on just one summer day during the month of August, a month of record heat (fig. 2.15). During the previous month, the ice loss was 197 billion tons or about 80 million Olympic-sized swimming pools. That same month, the Icelandic Okjokull Glacier was entirely lost to climate change.[29]

Climate Central says that glacier melt will cause flooding for millions of coastal dwellers around the world. In a mere three decades (30 years), sea level rise, says Climate Central, will affect *every* coastal nation. The greatest effects will be in Asia, in low-lying areas of mainland China, Bangladesh, India, Vietnam, Indonesia and Thailand, where most of the region's human populations reside. By 2050 in China, 93 million people could be living on land lower than the height of average yearly coastal flooding. In India, that number is 36 million. Meanwhile in Bangladesh and Vietnam, 73 million people face saltwater intrusion into lands currently extensively used for agriculture.[30]

Beyond Asia, land in 19 countries, including Nigeria, Brazil, Egypt, and the United Kingdom, that is now home to 1 million people could fall below the permanent sea level line by the end of the century, becoming permanently flooded. By then, the end of the century, at a global scale and at higher climate trend projections, land home to 640 million people today (about 10% of the world's population) could be threatened by chronic flooding or permanent inundation.[31]

What is certain is that as glaciers continue to melt around the world, sea level will continue rising, placing more and more people at risk or in states of dire emergency. These people will have to move in order to survive. That will greatly alter immigration patterns around the world, and put enormous strains on many already troubled countries and economies.

Most of the land of some island states around the world is already very near sea level. Consider Kiribati, Micronesia, Maldives, or Tuvalu. The inhabitants of these island countries may have no choice but to leave and settle elsewhere. Let me be clear, this is not merely about vulnerable neighborhoods or communities, but *entire countries* that simply have to move because of sea level rise. These countries are literally doing the global rounds at international climate meetings, asking for other countries to donate land to them so they can move their entire population!

That is about as dire as it gets for people facing the immediate impacts of climate change. These, even by name, are esoteric places of the world, and their predicament may not hit home for you. We tend to associate these natural catastrophes with poor or remote areas of the world, but as we see our climate evolve before our eyes, places closer to home are being affected. Cities like Amsterdam, Buenos Aires, New York, and Miami, largely inhabited areas that are more mainstream to global citizens, may too, sooner than we think (and maybe during our lifetimes, or already as in the Keys), have to consider moving people out of climate-vulnerable coastal or other low-lying areas. Floridians for example, could soon be suffering a similar fate, having to consider moving half of the state's population to drier land. Recurring floods in the US Midwest and in Europe are placing strains on local populations, causing them to consider seeking higher land.

Imagine that the place where you live will one day, during your lifetime, be underwater! Does this possibility sound remote? Take a look at the house in figure 2.16 in Chesapeake Bay. It has been engulfed by the sea.

Sea level is rising faster in the Chesapeake area than in any other part of North America's Atlantic Coast, by some 3.4 millimeters per year, which is twice the global rate.[32] It is estimated that the sea in the Chesapeake region, including areas at and around Washington, DC, could continue to rise significantly in the coming years. Benjamin DeJong of USGS and a group of scientists published a study in 2015 that indicates that rising seas could cause serious problems for the nation's capital:

> Several lines of evidence suggest that sea levels will rise more quickly in the Chesapeake Bay region. If this acceleration continues, it could induce an additional rise of 15 cm for the Chesapeake Bay and Washington D.C. areas by AD 2100. Recent evidence also confirms the instability of glaciers in West Antarctica, which has the potential to raise global sea levels significantly, particularly beyond AD 2100.

Figure 2.16 House overtaken by sea level rise in Chesapeake Bay.
Source: Courtesy of Flickr user baldeaglebluff

As global sea levels rise and the Chesapeake Bay region subsides, storm surges are projected to increase both in frequency and magnitude. Bridges, military facilities, national monuments, and portions of the rapid transit system would be flooded in Washington D.C., and [about] 70,000 residents would be impacted by a 0.4 m rise in sea level.[33]

In California, many coastal communities, such as the residents of Pacifica, are facing severe impacts from rising seas, especially on high tide days, and especially when these coincide with stormy weather. All along the California coastline, north and south, from Humboldt County in the north to San Francisco Bay, to Los Angeles County and San Diego County in the south, the impacts of coastal erosion, cliff collapses, and other sea intrusion related damage are felt. You *could* try to build levees, seawalls, or other infrastructure to withstand the oncoming sea. For San Francisco alone estimates put the infrastructure investments needed to hold back 2 meters (6.5 feet) of sea level rise at $450 billion, and for the Port of Los Angeles $9 billion, but it seems unlikely that we will be able to stop Mother Nature when glaciers fully melt.[34]

About 100 million people around the world live within the first meter (about three feet) from the current sea level. Meanwhile, a *single* piece of ice broken off of Thwaites Glacier in Antarctica in 2014, once eventually melted, could raise oceans globally by up to 2 feet. If you're at sea level already, that brings water up to your knees or thighs. Think back to the flooded neighborhood of the Florida Keys. If inundated by two feet of water, permanently, it's game over! As mentioned previously, this is not a future scenario: the Florida state government is already encouraging people to leave the Keys because of the climate crises already occurring and that will likely worsen in the coming years.

Other similar pieces of colossal ice masses are constantly breaking off of the frozen continent, including from other massive glaciers such as Pine Island Glacier. These giant glacier pieces turned icebergs could increase that sea level rise to nearly 4 feet (that's up to your neck). And that's just a handful of glaciers that are withering away at the South Pole.

In early 2019, when I first started preparing this manuscript, yet another piece of ice, twice the size of New York City, was beginning to break off of Antarctica from the Brunt Ice Shelf. Scientists did not know how long this piece of ice would take to finally break off and find its way into the ocean, commencing a progressive meltdown process, but the beginning of the collapse was already in motion.[35]

And on the eve of my first submission of this manuscript, in January 2020, Chris Mooney, the climate change beat reporter at the Washington Post, published an article about Thwaites Glacier, a glacier larger than the State of Pennsylvania found to be melting from beneath. The full collapse of Thwaites Glacier into the sea could result in over 10 feet of global sea level rise.[36]

Consider that *all* glaciers around the world are melting at accelerating rates. Alaskan glaciers, for example, are melting up to 100 times faster than scientists previously thought.[37] When you start to add how much each of them will contribute to sea level rise, the predicament is horrific. Once glaciers break off and turn into icebergs there is no turning back. Floating pieces of ice begin to melt faster and faster until they are fully consumed by the ocean. They cannot be reattached if the weather changes.

If all of the world's existing glaciers fully melted, we could see over 200 feet (60 meters) of sea level rise. As noted earlier, most of South Florida would become like the Keys, an archipelago of tiny islands, once high points of the mainland.

Benjamin Horton stated in a recent TED talk that:

> From the last glacial maximum 20,000 years ago, to today, temperatures have risen by around five degrees C, that has melted two-thirds of the ice on our planet, raising sea levels by 120 meters, over 350 feet, we have committed ourselves to a further 5 degrees C rise, unless we do something about climate change. How much of the ice do you think that we will melt. We have over 60 meters on our planet. I would say to you that we can afford maybe two or three percent at best.[38]

A recent study (in 2019) by Michael Zemp and other authors indicated that we have systematically underestimated the contribution of mountain glacier melt to sea level rise. Aside from Antarctica and the Greenland Ice Sheet, glaciers occupy an estimated 706,000 km^2 globally, with an estimated 170,000 km^3, or 0.4 meter of potential sea level rise equivalent.[39] His research suggests that things in mountain environments could actually be much worse.

According to statistics from 1961 to 2016 (a 55-year period), glacier melt in mountain areas has contributed as much melt as that of the Greenland Ice Sheet and clearly exceeds that from the Antarctic Ice Sheet (see fig. 2.16). It accounts for 25–30% of the total observed sea level rise. The article goes on to state that "present mass-loss rates indicate that glaciers could almost disappear in some mountain ranges in this century, while heavily glacierized regions will continue to contribute to sea-level rise beyond 2100."[40] Zemp also found that glacier melt is occurring in all of the planet's regions, with the fastest melt occurring in South America followed by the Caucasus, Central Europe, Alaska, Western Canada, and the United States.[41]

Current projections by the most advanced scientific community studying climate change are sobering. By 2010, the ocean will rise largely due to three global phenomena, all related to rising global temperatures. Melting glaciers, melting ice sheets (which are basically also melting glaciers), and the expanding ocean (because as global temperature rises and the ocean warms, that heat expands the water).

The subsequent sea level rise resulting from these global phenomena (even if we limit temperature increase to the Paris Agreement goal of +1.5°C) will be between 0.43 meters (1.4 feet) and 0.84 meters (2.75 feet). These numbers can be dizzying and confusing, but to gain perspective, if you already live at sea level, the water will be somewhere between your thighs and your neck.

After that (after 2100), the sea will simply continue to rise for centuries due to continuing ocean warming, which occurs at a slower pace than the warming of the atmosphere (it catches up eventually) and due to continued melt of glaciers around the world.[42] Oceans that warm also expand, and they are especially adept at absorbing heat, which is not good news. Almost all of the heat caused by humans since the industrial revolution lies already within the oceans, and is hidden away in their depths.[43] This is troubling since the warming we are experiencing is actually on the surface— much more is stored away and yet to come. In sum, the rising seas will stick around for thousands of years!

The warming oceans are also an enormous problem for glaciers and ice sheets that jut out from the mainland onto the ocean. Nerilie Abram, Matthew England, and Matt King report that the Antarctic ice sheet has a weak underbelly, as in some places, the South Pole ice sheets sit on ground that is below sea level, and in some cases, this ice comes into contact with warm ocean water (see fig. 2.17). Warm oceans do a good job of rapidly melting glaciers and destabilizing the ice sheet. The latest estimate, say the authors, is that 25% of the West Antarctic Ice Sheet is now unstable and that Antarctic ice loss has increased five-fold over the past quarter century. Antarctica has lost a staggering nearly 3 trillion tons of ice since 1992.[44]

Thwaites Glacier on the Western Antarctic Ice Sheet, on the Amundsen Sea, which is about the size of the State of Pennsylvania, is melting fast, faster than most glaciers on Earth, but not only that, it's collapsing in real time. It's called the "doomsday glacier" because it's melting so fast. The glacier's elevation falls by several feet each year. The entire region of West Antarctica, which includes Thwaites Glacier, Pine Island Glacier and many more, is actively melting off at alarming rates. All that melting feeds off of itself and creates … *even more melting*. Just the *meltwater* of Thwaites Glacier (not the glacier but the *melting of the glacier*), is roughly the size of Florida and accounts for 4% of sea level rise on an ongoing basis.

If Antarctica were *only* to lose Thwaites Glacier, sea levels would rise by about 2–3 feet, flooding cities like New York and completely drowning New Orleans, Venice, and many other coastal low-lying cities. But that of course is not what is happening. All the glaciers are melting, and that brings much more severe impacts to low-lying communities around the world.

The problem is that Thwaites Glacier is part of an ecosystem of glaciers in the West Antarctic that fit together like a continental-sized puzzle, and as we know, when we break off a piece of a puzzle, oftentimes many others come undone with it (fig. 2.18). The entire collapse of the West Antarctic Ice sheet could spur rapid ice melt causing some 10 feet (3 meters) of sea level rise.[45]

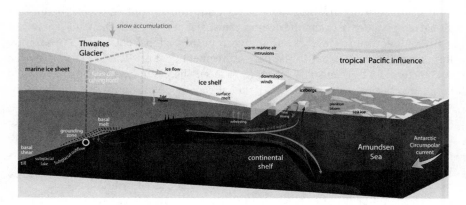

Figure 2.17 Thwaites Glacier and ocean model shows how warm water seeps underneath cold ice and accelerates warming.

Source: Credit: International Thwaites Glacier Collaboration

Figure 2.18 The West Antarctic Ice Sheet is a puzzle of rapidly melting and delicately positioned glaciers that, if released into to the oceans by melting and collapsing ice, could raise sea level by 10 feet (3 meters).
Source: NASA/GSFC/SVS

These calculations have a lot of uncertainty related to them, but for the most part, the uncertainty generally errs on the side of conservatism. What is certain is that in the last few decades, the melt, and the unexpected events of glacier breakdown around the world, have generally come *before* the dates expected in past scientific studies and projections, sometimes by several decades or even centuries.

Especially in the Antarctic and Arctic regions where ice sheets and glaciers are co-lossal, sudden collapse and break-off of large chunks of ice the size of entire countries can quickly change forecasts and projections, only accelerating the timing for impacts to sea level due to accelerated melt. At the time of the final edits to this chapter (April of 2021) yet another scientific article appeared and received global attention, alerting that Pine Island Glacier (one of the Antarctic glaciers making up the puzzle) showed clear signs of great instability, and that 3 distinct tipping points were emerging as a consequence of ocean-induced melt.[46]

Yet another report indicated that the glacier melt-off from the Greenland Ice Sheet was breaking new records, *again*. The 2019 summer shattered all records with 586 billion tons of ice melting—that's more than 140 trillion gallons of water. This is more than twice the yearly average melt, which has been about 259 billion tons of ice since 2003, and it blew out the previous record of 511 billion tons set in 2012. It's impor-tant to note that the northern areas of the planet warm more quickly than the rest. Greenland has warmed by 1.7°C (about 3°F) since 1991.[47]

Ingo Sasgen and other researchers who collected this data indicate that since the mid-1990s the Greenland Ice sheet has been steadily increasing its contribution to global glacier melt, graduating from being a modest contributor to nearly melting off a volume of ice almost equal to all other glaciers around the world combined! The reason, say the researchers: rising ambient temperatures in the Arctic, reduction of albedo, the migration of snow to higher altitudes, an increase in the total melt area, and cloud-radiative effects.[48]

Some researchers now think that the Greenland Ice Sheet has already melted beyond a point of no return. "The ice sheet is now in this new dynamic state, where even if went back to a climate that was more like what we had 20 or 30 years ago, we would still be pretty quickly losing mass," says Ian Howat, co-author of a study about the melting of the Greenland Ice Sheet, and professor at Ohio State University.[49]

The impacts of sea level rise are more immediate and evident for coastal dwellers. They will be flooded and will have to move or invest lots of money to stop the flooding (if that is even possible). The result, especially for poorer communities living in flood areas that are at risk due to climate change, is that they will have to leave or face dire outcomes for their homes, their agriculture, their livelihoods, and even in some cases, their very lives. Impacts from sea level rise will not only be related to flood damage, but to land erosion and salinization (rendering agricultural lands worthless).

At a national scale, for many countries, this phenomenon will cause substantial migration inland, putting political and economic pressure on other areas of the country or on other nations to receive fleeing populations.

Another option would be to try to *stop the ocean*, which seems a herculean and an almost irrational task. A few years after moving to Florida in 2015, I sat through a painful session with local mayors and policymakers in South Florida debating how they think they can stop the ocean by building walls and installing powerful water pumps to pull the water out of their streets when the floods come. I thought to myself, as the water pump proposal from one local mayor rang through the room, if the entire South Florida region is flooding, and you're getting water in your community (just like everyone else), where are you planning to pump this water? On your neighbor? I suppose this discussion *was* progress from just a few years earlier when the governor of Florida had prohibited members of his staff from even mentioning the term "climate change"![50]

I remember a Chinese fable from my childhood where two brothers go to the sea and one of them is able to swallow the ocean so that his brother can explore the sea floor. Before taking the big oceanic gulp, he warns his brother not to stray too far from the coast or be out for too long because he cannot hold the ocean in his mouth for very long. Eventually, he must release the water and to the detriment of his brother, who did not return, the flooding ensues. It seems to me that those that think they will be able to keep the incoming water from flooding their lands are like these two brothers, buying time to enjoy the scenery while they can, but regardless of their walls or pumps, eventually the water will come and we won't be able to stop it.

You can build levies, you can build sea walls, you can build higher even, but when the ocean rages and begins to swell by several feet at a time, there is no stopping it and it will be necessary to move people to other areas in a forced relocation.

For some countries with alternative land, this may be possible, for others, particularly for low-lying countries or islands whose entire territory is below or close to sea level, it will be deadly. They will simply cease to exist. This predicament may also only exacerbate already existing inequalities in our societies, with those living in wealthier neighborhoods, either because of their larger bank accounts or because they are able to influence and pressure policymakers, able to assure public works help them adapt to flooding, by building levies, installing better drainage systems, or building retention ponds, and other stormwater systems. But again, even this will seem minuscule when the ocean grows to unimaginable levels and surpasses or breaks the levies, as occurred in New Orleans with Hurricane Katrina.

I started this chapter with a reference to a news item that broke as I began the draft, the story about the flooded neighborhood of the Florida Keys. Alarmingly, by the time I finished the first draft of the chapter another more famous story had already broken into the mainstream news, the sudden flooding in Venice, Italy. And then, yet more environmental news came from my good friend Jared Blumenfeld's Podship Earth series—a story on king tides in California's Humboldt Bay and the risk sea level rise poses to this climate-vulnerable community as the seas invade once solid mainland.[51]

As many coastal economies struggle with having to confront the rising costs of dealing with climate change and related sea level rise, the future costs of both hedging your coasts for upcoming sea-level rise or cleaning up the aftermath when tragedy strikes, will be high and will only grow.

Projections for end-of-century scenarios suggest that sea level rise will cost societies tens of trillions of dollars in losses due to infrastructure destruction and economic activity loss each year and will displace some of the world's most economically vulnerable and disadvantaged communities. Even with extensive adaptation measures such as constructing levies and ocean walls, and even if such measures could stop the ocean (at least for a while), the costs of investments and damage could easily exceed trillions.[52]

K. M. Befus and a group of climate researchers recently published an article in the journal Nature Climate Change about sea level rise and climate change presenting startling data showing just what is in store for coastal communities in California. It's not only waves coming from the ocean that will invade land. Sea level changes can result in water coming up from beneath your feet via raising coastal water tables. The researchers indicate that over the next century, rising sea levels are predicted to cause widespread inundation of coastal terrestrial areas, wetland loss, and more severe nuisance flooding.

These impacts will not just affect coastal dwellers, but also people living inland who may see water seeping up in their gardens or from under their homes, possibly causing irreparable damage to buildings and other infrastructure such as roads and sidewalks. According to their calculations, a 1 meter (3 feet) sea level rise on the coast could have an impact up to 50–130 meters (165–500 feet) inland. Their calculations raise the risk for low-lying communities around the San Francisco Bay area and they warn that public officials should take note and start preparing to build irrigation systems to contain and remove floodwater before it's too late (see fig. 2.19).[53]

Even more alarmingly, they suggest that 13.8–43.9% of the areas defined as populated places along California's coastline are at risk of the hazards associated with emergent or shallow groundwater. With a 1 meter high sea level rise projection (3 feet), these risks grow by 1.1–6.4%.[54] They are especially concerning in regions like the San Francisco Bay Area where low-lying water tables will be greatly affected by rising tides. Other at-risk areas identified by the authors include the Port of Los Angeles, Santa Barbara, and the San Francisco Airport area (which is not in the city itself). There will also be impacts, say the authors, to roadways and to sewer and septic drainage systems, and there is the potential for soils further inland and currently above the water table to be contaminated by intruding saltwater as it reaches higher and higher.

Sea level and groundwater tables are intricately linked, as this study shows. A rise of one means the rise of the other. As the sea rises, the pressure exerted (the push) on groundwater increases and that water needs to go somewhere. It will work its way inland, into crevasses, gaps, and any space it can find, bringing saltwater into freshwater containment areas. This will not only result in flooding but also in saline intrusion

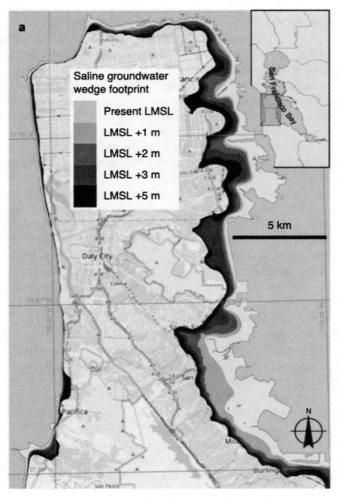

Figure 2.19 San Francisco and San Mateo County saline sea level intrusion into water table in multiple sea level rise scenarios. (LMSL = local mean sea level).
Source: Open Street Map Contributors

that will greatly impact delicate coastal ecosystems and areas that may be devoted to agriculture.

Past studies relating to sea level and California estimate that climate change driven overland flooding could threaten 600,000 people and $150 billion in infrastructure across the urbanized coast of California.[55] "We can build protections, walls, berms to prevent overland flooding, but that's not going to stop the groundwater," says Patrick Barnard, with the USGS Climate Impacts and Coastal Process Team. "It's this slow creep upward that gets into garages and foundations and roadbeds."[56]

California newspapers have been publishing stories about climate change and sea level rise to raise awareness of the issue among locals and convince them that climate change is not only real, it is happening now and it is happening faster than ever before, echoing the words that Governor Gavin Newsom recently proclaimed from the scene of unprecedented climate wildfires (see the end of Chapter 9 for excerpts from that moving speech).

Rosanna Xia of the LA Times warns Californians that they are not safe from sea level rise and that the water they should fear may not be the one coming overland. Seawater is pushing its way into rivers and deep aquifers, ruining the water supply for valuable farmland in Salinas Valley near Monterey, California. The same is happening under the Oxnard Plain and in Los Angeles County where water agencies are fighting salty sea intrusion into water supplies.

Xia points to a black and working class community in Marin City, nestled just above Sausalito, which is better known for its wealthy towns, pricey dining, fancy sailboat harbors, and some of the highest priced property in the world. The much less wealthy community of Marin City is not likely to feel sea level rise *overland* until the sea level marker reaches 3 feet (about 1 meter), but long before that, the water will have seeped in from the ground. "Basements will heave, brackish water could erode sewer pipes, toxic contaminants buried in the soil could bubble up and spread." One foot of flooding in Marin County (where Marin City is located) could produce increased flooding, and any pollution now contained in the soil could be pushed upward by incoming tides, says Kristina Hill of UC Berkeley. Hill estimates that twice as much land in the Bay Area could suffer from this underground intrusion of seawater as would be flooded by overland sea intrusion.[57]

Rosanna Xia's journalism is bringing forward an issue to Californians that very few people have really grasped—like the issue of invisible rock glaciers up in the Sierra Nevada, which we will talk more about in Chapter 8. The impacts to California could be much worse than those of the COVID-19 pandemic and the resulting collapse of the economy, which already seems monumental.

The USGS has done studies showing that $8 billion in California property could be underwater by 2050, with an additional $10 billion at risk during especially high tides. In less than 30 years, sea level increases of 10 inches combined with strong storms (that's the clincher, rising sea level + storms) could flood the homes of more than 150,000 people and cause $30 billion in damage. By 2100, when the seas could reach more than 6 feet higher than they do now (1.8 meters) the impacts could affect more than half a million people and cost the state $150 billion, or 6% of its GDP. These impacts, says Xia, could be worse than wildfires and earthquakes combined—massive natural and climate disasters that have already made California infamous.[58]

Over 600 million people live in coastal zones worldwide (that's at less than 10 meters elevation, or 33 feet), says Barnard, the author of the California coastal study, and that number is set to increase to more than 1 billion by 2050.[59] The risks and impacts of coastal flooding will likely surpass the trillion-dollar mark by the end of the 21st century, which will lead to great suffering, destabilization of local economies, and mass migration.

The authors project $150 billion in damages from coastal flooding by 2100. Sometimes such large numbers are dizzying and hard to truly grasp. To put this in perspective, consider the cost of other massive natural disasters or the size of some large public budgets:

US budget for Medicaid	$466 billion
Tohoku earthquake and tsunami	$325 billion
Return of historic 1860s flooding	$300 billion
California coastal flooding projection	**$150 billion**
2020 wildfire season (not over)	$150 billion
Hurricane Katrina	$127 billion
US budget for primary/secondary education	$79 billion
2017 California wildfires	$18 billion
1989 Loma Prieta earthquake	$10 billion

Across the country, the city of Boston, Massachusetts, is also preparing for the worst when it comes to sea level rise. Just during the 20th century, the seas surrounding Boston crept up by 9 inches. According to projections, as climate change accelerates, the pace of sea level rise does too. In Boston, a report commissioned by the city indicates a tripling of the rate of sea level rise, adding 8 inches to 2000 levels by 2030 and as much as 3 feet above 2013 levels by 2070. Boston's city government is also considering a more alarming scenario that projects a 7 foot 4 inch rise (2.2 meters) by 2100 if we cannot contain CO_2 emissions.[60]

Stefan Talke and a group of researchers at Portland State University, Oregon, draw a direct connection between sea level and historical flooding in Boston. They found that the sea has risen about 1 foot (28 centimeters) over the past two centuries, but the rise has been increasing since 1920. They can also link extreme tides to flooding in the city. The big problem, they say, and the historical data proves it, is when high tides coincide with big storms. They also note that some of the highest levels of flooding since European colonization occurred in 2018. After correlating Boston's flood risk with the seasons and the gravitational phenomenon related to tide shifts, they have found that flooding occurs fairly predictably over the decades. Their calculations and projections indicate that the city will likely face abrupt, elevated flood risk by the 2030s.[61]

In response, the city of Boston is raising streets, building berms, and mandating that new buildings in its harbor area include movable "aqua fences" to deflect incoming waves (you see these movable flood barriers pop up in Miami as well before storms), reports Steven Mufson of the Washington Post. The city mayor is spending millions to prepare the city for the sea level rise they expect from climate change.

The map in figure 2.20, taken from Climate Ready Boston datasets and reproduced in Mufson's article for the Washington Post, shows the alarming scenarios Boston will

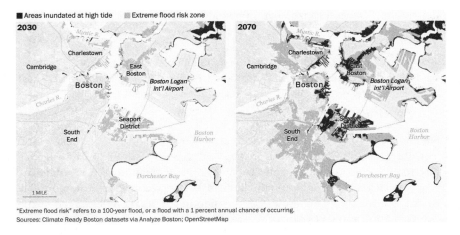

Figure 2.20 Extreme flood risk, Boston.
Source: Washington Post citing Climate Ready Boston datasets via Analyze Boston; OpenStreetMap

have to face in 2030 and 2070 (that's less than a decade away for the first projection) with rising seas. The vulnerable spots, visible on the map as dark shaded areas, are the commercial piers, the city's international airport, low-income neighborhoods, the South End, the New England Aquarium, and apartment buildings in the newly re-developed Seaport area.

The city and the metropolitan area are scrambling to design and implement climate resiliency projects that can handle the sea rise, says the Washington Post report from its Climate Solutions Series. Charlestown plans to raise Main Street by 2 feet (60 centimeters) to protect 250 homes and 60 businesses from flooding. The Massachusetts Bay Transportation Authority is building water-tight steel doors that can be closed near Fenway Park. It is strengthening the entrances to tunnels around the city to ensure water does not enter during future floods. The city is also studying natural land formations that can serve as water barriers in case of mass flooding. Many park areas are vulnerable to flooding because they have been built on land-fill. These floodable areas are then a climate vulnerability for adjacent communities as they too can flood due to their proximity. To avoid this, the city is planning to build berms as high as 10 feet tall and underground storage compartments to hold flood water. The city's bus depot in the Charlestown neighborhood is also in a re-curring flood zone. Future floods could eventually disrupt an important source of transportation for over 100,000 people who regularly utilize bus service in the metropolitan area.

Another risk, says Mufson, is that the city's sewer and drainage system is built for evacuating 4.8 inches of water in 24 hours. This drainage capacity is sufficient for about 90% of storms in an average year, but with climate change bringing on more rain, by 2100 rainfall could hit 6.65 inches in a day, overwhelming the city's drainage system. More to worry about.

Mufson points to an even greater concern, faced not only by Boston but also by many other cities: generally low-income, disadvantaged, and minority populations suffer the most from adverse environmental impacts. Boston is no different, says Mufson, and the rising seas are forecasted to be very discriminatory.

> Boston's low-income neighborhoods, where public housing projects were built on landfill, are particularly vulnerable to flooding. By the end of the century, a large part of the Dorchester neighborhood, which on its own would be the fourth-largest city in Massachusetts, could be underwater.[62]

As in many cities, city design, construction, roadways, buildings, and many aspects of city life were developed decades or even centuries ago. As our climate emergency begins to take front and center concern, new homes and buildings are built to codes that have been adapted to address climate change. They're more resilient, more able to withstand storms, and more protected from flooding. Migration of local populations to outlying areas also takes care of some climate risks in the future. But low-income communities and residents who do not have the financial resources to make the necessary investments to increase climate resiliency or to move to the more expensive and climate-safer suburbs cannot afford to improve their climate resilience.

Further south, 1,500 miles (2,400 kilometers) from Boston, the Union of Concerned Scientists (UCS) reports that the low-income localities of Opa-locka and Hialeah, in South Florida's Miami-Dade County, at only an average 6 feet above sea level and sitting on porous limestone (as does much of South Florida), face increasing risks from rising seas, rising water tables, contaminated wells, and severe inland flooding. The UCS study, from 2016, suggests that South Florida has already seen 8 inches of sea level rise over the last 100 years and 9–24 additional inches (23–61 centimeters) more are expected by 2060. The environmental justice dimension of sea level rise is also worryingly present in South Florida. With median household incomes at $29,961 for Hialeah and just $20,338 for Opa-loca, both cities are defined as low income. They are also home to predominantly minority populations: 66% African American and 35% Hispanic in Opa-locka; 95% Hispanic and 3% African American in Hialeah.[63]

Victims of past hurricanes in South Florida can attest to the severe damage and the recuperation time following hurricanes, which can be prolonged for communities of meager economic means. The UCS report shares the experience of a local Hialeah resident, Christie Diaz, who recalls her childhood experience of Hurricane Wilma in 2005, which aggravated the impacts of Hurricane Katrina just weeks earlier. Her home was without power for two weeks and she and her neighbors were dependent on deliveries of food from the government in the days following the back to back storms. Her parents could not get to work for days and she missed school for nearly a month. The elderly had the hardest time of all, recalls Diaz.[64]

Gilberto Turcios of Opa-locka also survived Hurricane Wilma and the many hurricanes since that now seem to be ever more frequent and worrisome. He is concerned about recurring flooding. After hurricanes past, he remembers trees fallen on homes, including on his aunt's home, which suffered serious damage. He recalls long lines of people waiting for food, water, and supplies from the Federal Emergency Management Agency (FEMA), the Red Cross, foreign embassies, churches, and other charity groups and drives.

"The government didn't do the right thing in poor communities," says Turcios, "they just did a quick fix and forgot about it. Over the last 20 years I've noticed flooding happening more often. I'm concerned because the more water tends to be around a home the more the water can crack the walls and weaken the base of the house. It's an economic loss, it hurts property values."[65]

And yet again, just as I was making the final edits to this chapter, more news from Hialeah and Opa-locka came in, this time regarding flooding from yet another storm, Tropical Storm Eta, which thankfully did not turn into a hurricane. Evgeniya Ignatushchenko hadn't even moved into her new first-floor apartment in Hialeah and it was already flooded. The parking lot was flooded, and surrounding streets looked like Venice, ironically, the name of her building complex.

"The floor was wet. I smelled mold. I could see that water had gotten inside because of the marks on the cardboard boxes I had left here. I've never seen anything like that. Last week the condominium association assured me that it was impossible for water to get inside, but it did."[66]

The rain fell just overnight onto Miami-Dade, dumping more than 16 inches (40 centimeters), overflowing canals and retention ponds, turning roads into rivers, and flooding neighborhoods. The problem is not so much that it's a lot of rain at once as much as it's that the stars aligned the wrong way for a neighborhood that usually stays dry. First, lots of rain, then lots of rain again, and then even more rain from yet another intense tropical storm. Place that sort of sustained water intensity on a 72-year-old drainage system designed to handle 6–8 inches a day (15–20 centimeters) and you're bound to have a problem. Even though the drainage system works just as it should, it simply cannot keep up with the increasing flows of water we're seeing in South Florida (for this storm, 14–18 inches or 35–45 centimeters). The results: 1 foot of water (30 centimeters) in your one-story apartment building that supposedly was impossible to flood .

Pablo Santos, a meteorologist at the Miami Office of the National Weather Service, could not drive out of his home with his Prius; it simply wouldn't get through the water.

"Living here 20 years, I have never seen the flooding that I saw." He said that the thing with Eta's flood was that the region had seen 30 inches of rain (76 cms) in one month during a season that is supposed to be a dry season. The area generally gets 60 inches of rain per year (152 cms). It got that much in a month. Then Eta hit, not a major hurricane but enough of a storm to cause havoc. "In effect, we have not had a dry season. The water table was already very high, in some areas right at ground level. All this rain comes and it has nowhere to go. Does it mean there's something wrong here? No, it means that every system has a limit."[67]

In my own experience in South Florida, a new arrival in 2015, one of the things that most surprises me is that *after* a hurricane you can drive around for miles along city streets and see enormous piles of fallen palm fronds, tree branches, or entire trees lining the streets waiting to be retrieved by the city. They can be there for weeks. So can the water.

The situation has not improved, and there isn't any hope that it will get better. Streets flood regularly in South Florida, from the expensive neighborhoods of South Beach to the low-income areas of Hialeah and Opa-locka, and the flooding, as in the Florida Keys, persists, sometimes for days, sometimes for weeks. Earlier, I mentioned my good friend Christian, the food delivery driver in South Beach. When it rains heavily for many days, he knows he will not be able to access certain neighborhoods so he won't even take calls requesting deliveries to those locations. Eventually, as the situation gets worse, he simply wont deliver there, ever.

A 2009 report[68] on sea level rise estimated that South Florida would get anywhere between 5 and 20 inches (12–50 centimeters) of sea level rise by 2060. An update to the report[69] upped that number to 17–31 inches (43–78 centimeters). Think about that, the drainage system of the Hialeah can handle 6–7 inches of rain in a day. If seas are going to rise by 17–38 inches, where does that leave South Florida residents on a dry day, much less on a rainy day? Neighborhoods that once remained dry will now flood, while neighborhoods that already are prone to flooding will get swamped.

Responding to climate change and sea level rise through public infrastructure investments is critical, but is not always easily accomplished. A case across the Atlantic, the city of Venice, Italy, is illustrative. We've all dreamed of sitting on a gondola and flowing through the liquid streets of Venice. I was lucky to enjoy such a passage when I was in college and studying in France. But that dream is slowly turning into an ugly nightmare. Venice today shares the fear of many coastal towns that must confront the reality of rising seas or perish. In 2019 a 6-foot high tide pushed by high winds (again, the combination of bad weather and extraordinarily high seas) submerged 80% of Venice, flooding homes, stores, restaurants, government buildings, and precious infrastructure, including the basilica in St. Mark's Square. Damage was estimated in excess of $1 billion. The city is having to take drastic and costly measures to try to keep the Adriatic Sea from completely submerging it on days with especially rough seas. An extensive mile-long submerged and movable floodgate system, with 78 gates that would rise from the bottom of the sea when needed, was designed precisely for this task … *17 years ago.* Nearly two decades later, with lots of money literally flowing under the bridge and a few corruption scandals hindering progress, this massive undertaking is still not complete. Meanwhile the water tidemarks on the basilica are rising.[70]

Further east, in Asia, Kotchakorn Voraakhom, a Thai landscape architect, designed a city park in the heart of Bangkok, Thailand, because her city is sinking by 1 centimeter per year, four times faster than the sea is rising. By 2030, she says in a moving TED talk, the city could be fully under the sea. She shares her architectural solutions to deal with recurring flooding in the Asian capital city, through a green roof park project which will help the city absorb water instead of succumbing to recurring flooding and imminent sea level rise. Such green spaces will not only help absorb water and help the city avoid flooding, it will also help alleviate the urban heat island effect by reflecting light (for more on the urban heat island effect, see Chapter 4).

The Thai architect tells how her country lived sustainably with its hydrology.

Before, the Thai, my people, we were adapted to the cycle of the wet and dry season, you could call us "amphibious" living both on land and on water. Flooding was happy event, when the water fertilized our land. But now flooding means disaster. In 2011 Thailand was hit by the most damaging and the most expensive flood disaster in our history. Flooding has turned central Thailand into an enormous lake. The water was overflowing from the North [the Himalayas], making its way into several provinces. Millions of my people including me and my family were displaced and homeless. Many were terrified of losing their homes and their belongings. Our modern infrastructure, especially our notion of fighting floods with concrete has made us extremely vulnerable to climate uncertainty. I cannot sit and wait as my city continues to sink.[71]

In 2018, UCS published a report on US coastal real estate values and the risks they face due to climate change. The report title is ominous, "Underwater."[72] It's all about rising seas and chronic floods. The cover photo is ironic, a sign that reads "Price Reduced: Waterfront" … is supposed to be on land, but water is covering half of the sign. Too late to sell it seems.

The UCS report focuses on 13,000 miles (21,000 kilometers) of coastland and the hundreds of thousands of buildings across the US that are situated dangerously at sea level, including schools, hospitals, churches, factories, homes, and businesses. Rising seas, says the report, will make many of these properties unusable and unlivable. Entire communities will simply cease to exist. The report also notes the somewhat ironic reality that property values have not reflected this real risk. It's as if the real estate market is sticking its head in the sand, simply barreling forward, like it did leading up to the mortgage crisis which ended in the explosion of the housing bubble of 2009. Real estate is turning a blind eye to what is really coming. The warning is very direct:

> "In the coming decades, the consequences of rising seas will strain many coastal real estate markets—abruptly or gradually, but some eventually to the point of collapse— with potential reverberations throughout the national economy. And with the inevitability of ever-higher seas, these are not devaluations from which damaged real estate markets will recover."[73]

UCS reports that 300,000 homes are at risk of chronic, disruptive flooding within the next 30 years with a cumulative risk of $136 billion by 2045. By the end of the 21st century, nearly 2.5 million residential and commercial properties, valued at $1.07 trillion, will be at risk of flooding. The report also stresses a very important point, that is worth repeating over and over: "These risks are masked by short-sighted government policies, market incentives, and public and private investments that prop up business-as-usual choices and fail to account for sea level rise".[74]

We are effectively building a *new* bubble, the climate real estate bubble.

Stephane Hallegates and his colleagues published a study for the OECD in 2013 announcing high risks for coastal dwellers from climate change and accompanying land subsidence (sinking). They estimate the risk for coastal flooding around the world will increase nine fold and that economic losses could be about $52 billion per year by 2050. Depending on what we do to avoid sea level rise between now and then, costs could be in the trillions per year. The study looked at 136 of the largest coastal cities, of which the most at risk were Guangzhou, Miami, New York, New Orleans, Mumbai, Nagoya, Tampa-St. Petersburg, Boston, Shenzen, Osaka-Kobe, and Vancouver. The countries most at risk from coastal flooding include the United States and China. The cities of Miami, New York, and New Orleans account for 31% of the losses across the 136 cities studied. Glacier melt and resulting sea level rise, again, are the lead culprits.[75]

I'd like to end this chapter with these final thoughts that I take from the various experts and studies that I have cited here. I hope that we will all step back and profoundly reflect on what is to come and take it very seriously, doing all that is within our means to change course, as this may be the difference between life and death for millions of people.

- Cities most at risk in the future are not necessarily those that are at risk today
- Port cities face the greatest risk
- Coastal defense actions are important but the associated investments in such defenses ironically increase the influx of population (the irony of taking action)
- If defenses fail, damage will be huge
- If defenses are weak and are breached, damages can be even worse
- Cities will need better crisis management and contingency plans
- Cities that have *not* faced flooding and storm surge in the past will be skeptical and dangerously resist preparing
- For smaller economies, a devastating flood can stall and ruin the economy
- Preparation and planning will save lives
- Ignoring the risks will be costly and will kill

3

Do You Drink Glacier Water? Probably

If I told you that your city was about to cut running water and that supermarkets would be running out of beverages, you'd probably waste little time before darting out to the nearest store (or several of them) and stocking up. I live in a hurricane-prone area and this happens every so often when we are on hurricane alert; it happened again during the COVID-19 pandemic. We all run to the supermarket to buy water before the stores are sold out. There's nothing worse than getting in late on the frenzy, showing up to the supermarket and finding that all of the shelves in the beverage section are empty. Your heart drops and you start imagining your terrible fate when the storm hits and you're stranded at home for several days with no power and no running water, *and* you have nothing to drink.

In the face of an imminent liquid shortage, you'd scramble to fill your garage with all the drinking water, juices, beer, wine, and any other drink of preference. Then you'd probably come to the realization that you couldn't buy a lifetime's supply of your favorite drinks and definitely not of water, and you certainly wouldn't have anywhere to put all that liquid if you could. You'd realize quickly that you needed a Plan B to do something to sustain your liquid intake for the long haul.

We all depend on the continuous, sustainable, and rationed delivery of water to survive day to day, and because it is so abundant and because society's institutions have resolved our need to obtain it in the right dosage over time, most of us take the availability of water for granted. For something so important that we never worry about, it takes only a small glitch in the delivery system to make us very unnerved about its sudden absence. Glaciers are the constant natural water delivery system that few people think about. Glaciers are one of Nature's key institutions that seamlessly take care of water delivery for us.

Glacier water, and water stored in subsurface frozen ground (permafrost), whether you know it or not, is likely an integral part of your water delivery system, and we've basically gotten the signal, worldwide, that it's running out. This is the hurricane about to happen that makes people run to the stores. It won't happen in a week, or a month, or even a year, and maybe not even in the next few decades (although in some places it very well might), but glaciers and permafrost are melting away fast. In some areas, they will disappear entirely and with them so will the natural water delivery systems that we are used to and that we depend on.

This won't mean that we won't have freshwater at all, since we still will have rain and snow, and those water sources will hopefully continue to provide us water for much of the year (although that source is also at risk as intense droughts happen more and more frequently). Water is a critical resource for our society, and its provision to agriculture, industry, and homes is a critical lifeline without which we cannot survive. The problem is that in dry months and especially during prolonged drought years, water shortages will become much more common and a normal part of life, and we'll all have to figure out, as a society, how to deal with *not* having water so readily available when we want it and how to distribute a resource that is becoming more and more difficult to harness and assure.

For communities that already have serious water supply issues—consider California for example—water availability and price are a serious a concern and make up an important variable that conditions how people define their lifestyles *and their budgets*. What most people do not realize is that water is expensive when you don't have easy access to it. Most people across the country probably don't think much about water price. Buying in bulk through water utilities—that's water that simply comes through the tap—for most people is fairly accessible. The average water bill across the United States is about $73 for about 12,000 gallons which is about how much an average sized home with four people might consume.[1] Paying $73 a month for a pretty large amount of water may seem reasonable for most. However, in a state like California, where water scarcity is already the norm due to severe and recurring drought periods, you'll find some of the most expensive water in the United States. In the Monterey Peninsula, for example, the average consumer takes in much less water (about 5,000 gallons per month) but pays a lot more for it. In Monterey 12,000 gallons would cost you more than triple the national average, or $240 per month.[2] Locals in Monterrey spend an average of $1,202 a year for 60,000 gallons. Add that to your monthly utility bills and all of a sudden, access to water becomes a significant portion of your budget, a luxury even.

Californians are obsessed, and rightfully so, with ensuring they conserve water to hold down their utility bill. They must ration everything from the water they use for showers, to dishwashing, to toilet flushing, to watering plants, and they usually cannot splurge on excesses such as carwashes, lawns, or refilling swimming pools. I remember growing up in California and my father placing bricks inside the water tank of the toilet bowl to reduce the amount of water per flush. In some cases, for some uses, excess water consumption is illegal.

We may think gas is expensive when we fill up our gas tanks. But, if in a sudden water shortage you had to buy freshwater at retail value for personal consumption, it would actually be much more expensive than gasoline! The average price of gasoline in the United States at the time I drafted this chapter was about $2.50 per gallon, or about $0.66 per liter. A small 20-ounce bottle of water at the same gasoline store down the street was about $1.45. That would be $9.28 per gallon! That's almost four times more expensive than gasoline! Consider that this comparison is taken at a time when there is no shortage of water. Imagine a time when there is a severe prolonged water shortage. You sometimes hear that in certain developing countries people have to walk for miles with buckets on their heads to get water for personal consumption. Such a day may not be so far off for more of the planet's population. Our buckets in the industrialized world will be our wallets. The price we may have to pay for rationed water could increase multifold! Californians are already there.

While glacier melt is important to the daily water supply of billions of people around the world, we don't really fathom just *how* important glaciers are to our hydrological intake. That's because generally we don't think much about where the tap water or the water in our toilets or from the hose in the garden actually comes from. We just turn the tap and out it pours.

Even if you are more water-conscious than the norm, you still may not realize that the water you consume may have, at some point up the provision chain, derived from glacier melt. We don't tend to associate glaciers with the water supply chain. We might guess that our water comes from a dam somewhere up in the mountains, and that may be correct, but we don't really think about where the water in the lake behind the dam came from. Sure, we'll presume that it came from the rivers upstream from the dam, and before that from the streams that together form the rivers. But where did the streams come from? Maybe . . . from glaciers.

Californians, who are very water-conscious, *do* have the general knowledge that water comes from the annual rain and snowfall, and each year they carefully monitor the rainfall and snowpack up in the Sierra Nevada Mountains to determine how much water they will get that year for the state's water supply. They monitor the snowpack because they know that the snowpack helps conserve water by slowing the water runoff (from the snowmelt) providing their streams, rivers, and dams water for longer periods than rain. But even that knowledge, of a very water-conscious population, is limited when it comes to glaciers. While much of California's drinking water indeed comes from rain and snowfall, a lot of it also comes from glaciers—more specifically from *rock glaciers* and other permafrost features of the periglacial environment (see fig. 3.1). That is something most Californians, even the more environmentally informed, simply know nothing about.

If you live in Louisiana, for instance, you don't really think about the fact that your tap water may come from frozen environments in the Rocky Mountains. In fact, most Louisianans probably have never gone to the Rocky Mountains. You, the reader, if you

Figure 3.1 California's periglacial environment.
This is the extensive frozen terrain of California's periglacial environment.
Note the small points on the mountaintops: these are rock glaciers. Also note that all waterways that come down the mountain (on both sides) derive from glacierized zones.
Source: Google Earth. Mapping: JDTaillant

know anything about US geography, probably didn't make that connection either. It is revealing to pull up Google Earth on your computer, find your home, and try to identify the canals and river systems that lead to your city. You may be surprised with what you find.

Louisianans probably presume that their tap water comes largely from the Mississippi Delta, and most of them *have* been to the delta. They are familiar with *that* water supply. But follow the Mississippi River upstream and you'll eventually get to St. Louis, where the Mississippi and Missouri rivers meet. Turn left, and follow the Missouri River upstream. Pass Kansas City and continue on to Sioux City, Iowa. Keep following the water! Head north to South Dakota. You could keep going to North Dakota or to the Canadian border, but already, all of that area in the Dakotas is in the eastern watershed of the Rocky Mountains of Wyoming and Montana, and yes, you guessed it, this is prime glacier country, and again, like California, this includes subsurface rock glaciers and permafrost terrains that also contribute to the hydrology of the Rocky Mountain basin.

That's how the water that you are drinking in Louisiana from the tap, derives in part from glaciers and other related glacier and permafrost features nearly 1,500 miles (about 2,500 kilometers) away! But for most, the relevance of the glacierized areas upstream from our homes, sometimes many hundreds or even thousands of miles away, is simply not a consideration or even in our frame of reference. We do not connect the dots to understand why glaciers are so important to our daily intake of freshwater.[3] Dahr Jamail, author of End of Ice, stresses the importance of glaciers and mountain ecosystems to our water supply: "It is clear that mountain ecosystems are highly sensitive to climate disruption, and those very ecosystems provide up to 85 percent of all the water humans need, not to mention other species. Globally [he reminds us] glaciers contain 69 percent of all the freshwater on the planet."[4]

Glaciers as a Source of Freshwater

We've already talked briefly about glaciers being a critical source of our planetary freshwater supply. The graph in figure 3.2, taken from the US Geological Survey's website, shows us the global distribution of water supply. From it we see that of the Earth's overall water supply, 97–98% is in the oceans (it's saltwater) and that only about 2.5% is freshwater. Of that relatively small freshwater supply, close to 70% is in glaciers (mostly in Antarctica and in Greenland), while about 30% is in the form of groundwater and 1.2% is surface or "other" freshwater. From these percentages we see that most of the world's water supply is in fact "in storage" in the form of ice, and not being used through the active water supply cycle. Water in ice form also performs other important functions such as reflecting solar rays and cooling our oceans (we'll talk more about that later in Chapter 4), but in terms of water supply, it's not the most relevant source. Limited portions of this stored water at a global scale are actually taken out of storage and transformed from ice to freshwater for our ecosystems to utilize.

Because of changing and rising ambient temperatures and because of the interaction of glaciers with their physical environment, glaciers, particularly mountain

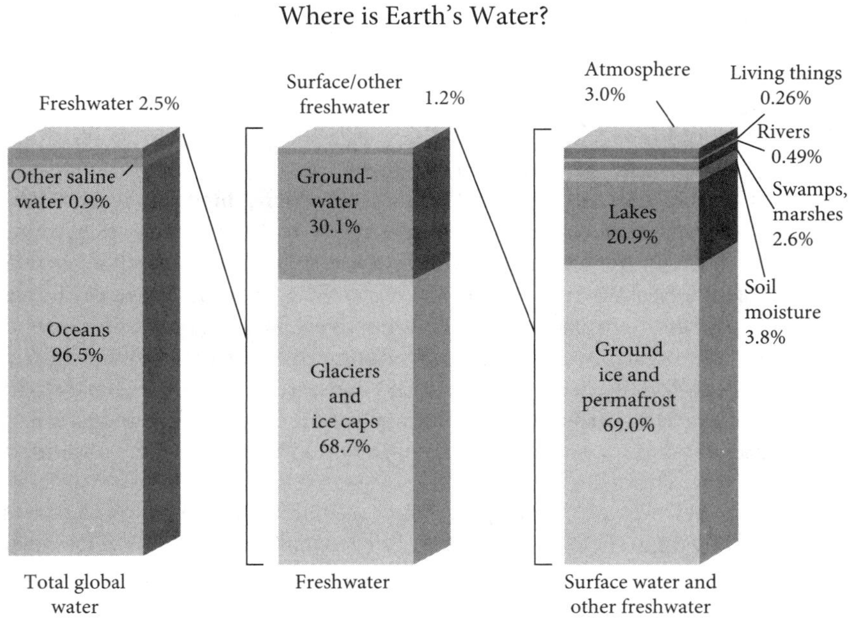

Where is Earth's Water?

Freshwater 2.5%

Surface/other freshwater   1.2%

Atmosphere 3.0%

Living things 0.26%

Other saline water 0.9%

Oceans 96.5%

Ground-water 30.1%

Glaciers and ice caps 68.7%

Lakes 20.9%

Rivers 0.49%

Swamps, marshes 2.6%

Soil moisture 3.8%

Ground ice and permafrost 69.0%

Total global water

Freshwater

Surface water and other freshwater

Figure 3.2 Where is Earth's water?

Source: Shiklomanov, Igor. World Fresh Water Resources. In Water in Crisis: A Guide to the World's Fresh Water Resources. Ed. Peter H. Gleick. Oxford University Press. 1993

glaciers, regularly melt off portions of their ice into their downstream glacier water basin. A glacier's surface may melt because the air around it is warm or because of heat at its base. It may also melt because of rainwater deposited at the surface, which is particularly effective in causing glacier melt. Or a glacier's surface may melt because of fine particulate matter from air pollution that finds its way onto the surface, causing it to darken and thereby increasing the glacier's capacity to absorb heat. (More about that in Chapter 4.)

As a glacier's surface and its internal ice melts and creates water flow, this water works its way down to the lower end of the glacier, and in many cases, glacier lakes are formed at the base. These lakes subsequently drain into lower water basins.

Whatever the reason for its melt, glacier melt induces water flow and water flow enters the downstream ecosystem through streams, rivers, and underground water networks, feeding plants, animals, people, farmland, and industry.

Glaciers and Subsurface Ice for Global Water Supply

The "other" category in the middle bar of the graph in figure 3.2 turns out to be quite significant. And while the amount of freshwater in all of the more well-known glaciers of the planet is very significant, most of that water is not as significant for human consumption and water supply needs as that much smaller 1.2% contained in the surface and *other* freshwater category. So let's look at that minuscule percentage that is so critical to human survival.

The bar on the far right of the graph reveals the percentages of fresh water supply that we see and can access on the surface of the Earth, or from what the graph calls "other." Of this, 21% is in lakes. We just mentioned that many of those lakes are fed by melting glaciers, so at least some of that supply is glacier meltwater. Other portions include seasonal snowmelt, rain, etc. Only 0.49% (that's less than 1%) of this freshwater is in our rivers, while 3.8% and 3% are in the form of soil moisture and in the atmosphere, respectively. Subsurface invisible ice and permafrost meanwhile, accounts for a whopping 69% of this remaining freshwater. This reserve is critical, just as glaciers are critical to provide basins with a steady flow of melting surface water. Permafrost nourishes surface and subsurface water at its lower fringes, continuously melting into the environment and into the downstream hydrology.

This reserve of frozen grounds high in water content includes rock glaciers, which we will talk about in more depth in Chapter 8. These are actually rivers of ice covered entirely by, and inter-mixed with, rocks and earth, holding vast amounts of subsurface water supply. Permafrost,[5] which is defined as ground that is permanently frozen for all or most of the year, can also contain ice in nooks and crannies beneath the surface of the Earth. As rain and snow falls on the surface, seeps into the ground between rocks, and freezes in the very cold subsurface temperatures, permafrost is formed. We may not realize it, but much of our water supply resides in these subsurface icy deposits. How big is this underground ice reserve?

If we consider this in terms of the Earth's total water supply, the percentages of these frozen underwater reserves seem small, but when we remember that most of the Earth's freshwater is locked away in Antarctica or in the Greenland Ice Sheet, those small percentages left for us to consume start to look a lot bigger. Society gets most of its freshwater for daily use from natural and artificial lake reservoirs, and compared to these surface water reserves, the volume of water stored in subsurface ice reserves, which continuously feed into the broader surface and groundwater supply, is comparatively significant. Ground ice and permafrost account for more than three times the water supply held in freshwater lakes. By comparison, the freshwater contained in subsurface ice is over 140 times greater than our global river water supply.[6] These significant subsurface ice reserves are especially important considering that they are commonly found in high mountain ecosystems upstream from significant portions of global populations. They are also feeding the lakes from which we are taking water. The dynamics of, and the relationships between, subterranean ice reserves and groundwater systems are mostly unknown and poorly understood as the technology to study and understand this system is woefully inadequate, but there is clearly an intricate hydrological relationship playing out beneath the ground between ice and the ecosystems below.

We'll focus more on subsurface ice resources in Chapter 8 when we look at invisible glaciers of the periglacial environment.

The Water Content of a Glacier

One of the underlying and major concerns (though not the only one) about glacier melt is that as glaciers melt, we lose the water they held in reserve. So a key question to answer is exactly *how much* water is in a glacier? The simple answer is *a lot!*

Let's look at a high mountain environment glacier, since this type of glacier is likely to be a significant source of water supply for a downstream community—unlike a colossal glacier in Antarctica that only provides freshwater to a few thousand scientists at any given time. Here we find that even very small high mountain glaciers can be significant in terms of the water they supply to their local environment. Precisely *how* they supply it is even more important.

There are for example, very small, thin elongated glaciers on Mt. Lebanon (in Lebanon ... *yes, there is glacier ice in the Middle East!*) (fig. 3.3). If we consider their total surface area, these small individual glaciers are about the size of a football field. This is generally the minimum threshold size for glaciologists to consider this a veritable "glacier." Some call glaciers that are comparatively small "glacierets" or ice patches. Many scientists that study large glaciers would not call these glaciers, but indeed, as we saw in the definition of a glacier in the Argentine Glacier Protection Act in Chapter 1, they are in fact *functioning* in many ways as the colossal glaciers of Alaska

Figure 3.3 Perennial ice resources (small glaciers or glacierets) near the town of Becharri on Mt. Lebanon in Lebanon (in some places above 3,000 meters) can hold significant volumes of water.

Source: Habeeb http://www.habeeb.com/images/lebanon.photos/Mount%20Lebanon/

or the Himalayas do, providing and rationing water for their ecosystems at some of the most critical moments of the year.

Surprisingly, if we were to tap these small glacierets for human consumption, we would discover that they can provide an entire family of four with water for drinking, bathing, cleaning, and for other basic household uses for an entire generation.

Let's do the math.

1 cubic meter (m³) of ice	=	approximately 1,000 liters of water[7] (3.8 liters is approximately 1 gallon)
1 small glacier (100 m²)	=	approximately 10,000 m³ of ice (that's 100 m × 100 m surface by meter deep)
	=	10,000,000 liters of water (that's 10 million liters of water)

1 person utilizes about 200 liters of water per day per person (drinking + household)
1 small glacier, hence, can provide 1 person with 50,000 days' worth of water (10,000,000/200)

That's 137 years' worth of water for just 1 person or 34 years for a family of 4 (137/4) In this example, we've presumed that the glacier is only 1 meter thick. But, even the smallest glacierets can be several meters thick, which would greatly increase the water supply stored in their interior. If the family of four in the example could tap this glacier for its water supply, it would have water for an entire century from this one small glacieret! Remember however that glaciers are not single-use, and in fact, their water supply is consistent and can provide a regeneration of water supply for many generations.

If we took a bigger glacier, say a glacier that was a mile or two long (1.6–3.2 km), three or four hundred meters wide (more typical of the size of a mountain glacier), and dozens of meters thick, we can see how the water supply of a single mountain glacier can be phenomenal, providing entire communities with water storage and rationed water supply for decades or even centuries.

Now let's consider a much larger glacier, one typical of the Himalayas or Alaska. The Yalung Glacier in Nepal, *a truly large glacier* (although not the largest), in the valley just under Mt. Kangchenjunga (the third highest peak on the planet) has a mind-boggling water volume (fig. 3.4).

Let's do the water supply math calculation for the Yalung Glacier.

1 cubic meter of ice	=	approximately 1,000 liters of water (3.8 liters is approximately 1 gallon)
1 large glacier		17 kilometers long, or 17,000 meters long by 700 meters wide by 100 meters thick (that's a conservative estimate on average thickness)
	=	approximately 1,190,000,000 m³ of ice (that's 1.2 billion, by the way)
	=	or 1.2 trillion liters of water!

The capital city of Dhaka, Bangladesh, which is downstream from the Yalung Glacier, is home to some 8.9 million people, who each consume about 100 liters of water per

Figure 3.4 The Yalung Glacier in Nepal, a key water supply source for people living in its water basin.
Source: Google Earth. GIS: 27 36 20.88 N, 88 02 57.48 E

day. Multiply it out, and we find that the water contained in the Yalung Glacier could provide water to the entire city of Dhaka for nearly four years![8] But again, the value of the Yalung Glacier is not for the short term and for single use, this glacier has been around for centuries, always regenerating its volume year after year, providing everyone downstream from it with a critical, sustained and carefully rationed water supply, all the time, never failing.

What's critical to understand in the case of the Nepalese (Himalayan) glaciers is that most of the world's population is centered in the water basins of these colossal ice reserves high up in the mountains. If they melt, all of a sudden, and disappear, which is actually happening, a whole lot of people could be out of water, just as happens every time we have a hurricane in South Florida and the water is cut off. And most of these people can't afford to run to the supermarket and buy water at retail value.

Glacier Water Basin Regulation

In terms of a glacier's water supply, of most importance is not *how much* water is stored in a glacier but rather how that water is dispensed rationally and sustainably over time. This is determined by the very nature of the glacier's environment (its *glaciosystem*) and the dynamics of interaction between its ice and the ambient temperature changes that are occurring at the glacier and to the glacier, year round.

Glaciers are extremely efficient at capturing precipitation (in the form of snow and rain) and holding on to it for when the ecosystem needs it most. Glaciers are creatively designed to store water for later use as opposed to having that water quickly run downstream in an initial flow into mountain rivers (when spring arrives). They are also specially adapted to hold back the evaporation of water, which occurs more rapidly when water is held in a dam. In this regard glaciers beat lakes at avoiding evaporation.

Glaciers delay water release into the environment, which as we can see is one of their most important functions. But here is the interesting thing, glaciers don't simply melt their water into the ecosystem and get smaller. If they did a glacier would only be good for a one-time use like plastic straws: you tap it for all of its water and then it's gone and you go on to the next one. Instead, Mother Nature was far wiser when she designed glaciers to create a recurring, renewable, and sustainable high mountain water tower. Glaciers, because of where they are located (in freezing environments) are good at recharging themselves with recurring snow and rain, so they always have the same amount of water in storage for constant use and reuse.

Glaciers melt into the basins below them when it is warm and then recharge from above when it is cold. In fact, they are continuously recharging. When glaciers are in equilibrium, they are constantly melting and recharging after each yearly snowfall and snowmelt cycle. That recharge can be between seasons or even in monthly or daily patterns according to the local weather cycles.

Glaciers capture water in the form of snow or rain at their higher elevations and turn this snow into ice, storing it in the glacier's body. Most of the glacier's volume is frozen solid because the local temperature around the glacier is at or below freezing for most of the year. Even if the surface of the glacier may be warmer on some days, weeks, or months, it's inner core is very cold. If you've ever visited a glacier in the summer, you know it can be quite warm at the glacier. In fact, last year I visited several glaciers in Alaska, and it was warmer at the glaciers for a few days during my visit than it was in my home state of Florida. But don't be confused, that only happens for a few months of the year. The rest of the time, the temperature at a glacier is low and usually below freezing.

Over time, the glacier ice accumulating from fallen snow and water converted to ice and entered into the glacier's volume reaches the point where the entire glacier starts to slowly flow down the mountain toward lower, warmer elevations. Glacier ice flows because overall the glacier is heavy, because it is on an incline, and because its base is wet, lubricating and facilitating the flow.

The very front (or lowest elevation area) of the glacier, called the *terminus*, is much lower on the mountain than the top (or root area) of the glacier. This is important because the lower you go on the mountain (and on the glacier) the warmer it is. Once the ice located in the area at the front of the glacier reaches the lower and warmer elevations, that region of the glacier begins to melt off slowly, feeding water into the ecosystem. Meanwhile, the higher elevations of the glacier are busy collecting more snow and converting it to ice.

This distinct spatial distribution of the functions of parts of the glacier in relation to temperature varies over time and is not always the same in the different areas of the glacier. In the wintertime, the warmer air is lower on the mountain, so the glacier ice is mostly protected—in fact, all of the glacier may be completely frozen. But in the summer, the warm air reaches higher onto the glacier surface where it begins to melt the ice. What's important for the glacier's overall survival and its sustainability is the *average* temperature throughout the year and over many years on the various elevations of the glacier.

Remember, a glacier can be a mile or even several miles long, so the higher up the mountain the colder the surface of the glacier will be. That's why for the most part, higher portions of the glacier are not melting (on average). It's the lower parts that lose more ice by converting it to water through melt. The water supplied by a glacier comes mostly from its warmer portions. This doesn't mean that upper portions can't melt away (they do on warm days). It simply means that on average, the higher portions remain mostly frozen while the lower portions, comparatively, melt and lose ice faster. In fact, whenever we refer to a glacier melting over time (years or decades), we always note in successive pictures taken of the glacier year to year, that the glacier is *receding* up the mountain. That's because the warmer areas of the mountain are constantly creeping up the mountain, as is the melting point where the glacier can no longer survive. The melting point is usually the lowest point of the glacier.

The glacier's lowest end will be where the average yearly temperature on the mountainside is no longer below freezing. This is the glacier's terminus area. We call this melting point area the Zero Degree Isotherm line (0° isotherm) since above that line, the air is, on a yearly average, *below* 0°C (or 32°F, the freezing point). Below that line on the mountain, the average yearly temperature is positive. So above the 0° isotherm, water freezes and stays frozen (necessary for glacier creation and glacier health), and below it water doesn't freeze permanently and any ice melts (the cause of the glacier's melting and poor glacier health). In most cases (with some exceptions we won't get into here), glaciers *cannot* be located below the 0° isotherm, at least for prolonged periods of time. If they are, then you get net glacier melt and in time, the glacier will disappear.

The glacier is a water storage machine continuously working to create water supply. It's constantly collecting snow and rain in its higher altitudes, converting it to ice, and sending it downhill until the point where the ice can no longer survive the warmer temperature. The glacier sheds the melting ice through ice melt during warmer days, weeks, months, or, as the case may be, years.

Let's consider snow and rainfall, and how it differs if it falls on warm land vs. a freezing glacier. When it rains, the precipitation falling on the surface of unglaciated portions of the Earth flows downhill immediately and fills the streams and rivers below. In the case of precipitation on a glacier (probably in the form of snow), the presence of sustainable ice helps retain the precipitation, by keeping it frozen for a time, effectively *slowing* the rain or snowfall runoff water flow so that there is a *slow but regular* flow of water to the rivers and streams at all times. If it weren't for glaciers, that original snowfall and rain would simply immediately enter the rivers and streams at the time it fell on the surface of the Earth, and we would lose out on the delayed release function provided by the glacier. If it weren't for glaciers, the winter snowfall would quickly melt when the temperature rose in the spring and summer, and then there would be no more water for the rest of the year, until the cycle repeated itself.

The glacier's water retention capacity and slow release into the ecosystem is what we refer to as "water basin regulation."

You can think of the glacier as a very large faucet in the mountain, opened ever so slightly to allow for a continuous flow of water into the lower ecosystem. You do the same thing with your tap water in your home that Mother Nature does with glaciers. While you have a steady supply of water at your tap, you don't leave the tap open, otherwise you'd quickly drain your water supply (or pay an enormous water bill at the end of the month). You may have inadvertently done this on occasion with your hot water supply, which is stored in a hot water heater. If you leave the hot water running, you'll quickly run out of water until the water heater can refill. Mother Nature stores and refills your water supply naturally, with glaciers!

Glacier Water Supply during Droughts

I mentioned earlier that glaciers provide water during extended drought years. This is a critical role that glaciers play during periods when snow and rainfall amounts are low. Californians are obsessed each year with the snowpack in the Sierra Nevada Mountains. Why? First of all because the snowpack melt fills their dams, but also because they need the snowpack to melt slowly over time. However, when the snow doesn't fall, or the rain fails to come, glaciers and ice-rich underground permafrost take on an even more important role.

In a particularly dry period, which might be a month, a year, or several years, meltwater from glaciers or subsurface ice runoff can become the only sustainable and constant source of water runoff feeding the lakes and rivers of high mountain ecosystems. And while the portion of glacier meltwater streaming into the hydrology of a state like California during a wet year might seem small or insignificant, during prolonged droughts like those that have recently been occurring in the state, and which are becoming the new norm under current climate change trends, they can be a significant percentage of total water supply.

Lake Evaporation vs. Glacier Water Storage

We mentioned above that glaciers are better than lakes at holding back evaporation. This is another important beneficial feature of the hydrological functioning of glaciers relative to water supply. Glaciers do a much better job than lakes and dams at retaining water. As Michael Baraer of McGill University in Canada explains, limiting evaporation is a key function of glaciers. Water in liquid form evaporates more quickly than water stored in ice, which sublimates off of the surface in the form of gas—similar to the steam you see evaporating from your lawn in the morning hours. When considering future water supply, the evaporation factor is critical. Faced with a choice between storing your winter snow in glaciers as opposed to creating dams to capture snow melt in the spring, in terms of evaporation you'd be much better off holding it in glaciers. So even if you thought that snowfall filling dams would take care of water storage, in fact, slowing that fill and dispensing it over time through glacier water capture is a much more efficient system. In other words, a hydrological system combining dams with glaciers is far more efficient than simply relying on dams. Glaciers are a great way to hold that snow in storage and they do a better job of not losing it through evaporation.

Building more dams to replace glaciers will not solve the problem of decreasing runoff that we face. Even if precipitation patterns are sustained in mountain environments, more water will occur in liquid form, says Richard Lovet of National Geographic who wrote an in-depth article on the water supply troubles caused by melting glaciers. In the article, Michael Baraer of McGill stressed: "Evaporation from reservoirs is much higher than sublimation [the conversion from ice directly into gas]. Dams will never, ever, replace the [natural] hydrological systems that are in place today."[9]

Global Dependence on Glacier Water

I've already mentioned that some 70–75% of the world's freshwater is held in glaciers, and while most of it is in the large ice sheets of Antarctica and Greenland, a very significant amount of that freshwater that is critical to the survival of entire populations is actually in mountain glaciers.

Glacier drainage basins, that is, water basins that are fed by high mountain glaciers around the world, are extensive, covering 26% of the global surface outside of Greenland and Antarctica, and hold approximately one third of the global population.[10]

The chart presented here is an excerpt (not exhaustive) of *some* US states and countries with high mountain environments whose populations depend, *at least in part*, on glacier water flow.

Mountain Range	Countries/States Fed by Glacier Meltwater
Rocky Mountains	Montana, Wyoming, North Dakota, South Dakota, Colorado, New Mexico, Nebraska, Oklahoma, Texas, Missouri, Arkansas, Louisiana
Cascades and Sierra Nevada, et.al	Washington, Oregon, California, Nevada, Arizona
Alaska Range	Alaska
Canadian Rocky Mountains	Canada
Andes Mountains	Colombia, Venezuela, Ecuador, Peru, Bolivia, Chile, Argentina, Brazil, Uruguay
Itzaccihuatl and Popocatepetl	Mexico
Spanish Pyrenees	Spain, France
European Alps	France, Italy, Austria, Germany, Switzerland, Slovenia
Scandinavian Mountains	Sweden, Norway, Finland

Mountain Range	Countries/States Fed by Glacier Meltwater
Mount Ararat	Turkey
Mount Aragats	Armenia
Caucasus Mountains	Georgia, Russia, Azerbaijan
Hindu Kush Mountains	India, Pakistan, Tajikistan, Afghanistan
Gissar Mountains	Uzbekistan, Tajikistan
Pamir Mountains	Kyrgyzstan
Tian Shan Mountains	Kyrgyzstan, China, Kazakhstan
Tavan Bogd Mountains	Russia, Mongolia, China
Himalaya Mountains	India, Pakistan, Nepal, China, Bhutan, Bangladesh, Myanmar, Thailand, Cambodia, Laos
Kilimanjaro	Tanzania, Kenya
Mount Ngaliema	Uganda, DR-Congo
Puncak Jaya	Papua

Large portions of North America depend on glaciers for land irrigation, including any place downstream from the Rocky Mountains in Canada (Alberta and British Colombia) and in the United States (Montana, Idaho, Colorado, Utah, North Dakota, South Dakota, Nebraska, Kansas, Oklahoma, Texas, New Mexico, and Arizona) or the high peaks and sierras of the West Coast including the Sierra Nevada and Cascade ranges (Washington, Oregon, California, and Nevada). And as we saw in the earlier example of Louisiana, even states that are very far from the glacierized mountains or cold mountain environments with permafrost are also likely obtaining at least part of their water supply from glacier- and permafrost-fed water basins.

Most South American countries are extremely dependent on glacier melt for water provision, particularly all along the Andes Mountains in Chile, Argentina, Peru, Bolivia, Ecuador, and to some extent Colombia and Venezuela. Argentina and Chile in particular have extensive frozen earth, permafrost, and rock glaciers. Large swaths of land, including prime agricultural lands in Chile and Argentina, rely heavily on glacier water and permafrost for irrigation. Even Brazil and Uruguay, which don't have *any* glaciers, depend on the flow of the Amazon river, and yes, you guessed it, if you follow the Amazon to its origins, it is born in glaciated regions and watersheds of the Peruvian Andes.

Michael Baraer of McGill notes that countries like Peru are already on the verge of water shortages caused by decreased glacier runoff. The Rio Santa in the Peruvian Andes, which has very large glaciers feeding water to millions of downstream communities, is already running low on glacier-fed water.[11]

Do Melting Glaciers Provide More or Less Water as They Melt?

As glaciers melt away and recede due to climate change, we might think that they release more water. This is true at the onset when, due to global warming, temperatures first rise and glaciers that had been stable begin to melt rapidly. At first, glacier runoff can grow significantly, up to about 50% of normal runoff.[12] This may be welcomed in some areas that need the water but the intensity of the melt runoff and the sheer volume can also cause severe or even deadly flooding and landslides. We'll talk more in Chapter 6 about what are known as Glacier Lake Outburst Floods (GLOFs) or as I refer to them, *glacier tsunamis*. Tragically, as I was editing this very chapter, a massive GLOF occurred in Uttarakhand, India, caused by a collapsing glacier (likely due to climate warming trends) and killed over a hundred people, destroying homes and infrastructure.[13]

In terms of long-term water supply, in a warming climate, over time (years, decades, or centuries) shrinking glaciers reduce in volume (vertically and spatially). If you were observing them from above over the years, you'd notice a receding of the ice volume up the mountain as well as a deflation of volume. This means there is less surface ice to melt and so after the initial increase in melt runoff, *peak melt-off* is reached. After that point, year after year and decade after decade, glacier melt runoff decreases even though the glacier itself is tragically bleeding out its meltwater. What this means for water supply for local ecosystems and downstream communities is critical, since as glaciers dwindle and eventually disappear, dry seasons or drought years will have less and less water from glacier runoff to cover downstream ecosystem needs.

Matthias Huss and Regine Hock, who study glacier melt runoff, indicate that in half of the 56 large-scale glacier runoff basins they studied, peak melt-off has already been reached, indicating that those basins are already losing water supply. Especially concerning are basins dominated by smaller glaciers such as in North America, Central Europe, and South America, where peak melt-off has also already been reached and where significant water supply reductions are expected in the coming decades. By the end of the 21st century, the authors note that 93% of the basins will experience seasonal glacier runoff reductions.[14]

Western and Eastern Europe also depend on glacier meltwater to feed their hydrological basins. Anyone who has traveled through the European Alps region, for instance, has undoubtedly seen raging waters traversing some of Europe's most important urban and agricultural centers. Countries like France, Switzerland, Austria, Italy, and Germany, among others, depend on glacier melt for their livelihoods. Additionally, countries such as Turkey, Lebanon, Georgia, Armenia, Azerbaijan, Bulgaria, and even Syria, have glacier or glacial ice (in glacierets, rock glaciers, and permafrost) that contribute water runoff to their regional hydrology.

In Central Asia, glacier meltwater provides some 70% of the area's major river systems during dry and hot summers for Kazakhstan, Kyrgyzstan, Tajikistan, Uzbekistan, and Turkmenistan.[15]

And then we get to the Himalayas. In glacier speak, the Himalayas are often referred to as the *Third Pole* because there is simply so much glacier ice in this high and cold mountainous region. The countries of this region and their water runoff basins are some of the most glacier-dependent nations on Earth and are also home to some of the world's poorest populations. India, Pakistan, Nepal, China, Bhutan, Bangladesh, Myanmar, Cambodia, Thailand, Vietnam, and Laos all depend on glaciers for their year-long hydrology, for agriculture and for human water consumption.

I lived in Cambodia for about two years in the 1990s and had to regularly cross the Mekong River to travel from Prey Veng Province (near the Vietnamese border) to Phnom Penh, Cambodia's capital. The flow of the Mekong is truly incredible, but so far from the Himalayas, most people don't associate the local water with mountain ecosystem hydrology. The vulnerability of Himalayan glacier melt, however, could kill this key lifeline river for so many Asian nations.

The glaciers from this region provide water into 10 major river basins and provide critical ecosystem services (water, food, and energy) that directly sustain the lives of 240 million people in the mountain regions, not to mention downstream communities, which are also in the hundreds of millions. Glaciers in the Tibetan Plateau, for example, hold the largest ice mass outside the polar regions. Many scientists today are studying the impacts of atmospheric contamination on these glaciers because they are so significant for such a large portion of the population. We'll talk about that in Chapter 4.

Glaciers in the Tibetan plateau area feed water to the Indus, Ganges, Brahmaputra, and other river systems, providing drinking water to over one billion people. Glacier melt provides up to two-thirds of the summer flow in the Ganges and half or more of the flow in other major rivers. In China, one-quarter of the population lives in western regions where glacial melt provides the main source of water during drier and warmer months.[16]

But if you consider the flow of water from this region into basins below, nearly 1.65 billion people live downstream of these glaciers, while 3 billion people depend on the food produced in these basins.[17] The links between glacier-fed rivers and agriculture in this part of the world is fundamental.

The UN's IPCC 2019 Report on Climate Change the Oceans and the Cryosphere states in relation to water supply:

> Melting glaciers can affect river runoff, and thus freshwater resources available to human communities, not only close to the glacier but also far from mountain areas. As glaciers shrink in response to a warmer climate, water is released from long-term glacial storage. At first, glacier runoff increases because the glacier melts faster and more water flows downhill from the glacier. However, there will be a turning point after several years or decades, often called "peak water," after which glacier runoff and hence its contributions to river flow downstream will decline.[18]

The report goes on to say, "glacier decline can change the timing in the year and day when most water is available in rivers that collect water from glaciers."

In early 2020, a group of scientists published what they call the Water Tower Index (WTI) analyzing the significance of high mountain environments, their climate vulnerability, and the effects of climate impacts on them and subsequently to society. They found that the most significant environments that comprise the WTI are also among the most vulnerable. According to the authors, 1.9 billion people living in or directly downstream from mountainous areas could be negatively impacted. The authors specifically refer to the mountain's buffering capacity for water supply, including its capacity to store water in the ground and in freezing environments where glaciers and permafrost are found. They also state that:

The vast majority of glaciers are losing mass, ... leading to future changes in the timing and magnitude of mountain water availability. [Further] the combination of cryosphere degradation and increases in climate extremes implies changing sediment loads affecting the quality of water supplied by mountains.[19]

We should also be concerned that as glaciers melt the water supply flowing from them might release certain toxic chemicals that are trapped in the ice, deposited over the years, decades and centuries by anthropogenic pollution that made its way to the glacier surface. The IPCC report mentions such risks:

Glacier decline can influence water quality by accelerating the release of ... legacy pollutants notably ... persistent organic pollutants (POPs), particularly polychlorinated biphenyls (PCBs) and dichlorodiphenyl-trichloroethane (DDT), polycyclic aromatic hydrocarbons, and heavy metals. ... mercury is of particular concern and an estimated 2.5 tonnes has been released by glaciers to downstream ecosystems across the Tibetan Plateau over the last 40 years. ... Both glacier erosion and atmospheric deposition contributed to the high rates of total mercury export found in a glacierized watershed in coastal Alaska.[20]

Glaciers and Biodiversity

Even small glaciers and their runoff can be extremely significant in providing water *and life* to local ecosystems just as alterations to glacier-melt runoff can cause enormous disturbances to river biodiversity. According to Alexander Milner, a professor of river ecosystems at the UK's University of Birmingham, the disappearance of a glacier can lead to a drastic drop in the biodiversity of the streams fed by that glacier. A 2012 report co-authored by Milner and published in Nature Climate Change found that the disappearance of glaciers could cause a drop of 11–38% in the number of species of macro-invertebrates in a river, and the loss of these little life forms in downstream river environments often results in the parallel collapse of whatever feeds on them, including fish, amphibians, and birds.[21] Milner went on in 2017 to publish a more in-depth look at dwindling glacier runoff and its impact to local water biodiversity in rivers, concluding:

Glaciers impart unique footprints on river flow at times when other water sources are low. Changes in river hydrology and morphology caused by climate-induced glacier loss are projected to be the greatest of any hydrological system, with major implications for riverine and near-shore marine environments. ... Glacier shrinkage will alter hydrological regimes, sediment transport, and biogeochemical and contaminant fluxes from rivers to oceans. This will profoundly influence the natural environment, including many facets of biodiversity, and the ecosystem services that glacier-fed rivers provide to humans, particularly provision of water for agriculture, hydropower, and consumption.[22]

Snowline Upward Creep: Another Glacier Enemy

I mentioned earlier how glaciers charge and recharge their ice volume each season by capturing snow and rainfall and converting it to ice and explained the importance of the 0°C isotherm line above which temperatures remain freezing for most of the year and how that line is slowly creeping up the mountain, crowding glaciers into tighter and tighter spaces. I've also stressed that ambient temperature is critical to establishing a glacier's equilibrium. But just as important is the *snowline*, that is, the actual place where snow starts to fall and settle on the ground as you go up the mountain. If you've ever driven up into the mountains when it's snowing, there is a moment when the snow appears on the surface of the mountain. That's the snowline at that given moment on that given day and time. This line varies year round, since temperatures vary, as does snowfall on different parts of the mountain from year to year.

The snowline in the middle of winter may be at the very bottom of a mountain or in the summer not exist at all. But where that line is year to year and especially where it is relative to existing glaciers or other forms of permanent ice (such as subsurface ice) is critical for the ice's survival, since glaciers and permafrost resources need cyclical snowfall to recharge after melt off. A sustained shift of the snowline in the right direction (down) can be a bonanza for a glacier's health, just as a sustained shift in the wrong direction (up) can mean its demise. If a glacier is located at 10,000 feet (3,000 meters) above sea level and the average snowline is at 5,000 feet (1,500 meters) it may be receiving snowfall for much of the snow season. However if the snowline begins to move upward and eventually reaches the glacier, it may be getting much less snow on average during the year. The average snowline may have moved up beyond the area where the glacier rests. That would be tragic for the glacier if the upward movement of the snowline were sustained.

In Yosemite, according to a 2008 paper from Bruce McGurk, hydrologist for the Hetch Hetchy Water and Power dam in California (nestled into the foot of the Sierra Nevada under Yosemite National Park), the snowline of the Hetch Hetchy watershed is expected to rise from 6,000 to 8,000 feet by the end of the century. Because of changes to the snowpack and the progressive rising of the snowline over time, California's snowpack is expected to decline by 25% statewide by 2050, according to the state's Department of Water Resources. This results in water rationing for urban areas around cities like San Francisco and Los Angeles and deteriorating quality of agricultural lands in places like California's Central Valley, which produces a quarter of the country's farm products. The Delta region around the San Joaquin and Sacramento Rivers provides water to some 25 million Californians.[23]

In California's few remaining glaciers, the amount of ice captured from each recurring snowfall could be substantially reduced over time, due simply to the fact that less snow is falling on the glaciers, which in turn means that the few remaining glaciers will have less capacity to regenerate their volume as they melt. In other words, the glaciers may be cold enough to regenerate (which with current climate trends this probably isn't the case), but if less snow is falling on the glaciers, irrespective of temperature, they will decrease in size over time as they melt off more snow than they receive each winter. When John Muir, naturalist and self-trained glaciologist, explored Yosemite, the surface of Lyell Glacier, California's largest glacier, was at least 100 feet above where it is today. It has completely deflated!

Another important consideration is the interrelationship between glaciers, snowfall, and the greater mountain hydrological ecosystem. When considering the role of the cryosphere in places like the Rocky Mountains, the Cascade Range of the Pacific Northwest, or of the Sierra Nevada Mountains in California we have to consider these hydrological systems as a whole. Water provision from these mountains (as in most glacierized high mountain ecosystems) clearly doesn't come entirely from glaciers. It *does* come from a complex and delicate hydrological ecosystem where glaciers are collaborating with other elements of the cryosphere, including snowpack, permafrost, rock glaciers, etc..

The snow that falls each season, on which glaciers thrive, covers the mountaintops and the forests, slowing down water flow that is critical to the local and broader ecosystem. That snow might otherwise fall as rain, creating a very different hydrological provision to the environment. This blanket of snow soaks the earth and keeps it moist for a much longer period of time than rain, which would quickly flow to the streams and rivers and out to sea. Glaciers (and frozen grounds in the earth that also harbor ice) are the tail-end of that slow flow system. Rain drains fast with very little water storage capacity. Snow drains more slowly and has more storage capacity (for a few months). Glacier ice (and subsurface permafrost) drains most slowly and has year-long storage and flow capacity. The entire mountain cryosphere—along with all of its pieces—is working to ensure that downstream ecosystems and populations that reside below are getting their water supply year-round.

Less snow also means fewer avalanches and this actually causes serious impacts to animals that depend on snow avalanches for their survival, says Dahr Jamail, author of the End of Ice. One would think that fewer avalanches might be a good thing—certainly skiers would say so! However, Mother Nature has a way of not wasting energy or effort, and as Jamail explains, even avalanches play a critical role in the cryosphere and in the ecosystem more broadly:

Snow avalanches disturb the mountain slopes. Their paths are ecologically valuable and are often referred to as "bear elevators" because bears use them to graze. Larger avalanches carry trees and soil to the valley bottoms, depositing them in the streams, which can rework the woody debris into trout habitats. Avalanches can also contribute microbes to the water, which go up the food chain to the fish. Fewer avalanches mean less food for fish to eat, which of course means fewer fish.

Without snow, the competitive relationship between animals are altered. Wolverines, lynx, and pine marten, which have adapted to the environment, are now dependent on the snow. For example, wolverines are only found in the boreal life zone, which is dominated by snow. Wolverine kits are denned in snow, and wolverines store food in snow as well. They are especially at risk because their capacity to adapt is relatively low.[24]

Conclusions

We generally do not think about where our water supply comes from. We just turn on the tap, or trigger our hoses, and the water flows. Much of the developing world today, however, struggles to secure water and is much more aware of water scarcity. Glaciers are not only an important water supply storage facility holding 70–75% of our world's freshwater, but the rationing function that glaciers provide is the key to guaranteeing water for billions of people around the world during drier months of the year, particularly during prolonged droughts.

In Asia, the threat of disappearing glaciers is one of the most frightening scenarios faced by billions of people, but similar impacts to other communities that live downstream from glacier basins around the world are just as concerning. Once headwater glaciers are gone, say Baiqing Xu and his colleagues, a dramatic decline in dry season water availability may ensue. And while total precipitation may increase (that is, more rain) in certain areas of the world due to the changing climate, the likely result of glacier loss will be heavier spring floods but reduced freshwater availability during dry seasons and during prolonged droughts. Over time, accelerated glacier melt will reduce overall water supply.

In Kazakhstan, the Tuyuksu Glacier, which is about a mile long and a mile and a half wide, is "melting like mad," report Henry Fountain and Ben Soloman of the New York Times, two climate change beat reporters that have been documenting climate change impacts across the globe. They interviewed Maria Shahgedanova, a glaciologist at the University of Reading in England, who studies the Tuyuksu Glacier to get a sense of what climate change is doing to the region's water supply. This glacier feeds into the water basin downstream, becoming the Little Almaty River, providing drinking water to the region's two million people and irrigation water for corn and other crops in the area. In just one year, Shahgedanova's measurements of the glacier indicated three feet of thinning. In the last six decades the glacier has lost about half a mile in length. The accelerated melting may not have reached the runoff peak (the point where the water derived from accelerating melting starts to decrease in volume), but Shahgedanova suggests that this peak may be around the corner in the next 10 or 20 years.[25]

Fountain and Solomon go on to point out in their introspection into glacier melt impacts, that across the Tibetan Plateau and in the Himalayan and Karakoram mountain ranges there are thousands of glaciers similar to the Tuyuksu and hundreds of millions more people along rivers like the Indus in Pakistan, the Ganges and Brahmaputra in India, the Yangtze and the Yellow River in China, and the Mekong in South East Asia, who rely on glacier meltwater for consumption and irrigation.

Scientists at NASA's Jet Propulsion Laboratory in the University of California at Irvine and the National Center for Atmospheric Research in Boulder, Colorado, are using advanced technology utilizing gravitational measurements on ocean waters to identify evidence of glacier retreat and related sea level rise. They stress that the evidence clearly shows that glaciers all over the world are melting rapidly and that while the impact of melting sea ice in Greenland and Antarctica on the world's oceans is well documented, in fact the largest contributors to sea level rise in the 20th Century were melting ice caps and glaciers located in seven other regions, including Alaska, the

Canadian Arctic Archipelago, the Southern Andes, High Mountain Asia, the Russian Arctic, Iceland, and the Norwegian archipelago Svalbard. In relation to water supply, the authors stress that

> this ice melt is accelerating, potentially affecting not just coastlines but agriculture and drinking water supplies in communities around the world. ... In the Andes Mountains in South America and in High Mountain Asia, glacier melt is a major source of drinking water and irrigation for several hundred million people. ... Stress on this resource could have far-reaching effects on economic activity and political stability.[26]

In the coming chapters we will look more closely at permafrost areas (the periglacial environment) and realize that not only glaciers of the cryosphere are contributing this critical function to our ecosystems, but that subsurface ice is also a key resource that provides water to downstream ecosystems in a similar manner to glaciers. These subsurface water towers of our cryosphere will survive when exposed surface glaciers fully melt, and consequently are just as important as water providers. We'll learn more about these fascinating "invisible" glaciers later.

4

Glaciers Are White, Oceans Are Blue, the Earth Is Warming, and So Are You!

Glaciers (and ice sheets[1]) cover 10% of the Earth's surface, mostly in the polar regions where ice covers the Arctic Ocean near the North Pole and most of the continent of Antarctica at the South Pole. The Greenland Ice Sheet has extensive glacier cover as do many mountain ranges that are home to smaller (but still significant) mountain glaciers scattered throughout most continents. Take a look at the images in figures 4.1 to 4.3, which show these regions and their glacier cover (or in the case of the Arctic, their *seasonal* ice cover). You get an immediate sense of the significance of white glacier ice cover to the Earth's surface. We notice right away that both of the poles have an extensive amount of white surface, while in the Himalaya region (considered by many the Earth's *third* pole because it has so much snow and glacier ice) there are comparatively more areas of browns, blues and darker hues. In the Himalayas the white surface cover, while important, is not nearly as extensive as it is at the North and South poles or in Greenland.

In terms of surface area, the polar glaciers and Arctic ice and the extensive glacier cover of the Greenland Ice Sheet clearly make up the bulk of the Earth's white ice cover, while the glaciers of mountainous regions are smaller and represent a significantly lower percentage of white surface cover. This is not to say that mountain glaciers are less important than polar ones; they're important for different reasons. But for albedo, as we will discuss in this chapter, surface area is critically important.

So if about 10% of the Earth's surface is covered with glacier ice or seasonal ice, that means that about 10% of the Earth's surface is white (or slightly off-white since glaciers are sometimes soiled and not completely white). As we consider the role of glaciers in our changing climate, this surface cover, while only 10%, is important because white surfaces reflect a significant amount of solar heat from the solar system back into space.

And while 10% may seem small compared to the 90% that is not ice, a simple example can show us how important that 10% can be. I went fishing for several hours recently on a Florida beach. To avoid sunburn, I tried to cover up. I was wearing a hat, dark sunglasses, a facemask bandana, and a light-colored long sleeved shirt to protect my skin from the sun. The shirt was my 1978 Argentine National Team soccer jersey, with a big "10" stamped in thick black numbering on the back. My routine was the usual, baiting my line, casting out to the ocean, walking around to collect sand fleas, moving back and forth to the water and into the surf, and back to my cooler to store the fish I caught (sorry, no pictures to prove it).

I spent the better part of the day on the beach in direct sunlight (mostly not walking back to my cooler). When the sun began to really hit hard near midday and in the early afternoon, I began to feel very warm, but most of the heat largely intensified on

Figure 4.1 Arctic (left) and Antarctic ice (right).
Source: NASA

Figure 4.2 The Greenland Ice Sheet.
Source: NASA

my back. And not only on my back, but especially on certain parts of my back. Clearly the big black numbers on my jersey were absorbing the heat, and my back near those numbers was burning up! I had to keep pulling the shirt off of my back to avoid the intense heat! The black numbers occupied much less than 10% of the surface area of my shirt, but boy was that minuscule dark color completely changing the temperature of my skin and my comfort level on the beach!

Figure 4.3 The Himalayas from space.
Source: NASA

What is at play here is the dynamics of light wave absorption and reflection. When we wash and dry cloths of different colors in our washers and dryers, all of the clothes, regardless of color, get as hot or as cold as the entire load. But when we take these clothes to the outdoors and expose them to sunlight, their reactions are different. Dark clothes warm more quickly than white clothes in the sunlight. Why?

The energy that is transmitted in light waves can be converted into heat and depending on the color of the light spectrum, this is done to varying degrees. A black object *absorbs* all wavelengths of light and converts them into heat, and as a result, the object gets warm, or *very* warm, as the case may be. A white object (such as a glacier), *reflects* all wavelengths of light, does *not* convert light into heat (or at least does it at a much lower intensity), and hence, does not warm as much as darker objects.

Lighter colored objects act more like white, while darker colored objects act more like black. So the dark blue of the oceans, for example, or the browns and greens of the Earth (as glaciers melt and expose these darker surfaces), will absorb more light and generate more heat than those same areas would if they were covered by white ice and snow. At the same time, as the planet warms and as glaciers melt away, reducing their surface cover, their "stay cool" capacity is greatly diminished.

The result: heat radiation from incoming sunlight will be *more* absorbed by those areas of the Earth with *less* glacier cover, thus increasing global warming. And just like the number ten stamped on my back warmed quickly and heated up my skin, the disappearing white cover of glaciers over the Earth's surface will lead to rapid surface and ambient warming where those glaciers once stood and where darker colors now prevail. In general, the Earth's surface and its atmosphere reflect about one third of incoming light, while two thirds is absorbed.[2] As glaciers continue to melt away and expose more and more of the Earth's surface, this percentage of light absorbed and heat emitted will increase.

So it is that we get to the issue of "the albedo effect," and as this chapter advances it will become clearer why *albedo* is so important to glaciers and to the Earth.

I like Guardian reporter Steven Poole's definition of the albedo effect. Poole researched the origins of the term and here's how he defines it:

> The term comes from the Latin *albus*, meaning white, from which we also get "albino" and "album" (which in ancient Rome could mean a blank tablet). In English it is first recorded, meaning "pure whiteness," in a letter by the astronomer royal John Flamsteed, complaining about Isaac Newton's *Optics:* "He calls the color of that representation of the sun which is made by the collection of his rays … whiteness … but 'tis far from *albedo.*
>
> From then on "albedo could be used medically (for an unusual whiteness of bodily fluids) or botanically (a fruit's pith). The modern astronomical use dates from 1860, and denotes the reflectivity of a planet or other object in space. Some experts suggest that we could nudge Earth's albedo back up again by painting all roofs white, which has to be worth a try." [3]

Not only do I like Poole's historical explanation of where the term albedo came from, I'd also like to make a parenthetical comment about his idea of painting roofs white. I realize this is a book about glaciers, but talking about roofs can help us understand in very practical terms what is going on with our glaciers, and why albedo is so important. It's also very important when we get to ideas about how to stop global warming and actually promote "global cooling" (see Chapter 9).

One of the things that I work on in my day job is promoting actions that help mitigate global climate change. Painting roofs white as a solution to climate change (or at least to deal with its impacts) is not as crazy as it sounds. Some people have even painted areas of mountains white to address the same problem.[4] I am currently engaged with a group called the Smart Surfaces Coalition[5] to address the *urban heat island effect*, which is a phenomenon that occurs in cities with extensive dark concrete jungles.

We know from experience that on very hot days non-reflective urban environments get very hot, much more so than shady, reflective, and tree-rich areas. Since the industrial revolution, cities have replaced natural tree and plant cover, including porous surfaces that offer cool shade and critical water absorption, with dark and impervious surfaces such as cement, dark tar rooftops, and dark pavement. This decreases a city's solar reflectivity, increases heat intensity as the sun's rays are absorbed rather than deflected, and provokes flooding due to increased runoff while lowering the soil's important water moisture.

Urban temperatures, due to the prevalence of heat-absorbing surfaces like dark cement and dark roofs or pavement, can be between 9°F and 15°F hotter than surrounding suburban or rural areas. During evening hours, these urban areas can be up to 22°F hotter than surrounding rural areas.[6] And while average city temperatures can differ substantially according to reflectivity, *site-specific* differences in intense urban heat islands are even more alarming. As heat is generated over dark-roofed homes and dark parking lots or pavement, the heat-absorbing infrastructure in select neighborhoods acts like a large oven. Roofs can account for about a third of urban surfaces and absorb about 80% of sunlight, converting the sun's rays into intense heat that lingers over buildings.[7] The difference in heat reflection between a black roof and a reflective roof, for example, is about 5% vs. 80% respectively. On a hot summer day with a temperature of 98°F (36°C), a black roof will be at 177°F (80.5°C), while a reflective lighter color roof will be more than 66°F cooler at 111°F (44°C)![8]

Take a look at Chicago's City Hall (fig. 4.4). The temperature on the hotter dark roof is over 150°F (65°C), while the garden roof (of the building on the left) is the coolest area in the image, at below 75°F (24°C). Hotter roofs heat up buildings (particularly the top floors). I am focused on this issue not only as a way to reduce global warming but also to reduce energy consumption by reducing the air conditioning loads required to cool our internal environments (something that would help slow climate change because we would utilize fewer polluting refrigerants that have very high global warming impact) (fig. 4.4).

So now let's get back to glaciers.

The National Snow and Ice Data Center (NSIDC) defines albedo more technically as:

A non-dimensional, unitless quantity that indicates how well a surface reflects solar energy. Albedo varies between 0 and 1. Albedo commonly refers to the "whiteness" of a surface, with 0 meaning black and 1 meaning white. A value of 0 means the surface is a "perfect absorber" that absorbs all incoming energy. Absorbed energy can be used to heat the surface or, when sea ice is present, melt the surface. A value of 1 means the surface is a "perfect reflector" that reflects all incoming energy.[9]

And you guessed it, because most glaciers are white, glaciers are close to 1 on the albedo range, while the *lack* of glaciers (a consequence of their meltdown and disappearance due to climate change) gets us closer to the 0 reflection value. Just-fallen

Figure 4.4 Urban heat island effect.
Infrared image comparing a dark roof with a garden roof at Chicago City Hall.

Source: ESRI: Mitigating Urban Heat Islands in the Twin Cities Metro Area; https://www.arcgis.com/apps/Cascade/index.html?appid=a26430a1dbdc4cb3a5e3f895461092f5

snow on glaciers is a pristine white color (high albedo close to 1), but as snow turns to ice, it interacts with the surrounding environment capturing dark particles of dust, smoke, etc., and begins to change colors. It darkens (lower albedo approaching 0).

Roberta Pirazzini of the Division of Atmospheric Scientists at the University of Helsinki is an expert on albedo in the Antarctic region and says that albedo changes in a place like Antarctica where the Earth has a lot of surface cover have intense impacts. "The amount of incoming shortwave radiation is very large in the summer over the Antarctic. Nevertheless, due to the high albedo the net shortwave radiation at the surface is small, and it is very sensitive to small changes in albedo: a decrease of 10% in the albedo can cause an increase of 50% in net shortwave radiation."[10] That's the same thing that happened to me on the beach with my jersey, even though the thick black numbers on my back weren't much of the surface area, they absorbed a lot of heat.

Dust in the air from the local environment, smoke particles (black carbon) brought by global winds from car, industry, and airplane exhaust, diesel emissions from ships carrying containers throughout the world or from cruise liners (including ones taking tourists to see glaciers), black smoke from forest fires, and ash and soot from volcano eruptions or from your fireplace or wood-burning stoves which are very common in the developing world, picked up by the global wind currents, all soil the surface of the hardening snow and the ice on glaciers and begin to darken the surface, altering glacier albedo.

Glaciers have a much higher albedo compared to the color of the Earth's surface or the oceans. According to the NSIDC, which offers an example of sea ice (not exactly the same as glacier ice ... but the example is nonetheless useful), bare sea ice has an albedo of about 0.5 to 0.7, which is fairly close to the reflectivity value of 1. This means that sea ice reflects about 50–70% of incoming energy in sunlight! Glaciers, and particularly snow-covered glaciers, that is, glacier ice that has newly fallen snow on the surface, can reach albedo levels as high as 0.9. This ice reflects a whopping 90% of incoming solar radiation. That's the sort of shirt I should have been wearing on my beach-fishing day!

By contrast, ocean blue has a much lower albedo of about 0.06. That means that the ocean only reflects about 6% of the incoming solar radiation and absorbs the rest—it's absorbing the other 94%! We can see how large areas of melting sea ice that begin to expose the ocean beneath can result in massive global warming! We will discuss ocean currents in more detail in Chapter 7, but it suffices to say here that as the ocean currents flow through the polar areas, water warmed due to albedo changes is picked up at the poles and distributed around the planet contributing to the destabilization of our ocean ecosystems.

We also need to consider, says the NSIDC, that melt ponds forming on sea ice when warm air from the atmosphere interacts with the ice have much lower albedo, in the 0.2–0.4 range, reflecting 20–40% of solar radiation, which in turn means that 60–80% of the energy from the sun is fully absorbed by these ponds (see fig. 4.5). This leads to even further melting of the ice at and around these ponds (increasing their extent), and as these ponds deepen, albedo can drop even more to 0.15 and the surface area of this ice can absorb much more solar energy. Glaciers, like sea ice, also often have melt ponds on their surface that act in very similar ways.

As is clearly evident from the photos in figures 4.1–4.3, the Earth's land is also much darker than white glaciers. As we fly over most any place on the planet, we see many shades of greys, browns, greens, yellows, blues, etc. As the Earth warms, glaciers that

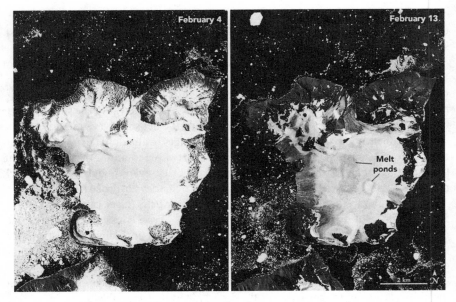

Figure 4.5 Glacier melt ponds on Eagle Island in Antarctica accelerating glacier melt.
Source: NASA

mostly cover the polar areas, the Greenland Ice Sheet, and mountain glaciers, retreat, and lands that were once covered by white reflective ice, where no organic matter could flourish, are exposed. As temperatures rise, these terrains become amenable to plant growth leading to further albedo (reflectivity) decreases.

Warmth-absorbing darker trees, bushes, tundra, and other plant matter begin to flourish where once only reflective white surfaces existed. Essentially, the tree line migrates north in the northern hemisphere, south in the southern hemisphere, and up mountain sides in high elevation areas, effectively covering the lighter *white* parts of the Earth in the polar regions and on mountain peaks. If glacier melt continues, these white areas will eventually disappear or be greatly diminished. It would be like adding thick black numbers to my soccer jersey while fishing in the sun at the beach. At some point, the heat becomes unbearable. Glacier and sea ice melt, hence, is causing an overall *increase* in the absorption of heat by the Earth due to an increase in overall darker surface area.

So what's causing the darkening of glacier surfaces that leads to these albedo changes?

One culprit we will discuss here is soot.

Soot

Soot is a carbon residue that results from the incomplete burning of organic matter. There are many sources of soot emissions, including for example, the burning of charcoal that you might use to make a barbeque, diesel-burning engines (mostly in trucks, tractors, ships, forklifts, and agricultural equipment), wood burning (such as from a forest fire like the recurring climate wildfires in California), or a wood-burning fireplace or stove (a common source of household energy in much of the developing world).

Soot also derives from a variety of industries, such as from energy produced from coal-burning, from the large engines that run natural gas compressors at oil and gas production locations, from the diesel trucks used to transport merchandise around the country, from commercial shipping that move millions of containers from port to port all over the world, or from the firing of artisanal bricks or ceramics in much of the developing world. Soot is produced in massive quantities from many different sources, and with the help of wind and global air currents works its way around the entire planet. If you live in an urban area, there is probably lots of soot in and around your home, mostly from car and truck emissions or from local industry. If you have a chimney, take a look up the flue and you're likely to see a thick layer of soot covering its internal walls. The inside surface of an outdoor grill, or the underside of a car near the exhaust pipe, will give you an idea of how soot quickly accumulates near an emission source.

But you don't have to go to specific soot-generating emission points to find it. Simply run a finger along almost any surface that is exposed to the elements, particularly in an urban setting. While some of the dust you'll pick up is also dust from soil, there is likely to be soot on the surface from the emissions around you. If you have tile floors, examine the grout between the tiles and you're likely to see soot. I recently had to take a wire brush and spend several hours on my knees to clean out soot from the tile grout in my home!

Soot can travel many miles, even thousands of miles, from its point of emission, be it out of the exhaust pipe of a truck traveling on a highway in Texas, in France, or in Mongolia, from a coal-burning power plant in South Africa or Wyoming or West Virginia, from artisanal brick kilns in India or Mexico, from container ships traveling through the Arctic, or from a coal or wood-burning stove in China, or from a barbeque grill in Argentina (something I miss dearly from my years living there). Once black carbon particles are picked up by the global winds, they can travel long distances and be deposited nearly anywhere on the planet. For glaciers, soot is a killer.

Because of the persistence of black carbon particles in the air throughout the Earth's atmosphere, their deposition on glacier surfaces is pretty common. We can see this sooty deposit on most glaciers, particularly at the end of the summer, when the seasonal snow has melted, and the local air has impacted and hardened the glacier surface. Glacier surfaces at the end of the summer are much darker than at mid-winter, when they are covered in pristine white snow and no soot at all is visible.

Throughout the year, particularly when it's not snowing, a thin layer of black carbon begins to accumulate on most glacier surfaces, darkening the surface slightly. The longer it goes without snowing, the darker the surface will get, sometimes reaching a thickness of several millimeters. The thin soot cover will eventually be buried into the ice as the ice hardens and as new snow falls onto the glacier surface in the following fall and winter. This alternating cycle of deposition of white powdery snow in the winter followed by dirty air soiling the glacier surface in the summer will continue year after year, to the point where if we were to take a vertical cut out of the glacier, exposing many layers of ice accumulated over many years, decades, or even centuries, we would see clear black lines concentrated and placed at intervals when the summer months peaked, separated by much whiter strips of ice when the clean, powdery, white winter snow accumulated on the surface. Over time, the surviving seasonal snow will turn to ice and harden inside the glacier, pressing the lines closer and closer together as the ice becomes more and more dense over the years, leaving behind a distinct striped interior with thin black lines representing peak soot accumulation during the summer and thicker white areas revealing winter or seasonal snow fall (see fig. 4.6).

Figure 4.6 Soot (black carbon) deposit on glaciers shows yearly snowfall, air contamination, and a glacier's age.

Source: Geoestudios

If we consider that snowfall cycles recur rather regularly over the years between winter and summer, it's easy to guess how you can tell a glacier's age. As you would with a tree by counting the layers of bark (or the rings of a tree trunk cut horizontally), which are laid down in yearly cycles, scientists count the soot lines to determine the glacier's age, as each black line represents a year of glacier life. They can also take core samples of these strips to analyze the particular year's global atmospheric contamination and the amount of snowfall registered each year for as many lines as their sample obtains. They can also identify prolonged periods of no or little snowfall as well as sudden snowstorms that may have occurred in a given year. They might also find that a volcano eruption occurred in a given year because they can actually find traces of the ash in the ice, which can usually be traced back to its source. Even volcanoes at the other end of the planet will leave their distinct trace in glacier ice. Any ash from a volcano that got into the global air currents could be deposited on a glacier surface thousands of miles away.

And while soot or ash deposits on glacier ice is a nifty tool to analyze our global atmospheric past, or decipher what was going on with our global climate in a given historical year, these deposits can also contribute to progressively higher and faster glacier melt rates year after year, particularly if prolonged periods of low or no snow fall occur.

We have extensive evidence to show that black soot is a principal cause of Himalayan glacier melt. One of the reasons that Asian glaciers are especially vulnerable to soot emissions is that urban life at the foot of the Himalayas is one of the most contaminated on the planet, as a consequence of soot-emitting sources such as diesel engines, coal-burning cook stoves, brick kilns and other industries. Remember, as more soot accumulates, the glacier gets darker, and albedo is decreased, resulting in greater heat absorption at the surface of the glacier, which in turn accelerates melting.

A group of scientists studying the Tibetan Plateau found that increased black carbon (soot) in snow increases the surface temperature by about 1°C, causing accelerated melt and reducing the spring snowpack more than CO_2-related increases do. This *increases* the rate of snowmelt in the late winter and early spring and leads to *decreases* in the late spring and early summer. You can imagine how these abrupt changes are altering the availability of glacier-fed water resources for farmers, when they need it most according to their traditional farming cycles.[11]

These scientists are alert to the impacts these changing patterns are likely to have on the entire Asian continent's weather patterns, as well as on water availability to ecosystems, to farmlands, and to human populations. Sometimes referred to as the Earth's third pole, because of the massive amount of glacier ice and snow that exists in the Himalayas, alterations to this part of Earth's delicate cryospheric ecosystem could be devastating for hundreds of millions *if not billions* of people. Some of the world's most important river systems are directly fed by Himalayan glaciers including the Yangtze in China, the Ganges in India, and the Mekong in Southeast Asia.

Biaqing Xu and his colleagues conclude from their study of soot in Tibetan glaciers that "black soot content is sufficient to affect the surface reflectivity of the glaciers and that the black soot amount has increased rapidly since the 1990s, coincidently with the accelerating glacier retreat and increasing industrial activity in South and East Asia." The study concludes that black carbon significantly affects the albedo of snow and ice, accelerating snow and ice melt in the springtime, at the start of the melt season. The glaciologists also stress that as concentrations of black soot increase, melt rates also increase.[12]

While climate temperature changes are probably the most influential factor in the current rate of glacier melt around the world, black soot is obviously an important and contributing factor that greatly leverages this accelerated temperature-related glacier melt phenomenon, magnifying glacier melt trends related to climate change.

A UN study dating to 2009 already highlighted the impacts of dust and soot on the Tibetan Plateau and Himalayan glaciers: "The black carbon deposition in Tibetan Himalayas is higher than other places in the northern hemisphere and has been a cause of significant albedo reduction in the Central Tibetan plateau".[13] The study finds that dust-induced albedo decrease could cause snowmelts as much as 30 days early, with increased evaporation due to early melt-out decreasing basin runoff by 5%. If you are a farmer that plants crops in a certain month because temperature conditions dictate when you should plant your crops, you're expecting rain to fall and river flows to increase at specific times during the season. Water arriving 30 days earlier than usual could mean the difference between a successful or a failed crop, just as the alteration of water flow quantities could also greatly change its accessibility and your capacity to manage your critical water supply.

Did Soot from the European Industrialization End the Little Ice Age?

An interesting article appearing in the Proceedings of the National Academy of Sciences (PNAS) argues that soot deposition on glaciers in the European Alps from the industrial revolution in the 1800s brought to a close Earth's so-called Little Ice Age[14] that took place from about 1500 to 1800 AD. The authors note, "the end of the Little Ice Age in the European Alps has long been a paradox to glaciology and climatology. Glaciers in the Alps began to retreat abruptly in the mid-19th century, but reconstruction of temperature and precipitation indicate that glaciers should have instead advanced into the 20th century."[15]

What was so potent that it could stop an ice age brought on by Mother Nature? The authors point to evidence that black carbon emissions deposited on snow increased quickly in the mid-19th century and argue that these deposits increased the sunlight absorbed by glacier surfaces, leading to accelerated warming and subsequent rapid glacier melt (more quickly than they could be replenished), causing glacier retreat, and reversing the naturally occurring Little Ice Age, which was recharging many of the Earth's mountain glaciers at the time.

As a result of the Little Ice Age, say Painter and his colleagues, glaciers in the European Alps increased in size between the end of the 13th century to the beginning of the 19th century. Beginning in 1865, say Painter and the other authors, they abruptly began to shrink. They contrast this tendency to recorded temperature readings for the Alps, which despite glacier shrinkage, were relatively cool between the end of the 19th century and the beginning of the 20th century (compared to the late 18th century and early 19th century). The authors stress that black carbon and mineral dust can significantly reduce albedo rates of snow and ice and result in great absorption of the sun's energy.

The Industrial Revolution in Western Europe, argue the authors, led to an increase in emissions of black carbon, starting in Great Britain in the mid-18th century, and spreading to France in the early 19th century and then to Germany and much of Western Europe by the mid-19th century.[16]

The Alps were surrounded by intensive industrialization and resulting black carbon emissions. Ice cores drilled into European Alp glaciers reveal that black carbon deposits coincide with these eras and the increased emissions from Europe, with increased concentrations starting at 1850, with general increases thereafter into the 20th century.

In much the same way as in Europe, the same phenomenon involving black carbon soot and glacier melt takes place in other parts of the world. My good friend and mentor, glaciologist Cedo Marangunic of Chile has noticed significant soot deposits on glaciers situated in the path of air currents that work their way up long mountain valleys and reach deep into the Central Andes. These air currents originate from one of the most contaminated metropolitan areas in the world, the Santiago metropolitan region of Chile. The Little Ice Age occurred later in the Central Andes than in Europe. It could very well be that dwindling glacier cover in the Andes, like in parts of Europe, may have a lot to do with local transport and industrial activity in the region that deposits soot on glacier surfaces.

Forest Fires

As noted earlier in this chapter, forest fires, which have been occurring more and more frequently and with much more intensity in recent years, in places like North America, are also a cause of concern for albedo reduction on glaciers and as a consequence, for the deterioration of glacier health.

In a study from 2017, scientists have been able to link black carbon deposition on Greenland's Ice Sheet in July and August of 2013 to large forest fires that summer in Canada.[17] If we consider that the melting of the Greenland Ice Sheet is one of the main contributors to sea level rise and that this ice sheet is also one of the world's most significant white surface areas, its accelerated reduction in size due to albedo reductions caused by forest fires is extremely alarming.

The study led by Thomas indicates that the Greenland Ice Sheet albedo has been falling since the 1990s. The authors' concern is that more wildfires in a world with a warming global climate can mean even more black carbon deposits on the ice sheet, further accelerating glacier melt. A new study published in 2020 by Into Sasgen and a group of climate scientists focuses on the rapidly accelerating glacier melt-off at the Greenland Ice Sheet, noting that Greenland has warmed by 1.7°C (about 3°F) since 1991.[18] Sasgen and his collaborators indicate that the volume of Greenland Ice sheet is melting almost as much as all other glaciers around the world combined. One of the key reasons for this is albedo changes. The phenomenal reduction of albedo capacity in the world's northern ice is drastically melting it. Other reasons for this melt include: rising ambient temperatures in the Arctic, the migration of snow to higher altitudes, an increase in the total melt area, and cloud-radiative effects.[19]

We are all painfully aware of the increase in wildfires, now called climate wildfires by some, in places like the Western United States, in Australia, and in the Amazon region. The impacts that these trends could, and will likely have, on the albedo of nearby glacier-rich regions (such as Alaska, Canada, the Sierra Nevada, the Rocky Mountains, Central Andes, and mountain glaciers in New Zealand) or even in ranges much further away are not necessarily being studied, but clearly have ominous consequences.

Ben Pelto, the Canadian scientist monitoring glaciers in British Colombia mentioned earlier, says that when it's fire season, you can taste the smoke in the water! Allison Dempster did a report for CBC News covering the issue of soot deposition on Canadian glaciers and interviewed Pelto and other scientists that are concerned with what they see as an alarming trend of soot deposits on glaciers accelerating glacier melt all over the continent (see fig. 4.7). Pelto noted that the summer of 2018 was the worst he has seen.

Hydrologist John Pomeroy studies glaciers in the Rocky Mountains (of Canada) and reports finding significant ash deposits on Canadian glaciers, probably blown from wildfires. Pomeroy notes that while glaciers generally absorb about 60% of the sun's rays, he's seen a steady increase in that number over recent years, from 70% to even 80% absorption.

Figure 4.7 Soot (black carbon) deposit on British Columbia Glaciers after wildfires.
Source: Ben Pelto, published by Allison Dempster, CBC

The mega climate wildfires that we have seen recently in Australia (2019) and the western United States (2020) will likely have enormous implications for soot deposition on the world's glaciers. At the time this book went to press, academic studies had not yet been conducted linking these mega climate wildfires to soot landing on glacier surfaces and accelerating melting, but surely the studies will come. We only need to look back a year earlier when the Australian continent was on fire and the impact this had on nearby glaciers in New Zealand. A headline in the Guardian in January 2020, not long after the Australian wildfires, stated: "New Zealand's glaciers turn brown from Australian bushfires' smoke, ash and dust ... snow-capped peaks and glaciers discolored ... from ash could accelerate glacial melting."[20]

Andrew Mackintosh, head of the School of Earth, Atmosphere and Environment at Monash University and former director of the Antarctic Research Center, indicated that glacier melt from Australia's 2019 fire season could increase glacier melt by 20–30% in New Zealand. New Zealand has some 3,000 glaciers and these have been steadily shrinking by up to a third since the 1970s. Predictions for New Zealand's glaciers are dire, with likely total wipeout by the end of the 21st century.[21]

The South American Andes is another region with extensive glacierized mountain ecosystems now threatened by ash and soot deposition from forest fires. Also appearing in the media around the time of the drafting of this chapter, extensive forest fires in the Brazilian Amazon and smoke plumes from human-induced fires in Bolivia and in Peru are sounding alarms for already vulnerable glaciers in countries such as Argentina, Bolivia, Ecuador, Peru, and Chile, which are in the vicinity of, and in areas affected by, air currents originating in places like the Brazilian Amazon. Between 2018 and 2019, fires in the Amazon region, due to increased temperature and clearing from forestry activity, increased by 70%. The smoke from these fires could be seen from space. South American fires emit approximately 800,000 tons of CO_2 into the atmosphere each year, which is about double Europe's annual emissions from energy use.[22] Soot is effectively choking nearby Andean glaciers.

Scientists in Brazil examining the impacts of recent Amazonian forest fires, as well as smoke from Peru and Bolivia, indicate that glaciers in Bolivia are being impacted by soot deposited onto glacier surfaces (with impacts also likely in other tropical Andes glacier regions such as Ecuador, Colombia, and Venezuela). They point out at that glacier runoff water is critical to communities along the Andes Mountains and that glacier melt runoff will be altered and land uses will change due to the changing albedo of Andean Glaciers as a result of soot deposition from Amazon forest fires.[23] Combined with extended droughts, the impacts of smoke and the deposition of soot from these fires on Central Andes glaciers could affect the downstream hydrology (as it is doing in the Himalayan and Tibetan Plateau regions) and the millions of people that depend on glacier-fed river systems up and down the Andes.

Mauri Pelto (Ben's father) who directs the North Cascades Glacier Climate Project at Nicholls College spoke recently to the publication *Glacier Hub* and reflected on the 2020 climate wildfires in California, Oregon, and Washington, along the west coast of the United States.[24] He complains of the difficulty of even getting to the glaciers in the area, due to low visibility and poor air quality caused by the smoke. "The smell was like being near a camp fire and the glacier melting was extreme." He points not only to the risks of albedo changes from ash and soot deposited on glaciers but also warming temperatures, which can interact to accelerate melting even further. Proximity to fires is a key factor, stresses Susan Kaspari, who also studies glaciers at Central Washington University, as ash is more likely to fall on glaciers closer to fires. What is also remarkable about the observations made by the glaciologist Pelto family is that glaciers are already so soiled by atmospheric soot that even these large wildfires are not necessarily contributing to "additional" albedo changes. They are already very dirty![25]

Volcanic Eruptions

There is yet another albedo reducer for glaciers (a completely natural one in this case): volcanoes. Volcanic eruptions result in the deposition of volcanic ash on glacier surfaces around the world. Anyone who has ever seen the volcanic ash plumes coming from erupting volcanoes knows how enormous they are (see the footnote if you haven't).[26] Once the ash from a volcano gets into the atmosphere, the global air currents circulate it over the entire planet. I remember being in South America when the Chaiten Volcano erupted in Chile. A few days later, flights were canceled as far away as New Zealand, 5,000 miles (8,000 kilometers), and across the Pacific Ocean because of ash in the air around the airport! Dark ash from volcanoes acts just like soot when deposited on glacier surfaces, reducing glacier reflectivity, increasing heat absorption, and hence accelerating glacier melt.

The Melting Arctic

Recent observations suggest that the Arctic is becoming a "new climate state."[27] We've discussed the differences between ice and glaciers in Chapter 1, and while this book is mostly about *melting glaciers*, and not *melting ice* generally, the reality is that vast areas of solid ice cover, such as occurs seasonally over the North Pole in the Arctic Ocean or off of some parts of Antarctica, are greatly affecting our Earth's cryosphere, and an impact on one of the resources is not only likely to have an impact on the other, but both are in a delicate balance and facing the same sources of impacts, and both are contributing to global climate change (albeit in different ways). This section will explore the dynamics of Arctic ice melt and its impacts on global climate change and on the vulnerability of glaciers and the cryosphere.

The Institute for Governance and Sustainable Development (IGSD) is headed by career climate activist, environmental policy expert, and good friend, Durwood Zaelke. He has brought together a group of renowned environmental leaders to concentrate their efforts on literally saving the planet by achieving the phaseout of *the most impacting pollutants* that are causing climate change, specifically *short-lived climate pollutants* such as methane, black carbon, tropospheric ozone, and hydrofluorocarbons (HFCs).[28]

One of IGSD's key focus areas is the study of albedo dynamics and glacier melt and their relationship to global climate change. Their research and findings indicate just how badly glacier melt is affecting albedo in the Arctic and how this is leading to accelerated climate change, which feeds back upon itself and only causes further glacier melt and more climate change—what is commonly referred to as a *feedback loop*. While Arctic ice is seasonal and hence not glacier ice, the dynamics of ice melt in the Arctic can help us understand how glacier melt occurs and the impacts it can have on our global climate. This section is the product of discussions with a handful of IGSD's leading climate researchers and climate activists to whom I am very thankful.

In their *Primer on Polar Warming*[29] (for those of you who would like to read more about climate change and the Arctic, I highly recommend it), IGSD reviews the startling scientific evidence showing just how fast our polar extremes are melting and the magnified contribution this accelerated melt and collapse of our polar ice has on overall climate change trends.

> The Polar Regions are particularly susceptible to self-reinforcing feedbacks and recent warming suggests we may be perilously close to crossing tipping points that could trigger the various tipping elements, shifting the poles—and the planet—into a new climate state.[30]

As global temperatures have risen, says IGSD, the Arctic has warmed by 2°C, twice the rate of the rest of the world. This is due to a "cascade of feedbacks" that collectively work to make warming in the Arctic far more intense than in other regions at lower latitudes. The reasons for this super-sized warming effect have to do with dynamics such as: albedo changes caused by loss of sea ice and decreased snow cover (together known as the ice-albedo feedback); increased water vapor in the Arctic atmosphere; altered cloud cover; added heat in the newly ice-free oceans; and a lowered rate of heat loss.

Temperatures in the Arctic from 2011 to 2015 were warmer than any year on record since instrumental recording began in the year 1900, and under medium to high greenhouse gas scenarios, says the IGSD report, the Arctic is projected to warm to an annual average temperature of 4–5°C above late 20th century values before mid-century (2050), which is twice what is projected for the northern hemisphere.

We keep hearing from climate scientists and in global climate news that 1.5°C warmer is the doomsday threshold, above which we should not go if we don't want to face drastic and likely irreversible climate catastrophe. According to the IPCC, 1.5°C of warming could be reached between 2030 and 2050, but we could actually cross the threshold sooner through natural climate variability, as early as 2026.

The problem with melting glaciers in the polar areas, and particularly in the Arctic, is that the Arctic-related climate-warming amplification can greatly alter large-scale atmospheric flow and contribute to extreme weather events beyond the Arctic region. This could increase climate uncertainty and alter weather events in lower latitudes including causing severe cold spells and other inclement weather. Arctic glacier ice imbalance can also affect accelerated glacier melting in the nearby Greenland Ice Sheet. Atmospheric changes in the Arctic region (caused by global warming) change cloud cover, which changes cloud formation patterns. Cloud cover can then shift to darker regions of the planet, which in turn changes albedo in those areas that now receive less sunlight, further warming Planet Earth. In other words, as clouds shift around due to the Arctic's changing climate, some areas like the Arctic will receive much more sunlight and will heat up more comparatively to other regions, generating vicious feedback loops for climate change that only get worse over time.

IGSD calls the Arctic region of the Earth the "wild card" of the climate system precisely because what is happening there is indeed *wild*. Climate change is faster and more severe in the Arctic than anywhere else on the planet. It's happening twice as fast! In the Arctic region, the reinforcing feedback loops, which self-generate greater and greater climate impacts and push our planet toward catastrophic tipping points that are irreversible on a human timescale, are extremely concerning.

The Arctic Ocean lost 40% of its ice cover between 1979 and 2011 and the record for the loss of ice is broken with every succeeding year. We hear about this all the time on our news streams. We may not pay much attention to it, but this is greatly altering our global ecosystem. Ice thickness has also been affected, and this is important because older, denser ice is more resistant to deterioration and climate change than newer, more fragile ice. The thickness of Arctic sea ice has declined by about 65% between 1975 and 2012.

This is indicative, says IGSD, of a "new ice regime" for the Arctic, which is more fragile than ever before. According to the National Snow and Ice Data Center (NSIDC), the Arctic sea ice cover continues to become younger, and as a consequence, thinner. NSIDC notes that nearly all of the oldest ice (more than 4 years old), which once made up 30% of Arctic ice, is gone.[31] It's the resistant thick ice that has been fighting against climate change. As this older resistant ice vanishes, the remaining new, thin ice will simply not be able to withstand the heat. It will break up faster and more abruptly than ever before. On February 22, 2021, an article ran on global newsfeeds indicating that the Arctic was now open for business (shipping) year round, even during winter. A large commercial vessel, sailed the northern sea route from Jiangsu China to Russia on the Arctic coast for the first time ever in the month of February, when usually, thick winter ice would not allow ships to sail during the winter months. The ice was gone.[32] And now, unless global leaders curtail shipping lanes through the Arctic in winter, the increased presence of shipping vessels in Arctic seas throughout the year, will result in the spewing of more and more soot on already vulnerable ice. Your newsfeeds about melting Arctic ice will just keep coming and they will be more and more alarming.

IGSD scientists sound a troubling alarm: the worst predictions are that the Arctic could become ice free within a decade. One of the features of Arctic warming that makes the predicament of this part of the cryosphere even more alarming is that there is little lag time between the warming climate and decreasing ice cover, that is, while in other parts of the Earth global warming may have a slow impact on glaciers, in the Arctic, this impact is not only big, it is immediate. We also know that anthropogenic climate change (the portion of global warming that is definitely caused by people and not by natural causes) is having a direct and accelerated effect on polar ice melt.

The graph in figure 4.8 is illustrative, plotting ice cover, albedo, and temperature. Note how albedo and ice cover (the black and green lines—the top two lines in (B) for those who only see the black and white text) move relatively in unison. They are declining over time. The red line (temperature—line rising left to right in (B)) is rapidly increasing. Here, we can see the clear relationship between albedo, ice cover, and temperature.

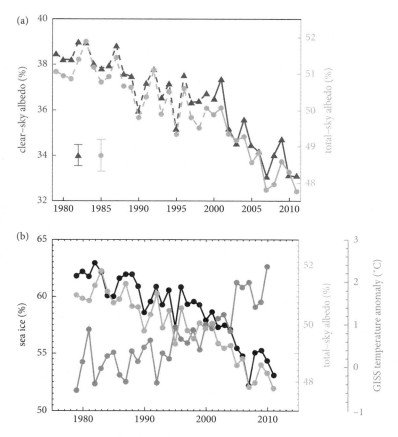

Figure 4.8 Temperature, albedo, and ice cover.
(a) Observed annual mean clear-sky (line starting on top in (a)) and all-sky planetary albedo (line starting on bottom in (a)) for the entire Arctic region. Solid lines are direct CERES observations, and dashed lines are estimates derived from sea ice observations. The error bars in the bottom left corner indicate the uncertainty in the pre-CERES clear-sky and all-sky albedo values. (b) All-sky albedo (line starting in middle in (b) as in (a) compared with annual mean observed sea ice (line starting on top in (b) area (as a fraction of the ocean in the Arctic region) and surface air temperature (rising left to right in (b)) averaged over the ocean in the Arctic region.
Source: Pistone, et al. PNAS. Vol. 111. No. 9. March 4, 2014

IGSD, in their Primer on Polar Warming, notes that the five years up to 2019 were the warmest in the Arctic, with 2018 being the *very warmest*. Average temperatures during 2015–2016 were about 3.5°C warmer than at the start of the 20th century, while high temperatures in 2016 were up to 15°C above normal. Meanwhile, average sea ice levels for the month of November have declined 5% per decade between 1978 and 2016. In December 2016, polar sea ice covered about 80% of the Arctic sea; it usually covers 95%.

On some occasions, rapid ice loss events occur, with massive amounts of ice suddenly melting away. In one case, a loss of 800,000 km^2 of ice was lost in merely seven days! That's about the size of the state of New South Wales in Australia, the entire country of Mozambique, or Pakistan. It's bigger than every US state except Alaska. Other severe rapid ice loss events occurred in 2007, 2014, and 2015 and are now being documented for 2019.

In April 2019, a record-low rapid ice loss event occurred in the Sea of Okhotsk (nestled between the north of Japan and Russia), when during the first half of the month, the region suddenly lost almost 50% of its ice. Temperatures in the area were critically high during the early part of April, especially over the East Siberian Sea (between Russia and the North Pole) and the Greenland Ice Sheet, where they averaged about 9 °C higher than usual.[33]

According to Ingo Sasgen of the Alfred Wegener Institute's Division of Glaciology in Germany and fellow arctic researchers, 2019 saw a return of rapid, and record-breaking, ice loss in Greenland. They indicate that between 2003 and 2016, the Greenland Ice Sheet was one of the largest contributors to sea level rise, as it lost about 255 gigatonnes of ice per year (that's a pile of ice about 87 kilometers tall over all of Central Park in New York city or about 100 billion Olympic sized swimming pools), while 2019 was marked by a record mass ice loss of about 532 gigatonnes (that's more than double the previous Central Park example or 213 billion Olympic sized swimming pools).[34] The graph in figure 4.9 plots arctic sea ice cover registered during the month of April for every year between 1980 to 2020, a 40-year span. The decline is alarming and undeniable.

The polar regions are clearly moving toward being ice-free during the summer months. This should be extremely worrying, in large part because of the changing

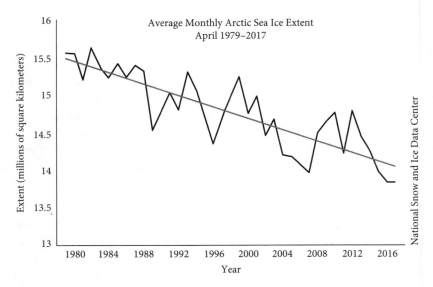

Figure 4.9 Average monthly Arctic sea ice extent, April 1979–2017.

Source: National Snow and Ice Data Center

albedo effect this will have and the resulting global warming effect on the planet. It will not only alter the local environment in the Arctic, it will also induce global warming trends generally across the planet. This will also increase the likelihood that glaciers around the world will melt faster, with likely local impacts to the nearby Greenland Ice Sheet as well as to permafrost (see Chapter 8).

By altering ocean temperatures and air currents, melting Arctic ice will destabilize weather systems, in part by causing havoc to the stability of oceans. Some of these impacts, say IGSD's scientists, may derive from thinning ice and resulting changes in wave conditions, which in addition to already rising ambient and ocean temperature, could induce more cyclones. The instability of the ocean can lead to ice breakup and collapse, while the stirring of warmer seawater enhances melting. Warming temperatures in the Arctic region are also concerning because they destabilize the important differences between the cold polar region and warmer equator. These natural (and welcome) temperature differences help generate differences in air pressure, which subsequently generates airflow between the different regions of Earth. Climate change is altering this natural airflow, which can result in the slowing and parking of storm systems in the mid-latitudes, thereby greatly increasing precipitation patterns. Are your storms lasting longer than usual? It could be because of warming in the Arctic!

Changing global wind patterns and temperatures can also destabilize weather and rain patterns throughout the Earth, including producing colossal *atmospheric rivers*. These intense changes in our weather patterns can bring on prolonged rain, which can at times be good for glaciers as they may bring more snowfall, but can also cause severe flooding. The National Oceanic and Atmospheric Organization (NOAA) defines atmospheric rivers as:

> Relatively long, narrow regions in the atmosphere—like rivers in the sky—that transport most of the water vapor outside of the tropics. These columns of vapor move with the weather, carrying an amount of water vapor roughly equivalent to the average flow of water at the mouth of the Mississippi River. Atmospheric rivers ... can create extreme rainfall and floods, often by stalling over watersheds vulnerable to flooding. These events can disrupt travel, induce mudslides and cause catastrophic damage to life and property.[35]

New studies are showing that atmospheric rivers will become 25% longer and wider, bringing heavier rains and stronger winds in atmospheric rivers appearing North America and Europe. The frequency (as well as the intensity) of atmospheric rivers, says NASA, will be nearly double.[36]

Through these atmospheric changes, alterations of the Arctic environment can cause extreme weather events in the mid-latitude regions, including severe cold spells and snowfall, dispelling the sometimes-heard erroneous idea that these cold weather events are evidence that the Earth's climate is not actually warming. Such extremely cold events are most likely in regions like North America and Eurasia, with the eastern third of the United States most affected. Other scientists, says IGSD, have found clear evidence that declining Arctic sea ice can be linked directly to intense drought cycles in places like California.

The Loss of Antarctica's Ice Cover

Meanwhile, at the opposite end of the planet, on the icy continent of Antarctica, things are no better than in the Arctic. The destabilization of the West Antarctic Ice Sheet threatens its rapid collapse into the ocean, which will not only result in sea level rise as melting colossal icebergs slip into the ocean, but will also greatly alter ocean temperature and decrease the continent's reflectivity.

Antarctica is really the last remaining massive permanent white and reflective surface area left on the planet (followed by Greenland). Antarctica is very sensitive to climate change and the changing environment of the region is referenced by many scientists as an "indicator" of climate change.[37] Arctic sea ice is cyclical, appearing in the winter but largely melting off in the summer. Antarctica has most of its white surface on the continent in the form of glacier ice, but also has significant amounts of sea ice. Both are in peril.

One of the main differences between ice near the North Pole and ice near the South Pole is territorial isolation. Since ice near the North Pole is surrounded by big continents, it is more exposed to warm air, while ice near the South Pole is mostly surrounded by big oceans and exposed to colder air, keeping it relatively cooler and safe (*or safer*) from global warming. That helps preserve glacier and sea ice. But it looks like things are becoming unstable in the south as well, and even the isolation that protects Antarctica's ice from melting off as fast as ice in the north may be showing signs of stress due to the southern encroachment of ocean and global warming.

Antarctica has a few characteristics that set it off from other parts of the Earth in terms of its glacier and ice cover. Stephanie, a scientific contributor to the American Museum of Natural History who works in Antarctica, offers a few insights about Antarctica to help us understand.[38]

First of all, Antarctica is at the extreme end of the planet and is very isolated from warm things (like air currents, cities, people, and industries) that are far away, which means it's very cold, in fact, it holds the world record for the coldest place on Earth, once hitting −89.5°C (−129°F).

Second, sunlight comes to the region at an angle (as opposed to head on at the Equator) so that it is less intense as it comes onto the surface.

Third, the extensive white surface of Antarctica gives it a very high albedo, so much of the sun's warming rays are reflected right back into space. In fact, the Antarctic continent gets about 60% of the warmth from the sun's rays as opposed to other areas that may get closer to 100%.

Finally, the Antarctic continent, as one of the coldest areas on Earth, as opposed to what you might imagine, is actually a desert—or well maybe, a "snow-desert" would be a better term. Despite the common misperception, it snows *very little* in Antarctica, less than an inch per year! In terms of building glaciers and ice cover, Antarctica is a pretty slow glacier-making machine. However, while little snow falls on the surface of the South Pole's continent, the snow that does fall is well preserved because of the extremely cold environment. This also means that it took a long time to build the continent's ice reserve.

Claire Parkinson of the Cryospheric Sciences Laboratory and of NASA's Goddard Space Flight Center indicates that Antarctic sea ice was steadily increasing over time (as has always been known) but then, suddenly it began to shrink. The reduction rates are phenomenal. A newly completed 40-year record of satellite observations of sea ice in the Antarctic region shows that a gradual increase of sea ice cover at the South Pole, that was slowly but steadily increasing until 2014, suddenly reversed course, and the melt-off rate was actually faster than at the North Pole, by a lot. Parkinson's research shows that the melt-off in Antarctic (South Pole) sea ice between 2014 and 2017, just four years, was comparable to Arctic (North Pole) melt-off that occurred over 34 years.[39] That's a seven fold melt-off difference in an incredibly reduced period.

In 2011 NASA scientists in Antarctica discovered a massive crack on the Pine Island Glacier near the Amundsen Sea, extending 19 miles (30 kilometers) and measuring 260 feet wide (80 meters) and 195 feet deep (60 meters). The portion of the glacier that is breaking off is about 350 mi^2 (900 km^2)—enough ice to fill the entire San Francisco Bay Area (see fig. 4.10). That was 2011. You can scour the internet and find numerous articles about pieces of ice breaking off of Pine Island Glacier, Thwaites Glacier, and all of the other glaciers of the West Antarctic Ice Sheet near Amundsen Bay. Another piece for example, 120 mi^2 in size (310 km^2) broke off in early 2020. It was 65°F (18°C) on the Antarctic Peninsula when the breakup occurred. Large calving events at Pine Island Glacier, according to NASA, have been occurring more frequently, including in 2001, 2007, 2013, 2015, 2017, 2018, and 2020. While calving of glaciers is normal, such large events and their frequency, are not. The primary driver of these events, according to researchers is

the influx of warm subsurface water into the Amundsen Sea Embayment, which is causing ice to melt from below. That, in turn, is related to shifting wind patterns that are pushing warm, deep ocean water onto the continental shelf. It's also in line with the bigger picture of climate change.[40]

Sea ice melt in the Antarctic is not uniform however. Different areas of the South Pole have different melt rates. The Amundsen Sea in West Antarctica, where some of the biggest ice sheet and glacier break-offs have been occurring, for example, has a 40-year negative trend in terms of sea ice cover and a reversal of trend to the upside since 2007. In four of the five regions where sea ice is present near the South Pole, sea ice cover is actually *growing*, notes Parkinson. These four regions acted in unison with the prolonged uptrend and then together fell off during the 2014–2017 period. The sudden sharp decreases in a short period of time warns us that while we may be in an uptrend in some places, an extremely accelerated melt-off event, such as occurred between 2014 and 2017, is still within the realm of possibilities.

"The Arctic has become a poster child for global warming, but the recent accelerated melt-offs in Antarctica have been far worse. All of us scientists were thinking eventually global warming is going to catch up in the Antarctic," Parkinson stated in a recent interview. Kaitlin Naughten of the British Antarctic Survey also focuses on Antarctic sea ice and she notes that glacier melt runoff coming from the Antarctic continent, the ozone hole above the South Pole, as well as ocean sea currents like El Niño directly affect ocean sea ice in the region.[41]

Figure 4.10 Colossal breakoff of Pine Island Glacier on the West Antarctic Ice Sheet, seven times the size of San Francisco, California
Source: NASA

As we can see from a look at our two (or three) whitest regions of the Earth, the North Pole, the South Pole, and the Greenland Ice Sheet, our planetary extremes are in extreme climate chaos.

We know clearly that the changing tone or *color* of Earth's surfaces, caused in large part by disappearing glacier surfaces and melting sea ice, determines how much sunlight is absorbed. The removal of glacier ice (the disappearance of glaciers) or the disappearance of sea ice effects whether or not sunlight is absorbed by the Earth adding to warming or reflected back into space causing cooling. Debris or dust—such as smoke from a forest fire, the ash of an erupting volcano, or emissions from vehicles and industry—circulated by global air streams and collecting on glacier surfaces make those glacier surfaces darker. And just as the color of our clothing affects our degree of warmth in the outdoors, this soiling warms glaciers and accelerates their melt, leading in turn to the acceleration of global warming.

These phenomena, which have to do with climate change and global warming caused by albedo changes, are occurring in parallel to a general trend of warming temperatures on the planet caused by yet other reasons. They are alarming. Our icy white glacier surfaces and the cold environments in which they thrive are an essential condition for stabilizing our climate. Yet, day in and day out, global warming is on a runaway path, and glaciers are suffering the consequence. The warmer they get, the warmer our Earth gets, and the warmer we get. It's a vicious cycle.

As I was writing that last paragraph I received yet another worrisome news article that showed that the temperature in Antarctica was the highest ever recorded since records began to be taken in 1961. By the time I edited the second version of this chapter a few months later, that record had been broken again.[42] In other news breaking during the very last revisions to the chapter (in April of 2021), ice melt at Pine Island Glacier seems to be reaching irreversible tipping points beyond which there may be no return.[43]

5

A Thawing Earth

As we have learned in previous chapters, glaciers have covered much of the Earth, *and then have not*, for millions of years, cycling in and out of existence, extending slowly over vast tracks of land and then retreating suddenly, as if they had never been.

As the Earth moves into an ice age, which in relatively recent Earth history has been occurring over periods of about 100,000 years, glaciers advance over organically active terrain covering existing vegetation. Covered by ice and unable to go through photosynthesis while temperatures plummet and prevent cell growth, this vegetation is destroyed and dies. The dead organic remains are then frozen, become trapped in the ice, and are conserved in this decaying state, for perhaps hundreds, thousands, or even millions of years.

Think of the time (hundreds of millions of years ago) when dinosaurs roamed the Earth, or more recently when woolly mammoths were a common sight (tens of thousands of years ago), when much of the planet was covered in greenery. Much organic matter that may have existed at that time was eventually covered by advancing glaciers and would have been buried deep under ice during glaciation periods. As warmer climate cycles return to the Earth, this frozen, stagnant *deteriorated* organic matter becomes newly exposed to the atmosphere and begins to release gas from the once putrefying organic matter. That gas is in part CO_2, but much of it is *methane gas*.

As if a CO_2 and methane emissions button were suddenly pressed, the decayed matter that emerges from its millenarian frozen sleep resumes its decaying cycle and begins to release these long-trapped gases during the ongoing (though delayed) rotting process (see fig. 5.1 of methane gas released from frozen grounds underneath Lake Baikal in Russia) . These gases, released as glaciers melt and retreat and expose frozen surfaces to warmer climates, or as frozen grounds beneath water bodies thaw, are greenhouse gases that causes global warming. One of the big concerns however is the methane gas release. The problem is that methane gas is dozens of times more potent than CO_2 in terms of its capacity to warm the planet. This is one of the ways melting glaciers (and in this specific case, thawing terrain in frozen environments) cause global warming.[1]

This frozen terrain located in some cases underneath, near or in the vicinity of glaciated areas (glaciers need not be present to have frozen grounds) is what we call permafrost: areas of ground that are permanently frozen. In particular, we define permafrost as grounds that have been frozen for at least two consecutive years and so have survived the warm months of two consecutive summers.

Neil Davis describes the Earth's permafrost features in the following manner:

Like many other words, "frozen" has more than one meaning so in addition to frozen ground we have things like frozen assets and people frozen in their tracks. Thus, frozen brings to mind both cold and immobility, but only the first really applies to frozen ground. Frozen ground definitely is mobile; it flows under load, it expands and contracts as its temperature changes, and its soil particles shift and deform and molecules of water move around between them.

Figure 5.1 Methane bubbling up from thawing organic matter beneath frozen Lake Baikal in Russia.
Source: Kristina Makeeva

In certain circumstances, freezing ground can swell up like a balloon, and when it thaws it can shrink to a fraction of its frozen size or slither away down a slope. Furthermore, the movements that accompany freezing and thawing within the ground can alter the makeup of a soil by reorienting its particles, compressing them, changing their shape, and sorting them according to size. So it is best not to think of freezing and frozen soil as a static entity. Albeit in slow motion, it churns and seethes; it is a thing alive that leads a most interesting life involving time scales ranging from hours to thousands of years.[2]

S. M. Govorushko, of the Pacific Geographical Institute in Vladivostok, Russia, has this to say about the distribution of permafrost around the planet (see fig. 5.2):

The permafrost area of the Earth is 38.15 million km², which corresponds to 25.6% [others such as Neil Davis suggest about 20%] of the land surface; 21.35 million km² fall in the northern hemisphere. Permafrost underlies 25.6% of the Earth's land area, including probably all of Antarctica, 99% of Greenland, 80% of Alaska, 61.5% of Russia, 55% of Canada and 20% of China. A substantial part of the present-day permafrost was inherited from the last glacial period, and now it is slowly thawing. ...

The maximum thickness of the permanently frozen ground is 1,493 m (4,898 ft.) in the northern Lena and Yana River basins in Siberia. In North America, the observed thickness of frozen rocks in northern Alaska reaches 740 m (2,428 ft.).

The largest area of Alpine permafrost, 1.5 million km² (580,000 mi²), exists in western China. Alpine permafrost in the contiguous United States is present on about 100,000 km² in mountainous areas of the west. It occurs at elevations as low as 2,500 m (8,202 ft.) in the northern states and at about 3,500 m (11,500 ft.) in Arizona.

In South America in the Andes along the Atacama Desert, permafrost begins at an altitude of 4,400 m (14,400 ft.) and is continuous above 5,600 m (18,400 ft.).

Large areas of permafrost also lie under the Arctic Ocean, and on the northern continental shelves of North America and Eurasia.[3]

Figure 5.2 Permafrost extent on the Earth.
Permafrost on the Earth is plotted in the image by an online permafrost zoning tool. The northern hemisphere has extensive permafrost regions. While the southern hemisphere is not shown, most of Antarctica is permafrost, while permafrost is also found in select areas along the Andes of South America, some high mountain environments of New Zealand, and a handful of islands in the southern hemisphere's oceans, such as South Georgia, and South Sandwich and the McDonald and Heard Islands (belonging to Australia), and a few last remaining enclaves on the African continent (Mount Kilimanjaro for example).
Key: Blue/Red Continuous—darker colors, Yellow Favorable—light colors, Green Uncertain—intermediate grey color.
Source: Google Earth. Permafrost Zoning Map: Stephan Gruber, University of Zurich

There is one tricky aspect to permafrost. Despite being defined as "permanently frozen ground," permafrost is not always *permanent*. Let me explain. Some permafrost *is* permanent, that is, always frozen, always, all the time. I stress that simply to distinguish permanent from "discontinuous" permafrost, which can be frozen most of the time, but sometimes melts. That is, it freezes, and then melts, and then re-freezes. Then there is "sporadic" permafrost, which as the qualitative word suggests, is occasional. Finally there is "isolated permafrost," which is even more sporadic.

Scientists classify various types of "permafrost" as follows:

Permafrost Categorization	
Continuous	frozen 90–100% of the time
Discontinuous	frozen 50–90% of the time
Sporadic	frozen 10–50% of the time
Isolated	frozen 0–10% of the time

Source: French, 2008. P. 94.

Curiously, even continuous and *permanently* frozen grounds (the 90–100% of the time category) aren't *always* frozen!

Permafrost terrain that cyclically melts is critically important for local water supplies, since the ground is capturing moisture in the form of snow or rain, freezing it during cold weather, and then melting it off back into the environment during warmer weather. This is a fantastic hydrological tool created by Mother Nature to ration water flow into downstream ecosystems. For this reason, the fact that permafrost isn't always frozen is also important, since it is the freeze-thaw cycles that help generate meltwater from the permafrost, otherwise it would simply be a solid block of ice, always.

In terms of CO_2 or methane emissions, we're not so concerned about permafrost that is sporadic or discontinuously freezing and thawing since we're probably not getting much emissions from organic material as it really doesn't have a chance to decay and generate large volumes of gas in such areas, and because even if it did, the recurring cycles of thaw and decay are already counted in the normal naturally-derived gas emissions contributions to the atmosphere. What we are most concerned with is the melting of deeper, permanently frozen ground, the ground in *very* cold environments that has indeed kept frozen over many centuries or millennia and that is holding a backstock of decayed organic material rich in gases, including very potent methane gas. That ground does have a chance of exposing new and very significant volumes of organic material that *was* buried way back in the past and which is not normally discharged into the atmosphere. It is the release of methane gas in this veritable "permanently" frozen permafrost that troubles us and that is such a nemesis for the climate today.

The Institute for Governance and Sustainable Development (IGSD) notes that 50% of the carbon stored in the world's soil is stored in Arctic permafrost and at current melt rates, vast amounts of soil carbon could be released into the atmosphere by 2050.

Alaskan soils have already been shown to be a carbon source from 2012–2014, and for the [periods of] October to December, CO_2 emissions rates increased 73% since

1975. [They point to the Arctic region] because the high-latitude regions have the largest stocks of soil carbon and the fastest rates of warming, the overwhelming majority of warming-induced soil [carbon] losses are likely to occur in Arctic and subarctic regions. Assuming 2°C of warming under business-as-usual, by 2050 soils would release 55 +/- 50Gt of carbon (equal to 200 billion tonnes of CO_2), increasing CO_2 concentration by 25ppm, equivalent to 12–17 percent of the expected emissions in 2050.[4]

Approximately 3.4 million km² (1.3 million mi²) of permafrost has already thawed in the 20th century, according to IGSD. In the northern hemisphere, permafrost has warmed by 2–3°C in the past two decades. Near-surface permafrost has warmed by more than 0.5°C in recent years. Globally, permafrost harbors about 1,700 gigatonnes of carbon, which is twice the amount of carbon currently in the atmosphere. Of this, stresses IGSD, 190 gigatonnes of this carbon lies at the top of the soil, within 30 centimeters (12 inches) of the surface, which is the most vulnerable to global warming, while more than half of all carbon in permafrost is in the top 3 meters (10 feet) of soil.

The Ice That Burns

Have you ever heard of ice that can catch on fire?[5] Well, you just may need your fire extinguisher on your next permafrost or glacier hike. As glaciers and permafrost melt, the small methane gas bubbles that were trapped in the ice thousands of years ago are compounded and pressed into small methane hydrates. These gas hydrates are ice-like crystalline molecular complexes formed from mixtures of water and methane gas, making this form of ice extremely flammable. You can bring a match up to a gas hydrate ice cube and light it and the ice will actually catch on fire. To be clear, it's the methane in the ice that is catching on fire and not the ice itself.

The oil and gas sector has expressed much interest in gas hydrates because this is actually the most abundant form of natural gas (and, for that matter, fossil fuel) on the planet. That's right, there is more fossil fuel trapped in permafrost than all of the other forms of fossil fuels combined, by a 10:1 ratio.

Figure 5.3 shows the global gas hydrate reserves that geologists believe could be tapped for fossil fuel energy, some of which have been verified, others are implied. The reason oil and gas companies don't exploit gas hydrates is that they don't know how to do it in a way that is economically efficient, otherwise I assure you they'd be churning up land and dynamiting glacier ice to try to get to it just as mining companies have done to get at gold in glacier and permafrost environments in places like the Central Andes and Kyrgyzstan. In fact, oil and gas companies *have* experimented with the extraction of gas hydrates from millenarian glaciers in certain areas of the Arctic, in Canada, China, and Russia, and even in the Andes.[6] In terms of the impacts on climate change and global warming, it would be better that they leave the gas hydrates in the ground.

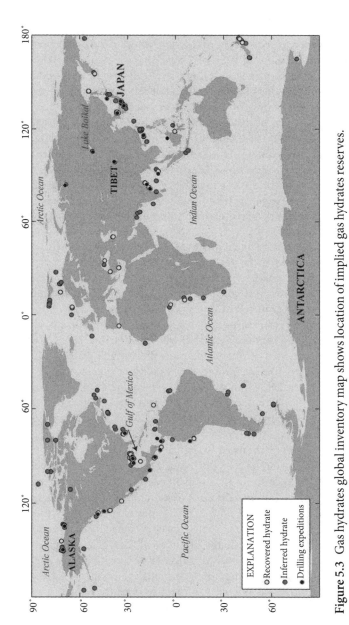

Figure 5.3 Gas hydrates global inventory map shows location of implied gas hydrates reserves.
Source: USGS

How Easy (or Complicated) Is It to Determine
Where the Ground Is Frozen?

The question, how easily can we determine where permafrost is, is key. It seems very important to know exactly where frozen ground exists to understand its dynamics and relevance, to ensure its protection where it can be protected, but also to ensure we are not building infrastructure on potentially unstable ground. By knowing where the Earth freezes and thaws cyclically, we can make determinations about many things. In terms of basic land and urban planning, such as constructing a highway or locating a home or office building, knowing where the Earth is permanently frozen can help us avoid costly mistakes. We don't want to build a road, or a city, over ground that is continuously expanding and contracting, bulging and collapsing, or that would be prone to contraction in a future with increased global warming. We also would want to know where our Earth is providing cyclical meltwater in freezing and thawing areas, in order to protect the hydrological value of that land and ensure that we use its hydrology efficiently where beneficial. Finally, it also behooves us to know where our future emissions may be coming from as permafrost around the world thaws.

In the past, determining frozen grounds was a painstaking task, involving long stints by geo-cryologists in barren permafrost lands. The weather was inhospitable, and the task arduous, involving digging and taking thermometer readings to determine temperature. Of course, any specific temperature reading was just that, one specific temperature reading at one moment in time. But determining permafrost existence requires monitoring ground temperature over time and at many sites, a seemingly impossible task if we consider that 20–25% of the world's land surface is permafrost of one type or another. Much of the data we did have was inferred based on findings noted in a variety of places. For example, if in a certain part of the Central Andes we found that at a dozen sites tested that permafrost exists at above 3,000 meters (about 9,800 feet) we might conclude that for the entire region there is permafrost at and above 3,000 meters. These were obviously rough calculations.

Another way to infer permafrost existence can be applied by observing specific elements of the terrain that help us distinguish the presence of permafrost, such as rock debris ordered into certain patterns by freeze-thaw dynamics. These patterns are produced by frost heave (an upward push by expanding, freezing water), the freeze-thaw cycles of the Earth, and the physical attributes of rocks interacting with gravity and the permafrost's dynamics or as rocks moved by permafrost dynamics flow down a slope and settle. The distinguishable presence of rock glaciers (glaciers that are a mix of rock debris and ice) covered by a thin layer of rock is also indicative of the presence of frozen grounds. When analyzing satellite imagery, active rock glaciers (rock glaciers that move—as opposed to inactive ones that don't) are easily identifiable and can be utilized to infer the presence of hydrologically rich frozen terrain. Inactive rock glaciers (rock glaciers that no longer show any physical displacement) are harder to identify through the use of satellite imagery but can still be distinguished by a trained eye.

Regardless, identifying permanently frozen grounds manually and on the ground is a difficult exercise. Today things are easier. With satellite technology, we can quickly ascertain the presence of frozen terrain from the comfort of our own homes. You can find rock glaciers, for instance, with your computer or even your smart phone if you know what you're looking for and follow the clues of rock debris flow in very cold environments. I will teach you to do that in Chapter 8.

The task of more effectively and more easily identifying vast areas of frozen grounds has been greatly simplified by modern science. Stephan Gruber of the University of Zurich in Switzerland has developed a free and easily utilized online remote sensing tool called the Global Permafrost Zonation Index Map, that can be used on Google Earth, to quickly identify places where the surface air maintains a yearly average temperature of 0°C.[7] While this is not exactly the same as measuring the ground for its surface temperature (or the deeper ground, which cannot be measured by the tool), it's pretty darn close. The tool can determine, with high accuracy, the presence of frozen grounds (at least at the surface) and essentially generates a global permafrost map.

You simply download a file to your computer, open it up in Google Earth, and you can see where the Earth is most cool. Simply visit the area you would like to look at, and the tool conveniently maps average surface air temperature from cold (freezing) to even colder (see fig. 5.4). Purple and red areas are the coldest (darkest and dark shade). Yellow areas are likely to be permafrost in favorable conditions (in the shade or on slopes facing the Earth's closest pole, for example), while green areas are areas of uncertainty—where we are not sure that there are perennially frozen grounds.

Figure 5.4 A permafrost zonation map developed at the University of Zurich allows us to easily identify frozen grounds on any area of the Earth.

Source: Google Earth. Permafrost Zoning Map: Stephan Gruber, University of Zurich

The Nature and Contents of Permafrost Terrains

While some permafrost may be short-lived, most permafrost has actually been around for a long time, perhaps for many hundreds or even thousands of years. The extent of permafrost regions in the Arctic for example is enormous—twice the size of the United States! Permafrost layers may be only a few feet thick or can reach thicknesses of nearly a mile.

There are lands and lakes or ponds including mountains, tundra, etc., in very cold areas like Siberia or the high central Andes Mountains and even in places in Alaska or the Rocky Mountains, where mean yearly temperatures are maintained at or below freezing. There is also frozen Earth beneath the water of the Arctic ocean in the deep ocean floor.

In these frozen terrains, any organic matter that may have grown in the past on these lands and in these deposits is dead (killed by freezing processes) and was trapped long ago in a permanently frozen state. It is believed that the carbon content of permafrost around the world is about *double* the amount of atmospheric carbon.[8]

As the Earth warms again (and unfortunately most of this frozen terrain is in the extreme polar regions where warming is occurring twice as fast as in the rest of the planet), this terrain thaws, and like organic matter trapped in glacier ice, as temperatures move above the freezing point, this frozen organic matter in the crust of the Earth resumes its decaying cycles that were halted centuries or millennia earlier. Subsequently, CO_2 and methane gases are released into the atmosphere from this newly exposed and decayed organic matter.

As the air warms above, frozen ground begins to melt from the top down, at a rate of a few dozen centimeters per year or more. Certain parts of the Siberian tundra, for example, where permafrost thaw is extensive, are melting at a rate of about 50 centimeters per summer (20 inches).[9]

Oleg Pokrovsky, Biogeochemist and Research Director of Tomsk State University in Russia, paints a troubling picture of what is to come:

> The effect of climate change on permafrost is strongly reciprocal; with the thawing of huge carbon stocks in frozen soils or methane hydrate release, these greenhouse gas components will be released into the atmosphere, thus further accentuating climate change, not only at high altitudes but also across the entire planet. The detailed mechanisms of a possible "permafrost time bomb" are unknown at present.[10]

Many have observed bubbles appearing in thawing lakes in vast tundra areas of Siberia or in parts of Canada and Alaska. These are methane bubbles, escaping from the once frozen lakebeds. Methane is quickly reaching the lake surface and being liberated into the Earth's atmosphere.

The release of methane occurs year round and mostly goes unnoticed with just small bubbles occasionally surfacing. However in the wintertime, lakes freeze over and emerging lakebed bubbles get trapped in the lake water, unable to escape the lake's ice cover. At the end of the winter, when the ice breaks up, the gas escapes suddenly and vast amounts of methane are released into the air. Put a match to methane escaping from a frozen lake, and you may cause an explosion![11]

A news story by the BCC showcased the work of Katey Walter of the University of Alaska at Fairbanks who studies methane release from permafrost areas. In the video you can see just how potent methane gas emissions are from escaping methane bubbles in a thawing lake.[12] Poke a hole in the ice in the winter, expose the water below, and you soon see shots of air coming up through the hole. Light a match to it and you literally get an icy explosion of gas and flame! (see fig. 5.5).

Dahr Jamail's book *End of Ice* includes an interview with Vladimir Romanovsky, a permafrost expert at the University of Alaska, Fairbanks, who has 40 years of data from local sources indicating that the permafrost temperature changes from Alaska's North Slope are some of the most dramatic temperature changes in the world. Over 35 years of measurements, the temperature readings 20 meters (66 feet) below the surface have increased by 3°C. At only one meter below the surface (3 feet), the temperature increase is an incredible 5°C.[13]

The impacts of permafrost thaw, especially from lakes, ponds, wetlands, and ocean bottoms, are particularly concerning in terms of global warming potential. The UN Intergovernmental Panel on Climate Change (IPCC) report, *The Ocean and*

Figure 5.5 Fire from a frozen lake due to the accumulation of methane gas. Scientist Katey Walter (left) and Louise Farquharson (right) of the University of Alaska Fairbanks and her colleague are thrown back from fire bursting out of a frozen lake, in a sudden release of methane gas from thawing permafrost. Source: Katey Walter Anthony, University of Alaska Fairbanks. Photo credit: Mark Thiessen.

Cryosphere in a Changing Climate, released in 2019, tells us just how significant this methane release is:

> Permafrost temperatures have increased to record high levels (1980s-present) … Arctic and boreal permafrost contain 1460–1600 Gt organic carbon, almost twice the carbon in the atmosphere. … Widespread permafrost thaw is projected for this century and beyond. By 2100, projected near-surface (within 3–4 meters) permafrost area shows a decrease of 24% (+/− 16%) … [leading] to the cumulative release of tens to hundreds of billions of tons (GtC) of permafrost carbon as CO_2, and methane to the atmosphere by 2100 with the potential to exacerbate climate change. … Methane contributes a small fraction of the total additional carbon release but is significant because of its higher warming potential. In many high mountain areas, glacier retreat and permafrost thaw are projected to further decrease the stability of slopes, and the number and area of glacier lakes will continue to increase.[14]

Recent studies suggest that an event 252 million years ago brought on the largest ever mass extinction in Earth's history—the Permian Extinction or "the Great Dying."[15] Five such mass extinctions have been identified, the Permian being the most recent. In that event, 90% of all living things were wiped out by runaway global warming, caused by a sudden emission of a large amount of CO_2 into the atmosphere (likely by volcanoes). The resulting warming melted permafrost, which emitted vast amounts of CO_2 and methane gases into the atmosphere to further leverage climate change.[16] The double-whammy of CO_2 and methane put the planet's climate over the top, making the ecosystems too warm and the oceans too acidic to sustain life. We're still a ways from the estimated average global temperature of 29°C reached at that time (we're at about 15°C now), however, our CO_2 emissions are through the roof, and our methane emissions are rapidly increasing.

This sudden CO_2 and methane increase could conceivably lead us into a sixth mass-extinction event. Some suggest that one is actually already underway and call it the *Holocene Extinction* (or the *Anthropocene Extinction*), referring to the epoch of the Quaternary Period we are now living in. The current rate of extinction of species is estimated to be between 100 and 1,000 times greater than the normal background extinction rate. Climate change, habitat loss, and the disturbance of delicate ecosystems are thought to be some of the primary causes of this possible new *sixth* mass extinction event. Humans are thought to be to blame, not only because of the high levels of CO_2 and methane gas we are releasing into the atmosphere, causing global warming, but also because we are considered to be a "global super-predator" of apex species, wiping out vast numbers of animals, fish, etc.[17]

The role of methane in the evolving theories of mass extinction, and the fact that the increase in emissions 250 million years ago which led to mass extinction came from melting permafrost, when considered in the context of what we know about climate change today, should definitely worry us. IGSD studies permafrost thaw and its impacts on the global climate and is especially concerned about widespread melting permafrost because it is very difficult to reconstitute permafrost terrain. Carbon and methane captured in permafrost, which is now being released because of global warming, operates on a millenarian time scale. Essentially, the melting permafrost, once it happens, is irreversible on human time scales. Basically, if the permafrost becomes unstable, it may result in a runaway situation, which when combined with our extremely high levels of CO_2 emissions, would lead to irreversible tipping points and rapidly escalating climate collapse.

In its Primer on Polar Warming, IGSD describes what it calls the permafrost carbon feedback:

> When permafrost thaws, organic matter is broken down into carbon dioxide (CO_2) and methane (CH_4) that is released into the atmosphere. There are four mechanisms through which permafrost thaws and releases carbon into the atmosphere: thickening of the active layer, talik formation, thermokarst development, and river and coastal erosion. The active layer is the upper layer of the soil that thaws and refreezes each year. Talik is an unfrozen layer of soil that has a high moisture content and temperature conducive to carbon release. Thermokarst lakes form when ice-rich soil thaws and the ground collapses, forming depressions that when filled with water are even more favorable for carbon release and additional permafrost thaw. Permafrost can release carbon as river discharge changes as well as along coastal boundaries from rising sea levels and wave and storm damage.[18]

Gases released from melting permafrost, emitted as CO_2 or methane, can have a colossal impact on our climate, says IGSD. The estimated impact varies according to different studies, ranging from 25% to 40% of total global warming. The difference will be determined by the type of soil and whether it is dry or moist.

Ira Leifer is an academic researcher specializing in subaquatic measurements of methane emissions from seabeds, particularly in the Arctic where most of these emissions occur. Dahr Jamail is concerned about permafrost melt in Alaska, a place he holds dearly as he has spent a considerable portion of his life there, and interviews Leifer to get his thoughts on the relevance of permafrost melt to methane release. Here is what Leifer had to say:

> Oceans have a thousand times the heat capacity of the atmosphere. Warming something in water at 200°F is far faster than warming it in your oven at 200°F. The East Siberian Sea is a huge shallow sea, so any methane there is going to get out. Same with terrestrial methane from permafrost.

Jamail echoes Leifer's concerns as he observes the warming of the Barents Sea, just north of the Atlantic, as well as temperature rises in the East Siberian Sea, which is likely in the next few decades. As temperatures continue to rise, once the methane in these seabeds is released, says Leifer, it will drive climate change further: "Methane pushes us over tipping points. If humans keep accelerating the rate methane is released, that ten-year timescale isn't a limit, because you keep hitting the accelerator further and further."[19]

Jamail's sobering conversation with Leifer steers to a "hot spot" of roughly a thousand square kilometers that contained a staggering 60 million methane bubble plumes. For comparison, the natural background rate of bubble plumes is in the tens of thousands for an area that size. Leifer sees the Barents as a warning of what will eventually happen across all the Arctic seas as the waters continue to warm and release more and more methane from the subsea permafrost.

Jamail also cites the findings of Natalia Shakhova, a former University of Alaska researcher and associate professor at the International Arctic Research Center, from the East Siberian Arctic Shelf (ESAS). This is the largest undersea-shelf in the world, spanning more than 2 million km², and holding 10–15% of the world's methane hydrates (see section on gas hydrates later in this chapter). The fear, she says, is that there could be a methane "bomb" waiting to explode under the ESAS that could result in the sudden and unparalleled release of gigantic volumes of methane gas, triggered by an earthquake or due to melting permafrost. This release could potentially emit 50 gigatons of methane, equivalent to 1,000 gigatons of carbon dioxide, increasing the modern atmospheric methane burden twelve-fold, with consequent catastrophic greenhouse warming.[20] Some scientists have questioned the magnitude and likelihood of this sudden methane release actually occurring and the suggested velocity and impacts to global warming of this scenario,[21] but certainly, the presence of massive methane volumes in the ESAS, given the observed rate of deterioration of permafrost around the world, is ominous. It could conceivably lead to another mass extinction.

Jonathan Watts of the Guardian ran an article recently with the ominous title of: "Arctic Methane Deposits Starting to Release". He calls attention to findings that indicate that a new source of greenhouse gases off of the coast of East Siberia has been triggered. Watts highlights recent measurements taken in October 2020 by an international expedition[22] on board a Russian research vessel that show elevated methane release from the Arctic Shelf. A Swedish scientist Örjan Gustafsson of Stockholm University on the vessel stated that the "East Siberian slope methane hydrate system has been perturbed and the process will be ongoing."[23]

Besides methane release, melting glaciers and thawing permafrost also create other types of problems. Both phenomena destabilize the Earth, ice (in the case of melting glaciers), and land (in the case of thawing permafrost). We talk more about such instabilities caused by melting glaciers and permafrost in Chapter 6. Thawing permafrost can also be devastating for the natural and built environment. Forests, including their trees and other plants, can sink suddenly because of collapsing ground underneath. The ground collapses as ice in the ground melts and the meltwater flows out of the area, leaving pockets of air behind. This causes the ground to destabilize and in some cases, collapse in on itself.

Anything that was once firmly rooted in the ground, such as trees, rocks or manmade infrastructure such as buildings or roads, can shift position because of thawing permafrost and can also collapse. Infrastructure such as homes, buildings, roads, bridges, etc., that may have been built when lands were permanently frozen, are directly affected by changing and thawing temperatures under the surface of the Earth. All are at risk of collapse as vulnerable permafrost areas melt, soils and rocks move, and grounds that used to hold everything together firmly, collapse. In these areas, roads may buckle due to melting permafrost terrain as ice in the ground under the roads melts and the water seeps out (see fig. 5.6) . These roads must be entirely resurfaced as the area warms up.

Figure 5.6 Thawing permafrost destroys Chief Eddie Hoffman Highway in Alaska. Chief Eddie Hoffman Highway, seen on June 28, 2017, is the main thoroughfare in Bethel and one of the few paved roads. It has become a roller coaster of a ride over the past couple of years. The state Department of Transportation is studying whether heaving from the thaw-freeze of permafrost is a factor in its deterioration.
Source: Lisa Demer/Alaska Dispatch News

As for the economic costs of melting frozen grounds, IGSD estimates that roughly four million people living in the Arctic region and approximately 70% of the current infrastructure located there are susceptible to being impacted by thawing permafrost. According to the environmental science and policy organization, a 2016 study calculated the economic costs of CO_2 and methane release from permafrost could amount to $43 trillion dollars over the next two centuries.[24]

Massive Sinkholes Caused by Thawing Permafrost

Another phenomenon spurred on by rapidly melting permafrost is the formation of sinkholes (or thermokarst).[25] In this case, frozen land held together by ice crystals beneath the surface rapidly melts and suddenly collapses, opening up colossal holes in the ground. These cryospheric sinkholes are sometimes found on glaciers, but now, as our Earth warms, we are starting to see them appear in land that has been frozen for hundreds or thousands of years. They form because ice has greater volume than water (it takes up more space). So, as subsurface ice melts, it leaves gaps in the soil and rock and unstable ground that can subside and collapse, leaving gaping holes at the surface. Water then runs to the bottom of these depressions, forming ponds and lakes.

Sergey Govorushko who studies melting permafrost tells us that "Thermokarst processes occur extensively in the depositional plains of northern Eurasia and North America. Countries with the most extensive development of thermokarst include Russia, Canada, and the United States."[26]

Merritt Turetsky of the University of Guelph and the University of Colorado, Boulder, and author of academic research on the cryosphere says that we are underestimating the impact of permafrost by about 50%:

> The amount of carbon coming off that very narrow amount of abrupt thaw in the landscape, . . . that small area, is still large enough to double the climate consequences and the permafrost carbon feedback. Where permafrost tends to be lake sediment or organic soils, the type of earth material that can hold a lot of water, these are like sponges on the landscape. When you have thaw, we see really dynamic and rapid changes. That's because frozen water takes up more space than liquid water. When permafrost thaws, it loses a good amount of its volume. Think of it like thawing ice cubes made of water and muck: If you defrost the tray, the greenery will sink to the bottom and settle. That's exactly what happens in these ecosystems when the permafrost has a lot of ice in it and it thaws. Whatever was at the surface just slumps right down to the bottom. So you get these pits on the land, sometimes meters deep. They're like sinkholes developing in the land. Essentially, we're taking *terra firma* and making it *terra soupy*. When you come back in, it's a lake and there's three meters of water at the surface. You have to probably say goodbye to your equipment.[27]

Matt Simon, author of the article just quoted, goes on to describe the findings by Turetsky and her colleagues as well as other scientists such as Christina Schaedel, a Northern Arizona University Biochemist and plant ecophysiologist. One of the most notable concerns is the effects of *rapidly* melting terrain as opposed to melt rates that might occur more naturally (or not at all). Considering that these areas are often rich in biological diversity, the melting of organic matter can also release significant amounts of CO_2 into the atmosphere.

In balanced ecosystems, the microbes that would otherwise feed off of "slowly" melting exposed permafrost, help digest nutrients, which produces CO_2. However, as climate change is accelerated beyond a *normal* pace, due to massive climate wildfires occurring for instance in Australia or the western United States or due to general global warming trends, these natural processes can be overwhelmed by the rapid pace of the degeneration and land collapse. This results in the exposure of greater volumes of deteriorating organic matter than the microbes can consume, resulting in intense emissions of CO_2 and methane. This is a dangerous feedback loop exacerbating emissions from permafrost thaw.

The Batagaika Crater in Siberia (fig. 5.7) is one of the most startling examples of the effects of thawing permafrost on the natural environment. The crater appeared suddenly, as previously frozen ground collapsed at the site, dropping nearly 300 feet (about 100 meters). The crater is about 1 kilometer long (0.6 miles) and 86 meters deep (282 feet). Katie Orlinsky, who documented the phenomenon for National Geographic, had this to say:

Figure 5.7 The Batagaika Crater in Siberia.
The Batagaika Crater in Siberia is an example of collapsing permafrost terrain, caused by a thawing environment.
Source: (Katie Orlinsky) National Geographic; see www.katieorlinskyphoto.com or Twitter: @ KatieOrlinsky

I had never really felt, in my bones, the weight of it all. I had never felt a sense of real fear, for myself, for those I love most, and for all of humanity. But that was what rushed over me when I finished working in Siberia. ... And for me, something has already changed. Standing inside the Batagaika Crater on my last shooting day, watching and listening to the earth as it tumbled down towards me, has affected me in ways I have a difficult time articulating. I wish I could just take everyone there, especially those in political power, to experience it first-hand.[28]

The Batagaika Crater in Siberia is one of dozens of sinkholes (or *thermokarst depressions*), as indicated above, formed by collapsing permafrost. These are often called "slumps." NASA writer Pola Lem suggests this is a case of a "megaslump." The exposed soil from the collapsing ground reveals 200,000 years of the Earth's geological and biological history, offering a rare glimpse into the past, a geo-historical reveal that is unfortunately becoming more and more common around the world.

The Siberian crater appeared in the 1960s, and has been growing since, a product of a number of local and global variables. Deforestation has removed much shade from the area, allowing the sun to penetrate and reach the ground, leading to a warming of the region's surface. The humidity in the air has also changed due to tree-cover loss, which has contributed to a warmer microclimate, which in turn affects the temperature of the soil. All of this coupled with rising global temperatures led to the progressively deteriorating state of the surrounding permafrost and to the Batagaika "megaslump."

Anton Troianovsky and Chris Mooney of the Washington Post did a special report on thawing permafrost in Siberia. They describe how melting earth is unearthing woolly mammoths which in turn is creating a new ivory trade. They describe the stench of decaying woolly mammoth bodies as putrid and filling the air, "decomposing after millennia entombed in a frozen purgatory. It smells like dead bodies."[29]

The authors describe the scene at Zyryanka, in Yukutia, Siberia, which has warmed more than many parts of the Earth by 3°C since pre-industrial times, triple the global average. Here is their description of events:

> The permafrost that once sustained farming—and upon which villages and cities are built—is in the midst of a great thaw, blanketing the region with swamps, lakes and odd bubbles of earth that render the land virtually useless. "The warming got in the way of our good life," said Alexander Fedorov, deputy director of the Melnikov Permafrost Institute. "With every year, things are getting worse." For the 5.4 million people who live in Russia's permafrost zone, the new climate has disrupted their homes and their livelihoods. Rivers are rising and running faster, and entire neighborhoods are falling into them. Arable land for farming has plummeted by more than half, to just 120,000 acres in 2017. In Yakutia, an area one-third the size of the United States, cattle and reindeer herding have plunged 20 percent as the animals increasingly battle to survive the warming climate's destruction of pastureland.
>
> Siberians who grew up learning to read nature's subtlest signals are being driven to migrate by a climate that they no longer understand. This migration from the countryside to cities and towns—also driven by factors such as low investment and spotty Internet—represent one of the most significant and little-noticed movements to date of climate refugees.[30]

The riverbeds are dwindling in Siberia, as permafrost melt makes once solid earth crumble away. Once stable grounds that used to be useful for farming, are now deteriorating wastelands of potholes, collapsed Earth, and puddles. The article goes on to highlight impacts to farming in the melting permafrost region. The area of cultivated land in the Yakutia has fallen by more than 50% since 1990, attributable to thawing permafrost and associated land deterioration. The region's cattle stock has shrunk by about 20%. Reindeer herds have also sharply declined.[31]

Permafrost Zombies

Thawing permafrost can also unleash real life zombies ... but let's not get carried away, these are not flesh-eating zombies in human form that walk around and try to break into your home to eat you. But they *are* in fact revived from the dead (the frozen state) and they *could* conceivably be flesh-eating!

We're talking about extinct disease-infested bacteria that have been frozen in permafrost terrain or in glacier ice for hundreds or thousands of years. With permafrost thawing, the bacteria are newly released into the surrounding environment after being extinct (or perhaps a better term, *dormant*) for hundreds, thousands, or even tens of thousands of years, if not more.

An article in Scientific American drew attention to cases of deadly disease impacts that have been widely reported in the media and allegedly related to ancient bacteria in Siberia.[32] A 12-year old boy died in the remote Siberian tundra and at least 20 others were diagnosed with a potentially deadly disease, while thousands of reindeer died. Authorities believe these were cases of newly released spores of *Bacillus anthracis* (anthrax) into nearby water and soil, and the culprit? Melting permafrost.

Think about people who died of terrible diseases centuries ago and were buried where land was permanently frozen. In a thawing Earth scenario, thawing corpses, many centuries old, could harbor diseases that may return to cause a health pandemic. While, fortunately, most microorganisms cannot survive the extreme cold of permafrost terrain, however, some could conceivably remain viable after many years, even centuries, in the freezing environment.

"Spores are extremely resistant and like seeds, can survive for longer than a century," says Jean-Michel Claverie, the head of the Mediterranean Institute of Microbiology in Aix-Marseille, France. Claverie also emphasizes that viruses that can infect humans, such as smallpox and the Spanish flu, could be preserved in permafrost.[33]

I should say, there is quite a lot of skepticism from scientists around the world around this idea that melting permafrost and retreating glaciers will awaken deadly diseases, and many experiments to revive these bacteria have failed, indicating that the hype and fear of zombie bacteria coming back from the dead may not be so warranted.

That said, one 2005 NASA-sponsored scientific study successfully revived bacteria from over 30,000 years ago that had been trapped in an Alaskan lake. "The microbes, called *Carnobacterium pleistocenium*, were frozen in the Pleistocene period, when woolly mammoths still roamed the Earth. Once the ice melted, they began swimming around, seemingly unaffected ... Once they were revived, the viruses quickly became infectious. [These bacteria] were discovered 100 feet underground in coastal tundra," reported Jasmin Fox-Skelly, in an in-depth article about ancient bacteria and viruses in permafrost, done for the BBC.[34]

Fox-Skelly provides evidence of a number of scientific efforts to study and attempt to revive these ancient bacteria. The COVID-19 coronavirus was spreading across China and the world as I wrote these lines in January and February 2020. My then 16-year old son, who is interested (as a 16-year-old would be) in anything to do with zombies, said to me, "Why would they do that? Isn't that dangerous?" He clearly has a point! One of the studies noted by Fox-Skelly recounts how scientists managed to revive an 8-million year old bacterium that had been lying dormant in the ice, beneath the surface of a glacier in the Beacon and Mullins valleys of Antarctica.

Claverie (whom I mentioned earlier) suggests that "viruses from the very first humans to populate the Arctic *could* emerge. We could even see viruses from long-extinct hominin species like Neanderthals and Denisovans, both of which settled in Siberia and were riddled with various viral diseases."[35] In 2017, in a Mexican mine, NASA scientists found 10,000–50,000-year-old microbes inside crystals. Once the bacteria were freed from the crystals they rapidly began multiplying.

Catherine La Farge, an evolutionary biologist, also studies thawing environments of the cryosphere. Daniel Ackerman reported on her work:

"You wouldn't assume that anything buried for hundreds of years would be viable," said La Farge, who researches moss in Canada. "The material had always been considered dead. But by seeing green tissue, I thought, 'well, that's pretty unusual.'". . . [La Farge] treated dozens of these samples with nutrient-rich soils in the lab. A third of the samples produced new shoots and leaves. "We were pretty blown away," said La Farge. The moss was resilient in hibernation as it came out of its centennial deep-freeze.[36]

Another biologist, Peter Convey from the British Antarctic Survey, also retrieved ancient moss and successfully revived it, this time, millenarian moss that was 1,500 years old and was buried more than 300 feet deep in the Antarctic subsurface.[37]

An article appearing in January 2019, by Colin Barras, tells of tiny animals dug up in a frozen lake in Antarctica where they were looking for signs of life.[38] These are pretty inhospitable environments, buried deep in frozen terrain where no life is supposed to survive. But biologists *have* previously discovered that crustacean eggs buried at the bottom of lakes for centuries can actually hatch and grow—leading to a new branch of biology called "resurrection ecology," says Barras.

Yet another scientist, Tatiana Vishnivetskaya, a microbiologist at the University of Tennessee, has also studied deeply frozen microbes of the Siberian Tundra. She has revived million-year-old-bacteria on a petri dish. I had to look at the number and check it several times. This is "million-year-old" bacteria, revived from the dead. So that's bacteria, but not quite zombie material. However, hold your hat for a moment and consider this. Later, Vishnivetskaya was able to locate and revive nematodes (worm-like animals), with a brain and nervous system, that were 41,000 years old—this would be the oldest living animal ever discovered on Earth.[39]

So the idea of reviving ancient diseases seems not so far-fetched.[40] There may be one case with good evidence of zombie diseases in permafrost now affecting modern humans. Zac Peterson recently carried out some permafrost melt investigations at an abandoned seal-hunting cabin near the town of Utqiagvik, Alaska, home to the Inupiat indigenous peoples. He was digging up perfectly preserved seals from frozen grounds. Hunters have been utilizing this location for seal-hunting for thousands of years. According to Peterson, "After kneeling in defrosted marine mammal goo . . . doctors treated me for a seal finger infection." Seal-finger is a bacterial infection that hunters contract from handling the body parts of seals.[41] This ancient seal infection transmission from animal to human clearly shows there are definitely things to worry about coming from thawing permafrost, maybe not zombies, but certainly some infectious diseases that should be extinct.

Until we have more data and clearer scientific results from analyzing bacteria in thawed grounds, zombie life unearthed from the depths of a Siberian crater may be more the subject of a fantastic plot for a new streaming Netflix series about the walking dead. What we should really think about, however, is how our thawing Earth and our warming climate is altering ecosystems around the world and setting the scene for accelerated global warming and potentially releasing disease vectors. Just as grapes can now be grown in places like Canada and the UK, thanks to climate change, so are favorable breeding grounds for disease-carrying mosquitos becoming more common around the world.

This is the real problem of the altering of our ecosystems by a warming climate. Whereas once glacier-covered lands or frozen Earth kept ancient remains and potent greenhouses gases from destabilizing our delicate global ecosystems, today melting glaciers are bringing a number of complications that destabilize our planetary ecological balance.

Thawing permafrost most importantly and most immediately presents a severe problem in terms of the release of methane gas, which has an impact on our climate and is many more times as powerful as CO_2. The release of this gas that is super potent for climate change could be devastating for an already stressed and warming climate, pushing us far beyond the point of no return and unleashing unstoppable severe weather events and patterns across the globe. This may already be happening, and as noted by IGSD, once the permafrost goes, we're not getting it back within any reasonable human timeframe. It's gone.

IGSD notes that during the Paleocene-Eocene Thermal Maximum (the hot years about 55 million years ago), Arctic temperatures were more than 10°C warmer than today, causing almost complete loss of permafrost.

> During the early Holocene (about 12,000 years ago), temperatures in the high latitudes were 2–4°C warmer, so a potential tipping point might exist between 4 and 10°C, which would be between IPCC scenarios. If temperatures increase 10°C, the amount of carbon released from the permafrost region could double.
>
> Failure to avoid permafrost thaw jeopardizes the likelihood that climate warming will stay below 2°C. Permafrost thaw has already led to the Arctic being an increasing source for carbon emissions. While the Arctic presently serves as a carbon sink, the region could become a carbon source as early as the mid 2020s from the permafrost carbon feedback. Carbon emissions from permafrost will continue for centuries, making the Arctic a long-term net carbon source by the end of the 23rd century.[42]

Thawing permafrost says IGSD, will result in a prolonged release of carbon and methane gas from permafrost that will compound projected global emissions of greenhouse gases, further contributing to present global warming trends. What we are sure of is that the thawing of our Earth and the resulting changes to our microclimates around the world are creating serious risks for our global ecological stability.

6
Run! The Mountain Is Coming!

In a seven-minute video filmed in May 2012 at Pokhara in the vicinity of Mount Machhapuchure and Annapurna of Nepal, you can witness the ferocious phenomenon that scientists call a Glacier Lake Outburst Flood (GLOF). I call them glacier tsunamis, which I think is more descriptive of their true ferocity and the damage these glacier melt induced floods can inflict.

Take a moment and watch the video (link in the footnote); it is truly frightening to imagine this glacier tsunami rushing down the mountain over inhabited lands.[1]

The video begins with a view of a tranquil mountain riverbed, mostly rocks with a small stream of water flowing down from the high Himalayan mountains. After about 20 seconds, you see the murky muddy front of the flash flood arriving from the glacier tsunami that has occurred many miles up the mountain and which is now growing in force. At 30 seconds, you hear the nervous chatter in Nepalese of the onlookers witnessing the sudden arrival of this curious slushy mix of water, mud, and small branches, which looks like lava flow running down the riverbed, in what is usually a tranquil riverside environment. In less than a minute, the rush of water fills the screen and is clearly gaining unstoppable and life-threatening power. If you had been on one side of the river, you'd suddenly realize you won't be able to cross it. Just after the first minute, at 1:09, you can already see Category 3 and 4 waves forming on a river that, only moments earlier, was but a trickle of water.

By this point you start to realize the severity of this event not only for downstream communities and anyone in the river or on its banks, but also for anyone that was unfortunate enough to have been upriver and already swallowed by the river's rage. Ten seconds later, at 1:25, you catch a glimpse of the deadly movement of water, which has picked up a full-sized tree that bobs up and down the river like a small branch and races downstream to demolish anything in its path. You realize that if this tree hits someone, it will kill. The next 20 seconds show more and more debris rushing in the unstoppable current, in what is now a raging river covering the full width of the visible riverbed in a monumental downflow. This is a glacier tsunami at full force.

The cameraman's chatter becomes increasingly nervous as more trees and debris rush down the river in waves that would make any effort at crossing impossible and most likely deadly. People can be heard screaming as the flood intensifies further and further as the seconds pass. We aren't yet 3 minutes into the filming and the situation is desperate.

Melting glaciers kill through these glacier tsunamis, destroying entire ecosystems and communities in minutes through sudden massive collapse, pushing ice, water, trees, mud, rocks, and anything else they can pick up down mountain valleys and taking out everything in their path. Thousands of people have died in glacier tsunamis. These ferocious events are generally caused by glacier instability or sometimes mountainside collapse due to thawing permafrost in very precarious environments, spurred on by altering climates that accelerate ice melt and make large glaciers highly unstable. These glacier tsunamis occur in places where unstable glaciers hang over lakes that formed when glaciers retreated in a warming climate. As the temperature rises further, glaciers melt off more and more water, filling the glacier lakes below them. The remaining glacier and its precarious ice hangs overhead, collapsing into the lake as its structure weakens. Sometimes the pieces of falling ice are so big that they cause deadly waves.

Most people living downstream from glacier-fed rivers may not actually understand what is triggering these floods, because for the most part, they likely live at quite a distance downstream from the source of the flood, and usually there's no one around to see the glacier ice fall and the glacier tsunami ensue. They may think it is a sudden rain storm, or flooding caused by high volumes of glacier melt, but in many cases, it is *glacier instability* interacting with glacier lakes below them that can trigger catastrophic events in high mountain environments. This is especially so for populated regions in hydrological basins near large glaciers and glacier lakes that are increasing in number and becoming unstable because of warming climates.

As I described glacier tsunamis to a good friend from Argentina, Viviana Krsticevic, she began to reminisce about a time in the early or mid 1980s when she was with her parents at the foot of the Andes in Argentina. A sudden river flood left their car stranded on a piece of highway that was, fortunately for them, above the flood. They ended up spending the night in the car, as they could not get out of the flooded area. I wondered if that could have been a glacier tsunami. I looked for past floods and related GLOFs in Argentina and found *many* historical recounts of sudden floods in the area, and many glaciologists that in time have associated these events with glacier vulnerability. The GLOF that Viviana remembered probably was a glacier tsunami. But no one living locally, including my friend, seems to recall these events as ever having been associated with glacier melt. They simply refer to them as floods (or *aluviones*, in Spanish) that just happened. They presumed it had rained heavily up in the mountains. No one thought of unstable ice. Today, glaciologists, with more and more information about ecosystem vulnerabilities caused by climate change, are connecting the dots.

We've learned in previous chapters that glaciers carve out channels beneath them down mountain valleys. As glaciers retreat in warming climates, they leave a big gap or hole in the ground, surrounded by rock debris moraines. Imagine pushing your fists and arms through sand at the beach. The sand piles up around your arms, forming a depression in the sand. If you pull your arms back, the next sea wave might be able to fill that depression with water, as if the sand were a natural dam. Well, glaciers generate a similar phenomenon but high up in the mountain, many thousands of feet in altitude. As glaciers dwindle in size, as they *retreat* up the mountainside because they are warming due to global warming, they leave massive holes in the ground they once occupied. Those holes get filled with their own meltwater, forming high altitude lakes that can be many hundreds, even thousands, of feet deep (see fig. 6.1 showing three such glacier lakes in the Sierra Blanca of Peru).

The problem with this phenomenon is that the original glacier, which may still be colossal and in many cases precariously dangling from a mountainside, now lies above the newly formed glacier lake formed at its base. As the glacier continues to warm, its

Figure 6.1 Three dangerous glacier lakes in the Andes Mountains of Peru.
Three glacier lakes in the Andes Mountains of Peru occupy the gaping holes formed by rock and earth moraines left as glaciers retreat up the mountainside. Dozens of looming unstable glaciers sit above the lakes, structurally weak from a warming climate. These melting glaciers can drop building-sized chunks of ice into the deep lake, causing massive waves that can break through moraines. Note the lake furthest left in the image, which shows evidence of one such past complete moraine-break in the front bottom portion of the lake's moraine. The glacier in the upper right portion of the image is 2.3 kilometers (1.4 miles) long and 1 kilometer (0.6 miles) wide—these glaciers are huge and so is the danger they pose! Huaraz, a town of 100,000 people, lies immediately below down the valley. In 1941 it was leveled by a glacier tsunami, which killed thousands of people following the moraine failure visible in this image.
Source: Google Earth. GIS: 9 23 43.51" S, 77 22 30.16 W

structure steadily weakens, and large pieces of glacier ice (called *seracs*) begin to fall off its retreating terminus into the glacier lake below. A piece of ice the size of a house or small building splashing into the water may cause a wave of 1.5 meters (5 feet) or 3 meters (10 feet) in height, but a piece larger than a very tall building or an entire city block can actually cause a wave tens of meters (dozens of feet) high.

As the image in figure 6.1 shows, such a large wave has the force of potentially breaking over or *even through* the glacier moraine (the rocks that contain the glacier lake)—you can see the moraine break at the outflow area of the lake (bottom left).

When this happens, we've got a problem! A wall of water, ice, and rocks can suddenly come crashing down the mountain at great speed, taking out everything in its path, and gathering up trees, boulders, mud, plants, and other debris, such as can be seen in the video of the Nepalese GLOF cited at the beginning of this chapter.

There are different types of potentially deadly events caused by glacier instability that can be induced by accelerated melt or other similar phenomenon. We've already spoken of the GLOFs or as I call them, glacier tsunamis. There are also *glacier surges*, which involve glaciers suddenly advancing at irregularly high speed. These glacier surges can rapidly accelerate the push of glacier ice through the surrounding environment causing the normal flow rates of large masses of ice to suddenly invade lands below with no warning. Then there is a *lahar*, which is a mudslide tsunami caused by the sudden eruption of a volcano, covered in glaciers and ash, that destabilizes the side of the volcano. Sudden excess meltwater from the heat of the eruption mixes with ice and ash, and a massive landslide ensues down the valley below.

One of the biggest and most-studied such events followed the eruption of Mount St. Helens, a stratovolcano, in Washington State in 1980.[2] It was the deadliest and most economically destructive volcano event in US history, killing 57 people, destroying 250 homes, 47 bridges, 24 kilometers (15 miles) of railways, and 298 kilometers (185 miles) of highways. Immediately after the mountain erupted, 400 meters (1,300 feet) of the top of the mountain was blown off, causing a massive landslide of glacier ice, earth, mud, ash, trees, and other natural debris. The blast left a new crater 1.6 kilometers (1 mile) across.[3] During the eruption, 70% of the glaciers on Mount St. Helens vanished. The ice of these glaciers was mixed into the massive deadly mudslides that ensued, reaching some 80 kilometers (50 miles) in extent. The volcanic blast and resulting lahar turned hundreds of square miles into wasteland.[4]

One of the curious facts following the Mount St. Helens eruption is that a *new* glacier was suddenly formed (now called Crater Glacier) that grew very quickly in size. It came together in the newly created "glaciosystem" from the entrapment of fallen ice and snow, subsequent snow avalanches from the new niche terrain formed in the crater, and from new snowfall cycles that have repeated each year since the blast. The glacier is already between 300 and 600 meters (1,000 to 2,000 feet) thick with ice only dating to 1980, which, geologically speaking is extremely rare for a glacier. This may be one of the Earth's youngest glaciers!

Such events as those described here, which have occurred over centuries and millennia, while common wherever large perennial ice accumulation occurs on mountains, have not been fully appreciated. Flash flooding in mountain regions may occur at far distances from glacier environments, and local communities may not attribute these events specifically to glacier instability. In a place like Alaska, with so many glaciers, and with climate change advancing so quickly, many times, people don't find out about these events. Mike Loso, a scientist working with the Wrangell-St. Elias National Park and Preserve says (in The End of Ice by Dahr Jamail), "All our parks are littered with landscapes that show that all kinds of crazy stuff is happening all the time, and only in rare instances do we get to witness or even know about it. These events are now completely common, but most of them are undocumented and unobserved."[5]

Michael Hambrey and Jürg Alean in their book *Glaciers* describe some of the most notorious glacier tsunamis, GLOFs, lahars, and surges that have occurred around the world.

In November 1985, for instance, the volcano Nevado de Ruiz in the South American Andes, in Tolima Colombia erupted. The snow and glacier ice trapped in the volcano's crater became unstable due to the sudden heat from the eruption. Snow, ice, and ash suddenly broke over the edge of the crater and raced downhill in one of the world's most deadly, known lahars. Unfortunately, the lahar headed straight for a densely populated region below at about 30 miles (50 kilometers) per hour, giving the residents little warning or preparation time. The entire town of Armero was completely buried in mud and more than 20,000 lives were lost to this tragic event.[6]

Half a million people live below Nevado de Ruiz, and as climate change continues to alter glacier environments, the risk of future and more deadly lahars looms for downstream communities living in Combeima, Chinchina, Coello-Toche, and Guali River valleys. Such is the case for many other volcanic areas around the world including in Mt. Etna (Italy), Mt. Mauna Loa and Mt. Kea (Hawaii), Popocatapetl (Mexico), Kamchatka Peninsula (Siberia), Mount St. Helens (Washington), Vatnajökull (Iceland), Ruapehu (New Zealand), and Kilimanjaro (Africa). The Pacific Rim of South America is especially prone to volcano eruptions. Many of the volcanoes along the Argentina-Chile border have craters filled with millenarian glaciers, snow, and ash. Every one of those is a glacier and mud tsunami of epic proportions waiting to happen.

On Christmas Eve, December 24, 1953, an eruption at Ruapehu, in Tangiwai in the Central North Island of New Zealand caused a lahar and GLOF that broke the glacier dam and sent a massive mudslide down the Whangaehu River. That lahar took out a railway bridge (with a train from Wellington to Auckland traversing it) and killed 151 people.[7]

Andreas Kellerer and his colleagues show that millions of cubic meters of ice can be suddenly released by changes in climatic conditions, in what are known as *glacier ice avalanches*. Peru has been particularly vulnerable to sudden veritable tsunamis of glacier ice and other debris running down mountainsides caused by accelerating climate change coupled with unstable ice hanging over glacier lakes formed by retreating glaciers.

In 1941 the town of Huaraz was flooded by ice, water, boulders, and debris in mere minutes from a glacier tsunami originating in a GLOF in the Cordillera Blanca Mountains just above the town. Six thousand people were buried alive in that incident (see fig. 6.2 showing the town of Huaraz relative to glacier lake location).

In 1962 at Mt. Huascaran, also in the Cordillera Blanca and the largest Peruvian peak at 6,768 meters (22,200 feet) as well as the most unstable of Peru's glaciated mountaintops, a large mass of ice broke off of the top of the mountain above the Rio Santa Valley of Ancash Province. An ice avalanche glacier tsunami ensued, traveling 16 kilometers (10 miles) and killing more than 4,000 people. This peak has been the source of other glacier tsunamis throughout the last century.

Not too long after this glacier tsunami in Ancash, there was another at Yungay when 50–70 million m³ of ice broke off of Mt. Huascaran, carrying away parts of the

Figure 6.2 Vulnerable communities downstream from glacier lakes at Huaraz, Peru. Google Earth image showing the vulnerability of communities downstream to dangerous glacier fed lakes and massive glaciers with deteriorating ice structures located along the mountaintop. Huaraz, a town of over 100,000 nestled at the foot of the Cordillera Blanca Mountains in Peru, has seen tragic loss of human life and total destruction of infrastructure in the past and is likely to face glacier tsunamis in the future.

Source: Google Earth. GIS: 9 30 38.73 S, 77 28 20.50 W

summit glacier, and rushed down the mountain to the town below. Over 18,000 perished in that disaster.[8]

Mark Carey of the University of Oregon has written a book called in the Shadow of Melting Glaciers, about glacier-related tragedies in the Andes Mountains. He suggests that "the historical effects of climate change and glaciers on society have been particularly powerful in the Peruvian Andes, where climate change has caused the world's most catastrophic glacier disasters of the last century."[9] According to Carey, nearly 25,000 people have died in Cordillera Blanca glacier disasters since 1941.[10]

Repeated floods have also occurred in Chile and in Argentina, all due to glaciers crumbling into precarious high mountain glacier lakes. Over the past three centuries, documented flash floods indicate recurring glacier lake outbursts at the foot of the Andes, running over towns that depend on glacier meltwater for agricultural production.

Glacier surges in the Mendoza valley (home to much of Argentina's wine industry) recur about every half-century, each time producing massive floods and destruction.

Both the Grande del Nevado Glacier and the Plomo Glacier, above the Tupungato River in Mendoza, have been rapidly melting off due to climate change and have experienced sudden surges in the last several centuries. In 1788 and then in 1900, in 1926 and again in 1934 the Grande del Nevado surged forward suddenly, producing catastrophic flooding in the valley below. The Plomo produced its own massive flooding event in 1985, following the damming up of river water in a temporary obstructive glacier lake that formed at the foot of the glacier.[11]

One of the oldest recorded glacier ice tsunamis took place in 1597 at the Balmen Glacier in Eggen, Switzerland, burying 81 people. Fortunately, most of the inhabitants of the town were off working in the alpine meadows, otherwise the death toll would have been much higher. The village, then situated near the Simplon Pass in the Canton of Valais in Switzerland, no longer exists, and neither does the glacier, note Hambrey and Alean. The town was wiped off the map by this glacier tsunami.[12]

In 1895, a glacier on the face of the Altels mountain in Switzerland dropped four million cubic meters of ice, which raced down the mountainside at 400 kph (250 mph) killing sheepherders and their livestock. The collapse piled up ice across the valley floor more than 400 meters high (1,300 feet).

In 1965, during the construction of a hydro-electric power plant at Mattmark in the Saas Valley, Switzerland, a mass of ice about 1 million m^3 large broke from the Allaling Glacier and, within seconds, swept over the construction site, burying workers alive and killing 88 people.[13]

Photographer James Whitlow Delano, who established the Everyday Climate Change Instagram feed to draw attention to climate change vulnerabilities, recently traveled to the Valle d'Aosta in the Italian Alps to focus on the plight of receding glaciers, specifically the Brenva Glacier, which has receded and reduced in size substantially. Delano notes that between 1960 and 2017 the Alpine snow season in the region of the Brenva has shortened by 38 days on average. It starts 12 days later and closes 26 days earlier. The winter of 2015–2016 was the warmest on record, with only 20% of the normal snowfall.

Less snow means less nourishment for Europe's Alpine glaciers. This in turn means that without the accumulation of snow year to year, their melt will exceed their recuperation, and thus they are dwindling in size and receding. This also means a weakening of the glaciers' internal structure. The Brenva Glacier, reports Delano, is now a highly unstable ice body that is subject to avalanches and rock slides, placing communities below on alert for danger.

But downstream communities do not face physical danger only. Snow that feeds the glacier will be replaced by rain. In contrast to snow, which is accumulated, stored and released slowly by the glacier, rain drains quickly and moves through the ecosystem. Delano vividly describes the situation caused by melting glaciers in the Italian Alps.

Receding glaciers mean less water stored up to feed rivers, especially in times of summer drought. Farmers and livestock pastoralists will find less grass in high meadows in summer to fatten up cows to produce milk for cheese.

Valdostani (the local inhabitants) are keenly aware that global warming means a shorter ski season and a greater dependence on costly man-made snow to keep that important regional industry going. Bars, restaurants, hotels, ski schools and rental shops are all negatively affected by a snow season that is 38 days on average shorter.

Pastoralists are also feeling the effects of less green pasture in the drier, hotter summer months, but millions of people, dependent on the Po, Rhone, Rhine and Danube rivers that are either born or fed by rivers sourced in the Alps, will be affected, hundreds of kilometers away, by an Alps with fewer glaciers.[14]

Nepal is one place that is very vulnerable to glacier tsunami activity, due to the high number of glaciers present along the Himalaya mountains, which are melting fast. There are an estimated 2,323 lakes in Nepal and 3,252 glaciers. Many thousands of small glacier lakes are forming through the Himalayas as glaciers retreat, all potentially increasing the risk of glacier tsunamis. The video link at the beginning of this chapter is an example of a glacier tsunami from Nepal. More are coming.

In August 1985 the Dig Tsho glacial lake in eastern Nepal was the site of one of these catastrophic events. A sudden ice avalanche produced a wave over 5 meters high (17 feet), which overtook the glacier moraine that was containing the lake. The lake, which was 18 meters deep (60 feet), drained almost completely and within 6 hours. The resulting glacier tsunami took out homes, bridges, agricultural land, and a newly constructed power plant, causing about $1.5 million USD in damage. Remarkably, very few people lost their lives, as there was a local festival at the precise moment of the outburst that had lured people away from the danger.[15]

The International Center for Integrated Mountain Development (or ICIMOD, as most people know it), studies glacier dynamics, vulnerability, and impacts in the Himalayas. They recently published a new inventory and report on the dangers posed by glacier lakes and the vulnerability of unstable high mountain glaciers due to global warming. They studied 3,624 glacier lakes in three basins, 2,070 of which are in Nepal, 1,509 in the Autonomous Region of Tibet, China, and 45 in India. Of these, 1,410 are already large enough to potentially cause serious glacier tsunami impacts. Of these, 47 are potentially dangerous due to witnessed instability, particularly of the glaciers above them. ICIMOD is calling for urgent action by public officials to try to reduce the potential damage, loss of life, and harm glacier tsunamis at these glacier lake sites could cause to the environment and to people, as well as likely property damage that would result to downstream communities.[16]

C. R. Stokes, a glaciologist of the Department of Geography at the Durham University in the UK, and his colleagues examine glacier retreat and glacier lake development in the Caucasus Mountains of Russia. They note a 57% increase from 1985 to 2000 of glacier lakes in this range, caused by retreating glaciers and accelerated melt and glacier thinning.[17]

In 2015, one of the biggest glacier tsunamis ever recorded was caused by a melting glacier, due to the weakening of mountainside walls after a glacier retreated as a result of global warming. The event occurred at the site of a southeastern Alaskan fjord, following a sudden tidal wave caused by falling debris. The wave reached nearly 200 meters (600 feet) in height. Imagine standing under a monstrous ice wave that's two football fields high! The source was 180 million tons of rock that crashed into the waters of the Taan Fjord below. The fjord's rock walls used to be contained under solid ice.

Now exposed to the environment due to the glacier's retreat, they are warming, and the thawing process is weakening them. Scientists studying the event call it "hazard occasioned by climate change." Curiously, 40 years ago, the fjord didn't exist as it was completely filled with ice. The gap left by the retreating glacier is now the source of catastrophic risk from sudden outbursts of glacier ice, water and debris brought on by a glacier tsunami and related flooding.[18]

In 2016, the following year, also in Alaska, and again attributed to thawing permafrost terrain that is holding mountains together, parts of a 4,000 foot high mountain collapsed in Glacier Bay National Park in a massive landslide that spread debris for miles across the Lamplugh Glacier below. The rockslide slid 5 kilometers (3.1 miles) from the base of the slope (see fig. 6.3).[19]

In 2017, this time on the west coast of Greenland in the Karrat Fjord, one of the biggest recorded tsunamis in history was set off by unstable permafrost lands and caused a wave 100 meters high (330 feet), devastating the remote village of Nuugaatsiaq. The tsunami was massive because of the sheer size of the collapse. Enormous boulders broke off of the side of the mountain fjord plummeting 1,000 meters (nearly 3,000 feet) to the glacier and glacier-lake below. This mountainside collapse was so large that it generated a 4.1 magnitude earthquake. The falling boulders were likely the size of buildings and on impact broke off colossal pieces of ice, causing the tsunami that ensued. The force of the impact was so powerful that the resulting wave reached the small town, 20 kilometers away (12.5 miles), killing 4 people and washing away 11 homes.[20]

Figure 6.3 Massive rockslide in 2015 in Southeastern Alaska caused by thawing permafrost, which slid 5 kilometers (3.1 miles) from the base of the slope.
Source: Paul Swanstrom/Mountain Flying Service

Surging glaciers (glaciers that suddenly and abruptly move forward at high velocity) can also cause havoc to local environments and can be deadly for flora, fauna, and people living in nearby areas. In what Hambrey and Alean call "a rare but spectacular type of ice movement," in 1986 the Hubbard Glacier (one of the longest in North America) suddenly exploded in velocity as the tributary Valerie Glacier surged onto it. The speed of the surge reached almost 40 meters (131 feet) a day (which is a lot for a glacier). This surge advanced over the Russell Fjord and over a forested island, smashing trees and creating a huge lake 50 kilometers long (31 miles).

Marine mammals (including seals and porpoise) became trapped, bird nests and chicks were destroyed, and other flora and fauna suffered fatal impacts. The newly formed lake reached 30 meters (nearly 100 feet) above sea level and began to threaten the village of Yakutat. When the lake finally gave and broke out into a GLOF, its flow was estimated at 35 times the flow of Niagara Falls. Hambrey and Alean describe the chilling scene, which they witnessed:

> Standing on the shore of this lake in July 1986 was an unforgettable experience. The jagged wall of ice of Hubbard Glacier constantly sent large masses of ice crashing in to the lake, the initial fracture roaring like thunder crack and the break-up a muffled explosion. Small icebergs littered the lake surface in front of us and bobbed gently among the drowning trees. Agitated birds fluttered frantically back and forth, seeking their flooded nests. All the while the water crept gradually up the flat plain on which we were standing.[21]

Just as this book went to press, the Muldrow Glacier in Alaska was surging, 100 times faster than normal, some 90 feet per day (27.5 meters). Muldrow is 1,500 feet thick (460 meters) and a mile and a half wide (2.4 kilometers).[22]

We now turn to look at the ocean currents, the jet stream, and polar bears.

7

Ocean Currents, Jet Streams, and Polar Bears

We've talked so far about many aspects of our society affected by melting glaciers and thawing permafrost. We've read about rising sea levels, altered and dwindling water supply, solar reflectivity and absorbed heat, destabilized ground, methane emissions from permafrost thaw, and deadly glacier tsunamis. But the impacts of melting glaciers on the planetary ecosystem do not stop there.

Glacier melt still has further effects on our Earth that profoundly impact our global ecosystems, our atmosphere, our climate, and our marine environments, all of which have consequences for human life and natural habitats. In this chapter we will examine four of these dimensions where glacier melt is wreaking havoc on global climate and other environmental ecosystems.

These are:

The Warming and Altering of Ocean Currents and the Global Jet Stream
The Surface Rebound Effect (or "Popping Earth")
Energy Production
Rivers, Animals, and Human Habitat

Let's take them in that order.

The Warming and Altering of Ocean Currents, the Jet Stream and Rivers

Ocean Currents

The latest (2019) IPCC report on Oceans and the Cryosphere indicates that over the last two decades the temperature at the Arctic has likely increased by more than double the global average. This increase in air temperature leads to localized, accelerated glacier melt in the Arctic region (we focused on that in Chapter 4), which in turn sends more and more glacier meltwater into the oceans. The sea ice albedo feedback (increased air temperature reducing sea ice cover, allowing more energy to be absorbed at the surface, fostering more melt), according to this report, is a key driver of sea ice loss in the region.[1]

We saw in Chapter 4, on albedo, how the large ice sheets of the Antarctic and Greenland as well as glaciers in places like Alaska that reach to the sea reflect enormous amounts of heat back into space. There are two sides to this coin, both of which are important. Reflecting heat back into space is clearly one. The other is the cooling effect below that is generated by these large ice sheets.

The cool oceans of the polar regions also provide an important ecosystem balance for ice-associated habitat.[2]

Additionally, large ice sheets in the polar regions act as thermal insulators for the oceans, separating solar heat from ocean waters. The ice essentially helps keep the polar oceans cool, an ice function that is fundamental for the role water currents play in cooling oceans not only at the poles but all over the Earth. As ice sheets in the polar regions of the Arctic melt and expose the oceans, two dynamics enter into play. As we saw in Chapter 4, the first is the albedo changes of the surface of the Earth that go from reflective white to darker ocean blue. This results in the ocean surface *absorbing* more heat instead of reflecting it back into space. Second, the uncovering of the ocean also causes the release of ocean heat, warming the lower atmosphere, which also provokes accelerated ice melt.[3] This produces a vicious circle and feedback loop of global warming.

As global warming heats the polar ice, ice sheets and glaciers also pump billions of liters of cold and fresh glacier water into the sea. As these ice sheets and glaciers melt away (particularly in the Arctic, less so in Antarctica) not only is the sea level rising at an accelerated rate (sea levels rise from glacier melt, but not from Arctic sea ice) but ocean temperatures are being altered drastically. The Earth benefits from a circulation of ocean currents that takes warm water from the middle swath of the Earth (oceans located nearer to the equator), run it through the colder waters of the polar regions to cool it off, and then recirculate it through the middle latitudes.

Warmer water runs toward the poles along the top of the oceans, while colder deepwater currents bring the cooler water back to the equatorial regions. The dropping of cooler ocean water at the poles is driven by differences in the ocean's water's density, which is controlled by temperature (thermo) and salinity (haline), in a process called Thermohaline Circulation (THC). In the polar regions, the water gets very cold, with much of it turning into ice. This separates freshwater from salt, making the Arctic sea very salty. That makes the remaining water denser and causes it to sink, pulling more surface water downward and initiating a mixing process that feeds into the deep ocean currents driving the global ocean currents.[4]

This circulation of water from the equator to the pole is also referred to as the Atlantic Meridional Overturning Circulation (AMOC). The AMOC is characterized by a northward flow of an upper layer of salty, warm water and a returning southward flow of cooler, deep water that occurs after the THC mixing just described.[5] The THC and the AMOC, hence, work together to mix ocean water and help it develop harmonized and ecologically balanced marine environments across the planet, keeping temperatures in each part of the world in equilibrium (fig. 7.1).

Think of the polar region oceans like the radiator of a car. Too much heat could destroy the motor if it heats up uncontrollably. The way we solve this is to circulate liquids that don't heat up as much (the role of antifreeze) from hotter parts of the motor (think of these as the oceans in the equatorial regions) through the radiator (the North and South Poles), which is being cooled by a fan (the glaciers and the sea ice). As hot liquid runs by and through the radiator, the entire system is conveniently cooled.

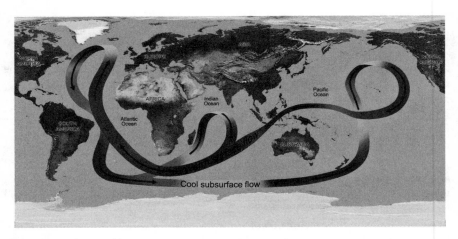

Figure 7.1 The Earth's ocean currents.
The Earth's ocean currents take water from the warmer middle latitudes of the Earth and run them through the cooler polar regions, cooling off the water and creating a more balanced and habitable ocean and terrestrial environment.
Source: NASA

This THC/AMOC cycle serves the fundamental purpose of maintaining a stable temperature balance in the world's oceans. Ocean ecosystems have adapted to this balanced cooling process and the resulting hydrological equilibrium. Today they need each other to maintain that balance. With global warming, this equilibrium is running out of sync. The polar oceans are, according to the IPCC, among the most rapidly changing oceans of the world, with consequences for global-scale storage and cycling of heat, carbon, and other climactically and ecologically important properties.[6] As the melting of glaciers and sea ice accelerates, more freshwater is flowing into the seas, disturbing the salinity levels, while the sea itself is warming due to global temperature increases and ocean absorption of solar heat. This greatly alters the delicate marine ecosystem and destabilizes the cooling balance and temperature control function of the recurring global sea currents and THC/AMOC cycles. It's as if we suddenly lost our car's radiator and the fluids started circulating without cooling. In just a few miles, everything would heat up and we would be in dire trouble. The engine would overheat, stall, and the car cease to work. That's what is happening to our oceans. You can fix the radiator of a car, put in new coolant, and get back on the road with some delay, but I don't know of any polar cooling garages out there to fix our THC/AMOC cycle.

A team of scientists from the University of Sheffield and Bangor University, including Dr. Clare Green, found in 2011 that the collapse of glacier ice into the ocean has significant impact on water circulation in both the Atlantic and Pacific oceans (especially in the Atlantic), triggering significant changes in global ocean current patterns, particularly in the North Pacific.[7]

Changing ocean temperatures and currents can have extremely disruptive consequences for the ocean's ecological balance. The sudden entry of excessive freshwater (from melting sea ice) into the ocean and the mixing of this new freshwater and salt-water, each at varying temperatures, and occurring at different strata of the ocean in a global churning process (the THC), can change the temperature, salinity and acidity balance of the water. This can lead to massive impacts on marine life and ocean eco-systems. A new study from NASA shows that:

> a major ocean current in the Arctic is faster and more turbulent as a result of rapid sea ice melt. ... The Current is part of a delicate Arctic environment that is now flooded with fresh water, an effect of human-caused climate change. Using 12 years of satellite data, scientists have measured how this circular current, called the Beaufort Gyre, has precariously balanced an influx of unprecedented amounts of cold, fresh water—a change that could alter the current in the Atlantic Ocean.[8]

Mattias Green, senior research fellow at Bangor University, highlights this concern, suggesting that:

> Freshwater entering the ocean from melting ice-sheets can weaken the climate con-trolling part of the large scale ocean circulation, with dramatic climate change as a consequence. During the period of our study, the global temperatures dropped by up to two degrees over a few centuries, but changes were not uniform over the planet, and it took a long time for the climate to recover after the ice sheets had melted completely.
>
> The team [of scientists] argues that it is not only the volume of freshwater being released from the melting ice-sheet which is important but also the state of the fresh-water: icebergs act to reduce the ocean circulation less than meltwater, but the effects of icebergs last for longer periods of time.
>
> This can be compared to the difference between adding very cold water to your drink and adding an ice cube or two. With meltwater—similar to adding water to your drink—the water spreads out quickly and has an immediate effect, but it is also absorbed quickly into the rest of the ocean. In a similar way to your ice cube, the ice-bergs drift along and melt more slowly. This means the immediate impact is weaker, but they are there for a longer time and they distribute the water over a larger area.[9]

In addition to temperature changes in currents, we mentioned earlier the importance of ice sheet and glacier melt into the oceans and the consequent altering of salinity (the amount of saltwater vs. freshwater in the oceans). This has a direct effect on the sustainability of the oceans' ecosystems, as well as their capacity to avoid freezing. Any and all changes to the level of salinity in the oceans could *and does* have drastic effects on natural ecosystem cycles for wildlife that depend on ecosystem stability.

The 2019 IPCC report indicates that, "salinity is the dominant determinant of polar ocean density, and exerts major controls on stratification, circulation and mixing [of the oceans]. Salinity changes are induced by freshwater runoff to the ocean (rivers and land ice), net precipitation, sea ice, and advection of mid-latitude waters".[10] The Beaufort Gyre (the ocean current in the polar region) for example, evidences a near 40% increase in freshwater intake between 2013 and 2017. An increase in freshwater entry has also been registered in the Bering Strait, as well as in rivers of Greenland.

The Jet Stream

In parallel to our changing ocean temperatures and water flows, we also see an emerging pattern change in our global stratospheric winds, known more simply as the Jet Stream (fig. 7.2). NASA calls it the "aerial superhighway."

The Jet Stream is a "river of air" that circulates around our planet (in middle latitudes between 30 and 60 degrees, and mostly from west to east) as strong winds influenced by the temperature at the polar caps. Pilots utilize the Jet Stream in the high skies to conserve fuel as their aircrafts are carried on the winds of the stream, which can reach velocities of up to 500 kph (310 mph)!

This stream of air shapes our regional climates, as cold air from the polar regions mixes with warm air elsewhere. A steady flow of cold air, particularly during the wintertime, generates predictable weather patterns around the globe that contribute to the seasonal cycles of our natural ecosystems. The warming of the polar ice caps due to climate change and to the various climate feedback loops occurring in the Arctic due to melting glaciers is affecting and destabilizing this pattern. As glacier ice melts and ice sheets retreat, the ocean is exposed, the ocean surface warms, and the ocean subsequently absorbs additional heat, leading to further acceleration of the glacier melting process. The ocean in turn sucks in more cool air from the atmosphere and warms it on the surface.

The temperature contrasts between cool Arctic air and other air in various regions of the planet is being altered. This contrast in air temperatures is what drives the global Jet Stream. Disturbing this balance of air flow is like taking a single cylinder out of a motor and replacing it with a different sized cylinder, throwing the entire motor out

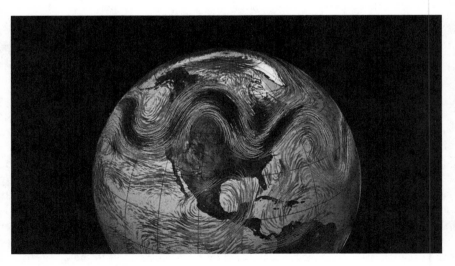

Figure 7.2 The Earth's Jet Stream.
The Earth's Jet Stream, or as NASA calls it, Aerial Superhighway, moves air streams from warmer areas near the Equator through cooler polar regions.
Source: NASA

of sync. This cycle alteration directly affects the stability of the air currents, weakening the Jet Stream winds, and creating wobbling and unpredictable wind patterns around the planet.

The temperature changes in the air generate less steady airflow and thus more unpredictable, irregular seasonal alterations.[11] This is part of the reason why we are seeing so many unusual weather events, cold weather in summer, hot weather in winter, a very hot day after a very cold day, etc., etc. Some people think that an extremely cold day or an especially cold spell of several days in winter or even during a usually warm month suggests that our planet is not actually warming, and yet reality is precisely the opposite. A very cold spell in an unusually warm month or even a particularly harsh winter may be indicating that our planet's climate is frightfully out of sync and off-course!

I recently traveled to Alaska excited to see glaciers and glacier ecosystems. I planned a visit to Denali National Park with my family and took advantage of the trip to also visit a number of glaciers around the state. I live in Florida and just days before traveling I checked the weather and was startled to find that during the week prior to my trip, it was warmer in Alaska than in South Florida! Wildfires were out of control throughout Alaska during this very unusual and hot spell.

Dr. Jennifer Francis, a senior scientist at the Woods Hole Research Center, described the phenomenon of a changing Jet Stream in simple terms:

> Because the Arctic is warming faster, the temperature difference is smaller and there's less "fuel" driving the jet stream. A weaker jet stream tends to meander north and south more in a wavy path. Those waves are what create the weather systems we feel on the surface, and when the waves get bigger, the weather patterns tend to move more slowly. Stuck weather patterns can result in extreme weather events, such as droughts, long heat waves, prolonged stormy periods and persistent cold spells, all of which are occurring more often now.[12]

According to Francis, who has worked at the Department of Marine and Coastal Sciences at Rutgers University for over two decades, the deterioration of the ice in the Arctic accounts for some 25% of global average warming temperatures. The vicious cycles of warming causes the Arctic to heat up faster than any place on Earth, and the subsequent melting of the sea ice causes warming to intensify across the planet.[13]

Tim Radford of the Climate News Network interviewed numerous scientists studying Jet Stream phenomena related to melting sea and glacier ice and subsequent warming temperature trends in the Arctic. He describes the change in the Jet Stream patterns as "dramatic waves." The "twists" in the direction of the Jet Stream that used to keep mainly to west-east directions are invading new territory, sending blasts of Arctic air into regions of the planet that would otherwise have more mild winters, says Radford. The changes to and instability of the Jet Stream are linked to prolonged droughts, extreme heat events, and other climate anomalies that we are seeing all over the world.[14]

Angela Fritz, an atmospheric scientist at Weather Underground, talks about "The Great Drought" in the United States in 2012:

The Great Drought of 2012 was as great as the dust bowl, and at one point 62% of the United States was in more than a moderate drought. ... We're looking at multi-year droughts now. This isn't just a single year event. The drought and the heat amplify each other." [Citing Jennifer Francis at Rutgers, Fritz points to the greening of the Arctic and the continental areas around the arctic, which could be changing how the Jet Stream pattern works in effect.]

It's slowing it down, it's making it more amplified. We are seeing more extreme weather and we are likely to see *more* extreme weather. And it will probably last longer. The Jet Stream exists because the tropics are warm and the Arctic is cold. It is strengthened by the difference between those two things. The warmer the tropics, the colder the Arctic, the stronger the Jet Stream is. When you make the Arctic warmer but the tropics don't warm as much, you're weakening the Jet Stream, making it slower, and then you're making the ridges and the troughs longer as well. So you're getting those extremes and you're getting longer extremes.[15]

The Surface Rebound Effect

Glaciers Are Heavy, Very Heavy!

Ever since the late 1800s when scientists really began considering the enormous role glaciers have played not only in our environments, but in shaping the land and its physical attributes, we've noticed that glaciers have a drastic effect on the Earth's crust.

When glaciers grow and expand over the Earth's surface, the sheer weight of the ice pushes the surface of the Earth that is underneath the ice, downward. Concurrently, at the edges around the ice, the ground is generally pushed upward, creating edges around the ice sheet called "forebulges."

Underneath the glacier, the ground can sink by many meters due to the weight placed upon it by the massive ice body. As glaciers retreat during great glacial melt, they expose previously compressed earth, and just as expected, and because the surface is elastic and wants to return to times past, the ground pops up to regain its former place.

Land swelling due to receding ice can be very noticeable in a short period of time, ranging from many centimeters per year to several meters per century. While this may seem small, it can have profound effects on local ecosystems and coastlines.

Jenny Chen of Smithsonian Magazine offers a very descriptive explanation of how this works:

During the glacier heyday 19,000 years ago the Earth groaned under the weight of heavy ice sheets thousands of feet thick. These enormous hunks of frozen water pressed down on the Earth's surface, displacing crustal rock and causing malleable mantle substance underneath to deform and flow out, changing the Earth's shape— the same way your bottom makes a depression on a couch if you sit on it long enough. Some estimates suggest that an ice sheet about half a mile thick could cause a depression 900 feet deep—about the height of an 83-story building.

The displaced mantle flows into areas surrounding the ice sheet, causing that land to rise up, the way stuffing inside a couch will bunch up around your weight. These areas, called "forebulges," can be quite small, but can also reach more than 300 feet high. As prehistoric ice sheets began to melt around 11,700 years ago, all this changed. The surface began to spring back, allowing more space for the mantle to flow back in. That caused land that had previously been weighed down, like Glacier Bay Park in Alaska and the Hudson Bay in Canada, to rise up. Places that were once forebulges are now sinking, since they are no longer being pushed up by nearby ice sheets.

For example, as Scotland rebounds, England sinks approximately seven-tenths of an inch into the North Sea each year. Similarly, as Canada rebounds about four inches each decade, the eastern coast of the U.S. sinks at a rate of approximately three-tenths of an inch each year—more than half the rate of current global sea level rise. A study published in 2015 predicted that Washington, D.C. would drop by six or more inches in the next century due to forebulge collapse, which might put the nation's monuments and military installations at risk. [16]

Land is popping up and deflating all over the world due to accelerated glacier melt. This is something of an anomaly in a time of general sea level rise. In most parts of the world, rising sea level means that the sea will come in over the land. For coastlines that were recently covered by glaciers, the dynamics of sea level change may be different. In places like Alaska where glaciers were once extremely large and very extensive and reached much of the coastline, the effects of the land popping up can be very significant.

Cornelia Dean writes about parts of Juneau, Alaska, where the retreat of glaciers is expanding coastal territories. "Morgan DeBoar, a property owner, opened a nine-hole golf course at the mouth of Glacier Bay in 1998, on land that was underwater when his family first settled there 50 years ago. . . . Now with the high tide line receding even further, he is contemplating adding another nine holes." [17]

And while for some, like Mr. DeBoar, newly surfacing land can be a positive, for others, the retreat of the sea can pose important ecological challenges. As the sea retreats, so does the water table, drying up land that may have had much higher humidity. Wetlands dry up and ecosystems are radically changed. If you had good farmland with just the right humidity, you might find that the land has become worthless due to retreating water tables.

These alterations in elevation due to expanding land and retreating glaciers, while they may be beneficial to some (like the golf course owner above) obviously pose serious threats to ecosystems, altering land humidity and destroying local habitat. Land can dry up where mangroves once existed, forests can suddenly pop up where salmon once swam, and coastlines can be entirely redrawn from one decade to the next.

Energy Production Affected by Melting Glaciers

Glaciers naturally or artificially melt in cycles. When those glaciers are in high mountains (such as in the Rocky Mountains, in the Andes of South America, in the Himalayas, or in the European Alps—just to name a few examples), that melting creates rivers, and sometimes very fast flowing rivers, and that in turn is a great opportunity to generate electricity.

Henry Fountain and Ben Solomon of the New York Times wrote a special report, which I highly recommend for its great information and stellar visuals. The report focuses on the energy opportunities created *and hindered* by glacier melt. "Where Glaciers Melt Away, Switzerland Sees Opportunity" is the name of the Times report.[18]

They begin their article with a 500-foot tightrope footbridge dangling 300 feet in the air over a glacier lake beneath the Trift Glacier in Switzerland. The glacier lake and the bridge did not exist a mere two decades ago. The cause, climate change. The glacier is rapidly melting and the water is running, forming the lake below.

This may seem bountiful for energy production, utilizing rushing river water to generate electricity. However, dwindling glacier size and eventual permanent glacier destruction could mean less water flowing down some of Europe's most prominent mountain rivers, so in time, there will actually be less water available to produce much needed energy.

In Switzerland, about half of the country's power is river-flow-generated electricity that is directly related to glacier melt. In the United States, about 7% of electricity comes from hydropower. In Switzerland, where climate change is advancing at a very accelerated pace, most glaciers are retreating and most will vanish by 2090, notes the New York Times article. Power plants were not designed with climate change in mind, and thus the impacts to energy and water supply due to shrinking glaciers could be significant.

The Aletsch Glacier, which is also in Switzerland, is more than 22 kilometers (14 miles) long and up to 900 meters (3,000 feet) thick, note the authors. Aletsch could lose 90% of its ice by 2100. That could be devastating. And while the future supply of glacier meltwater for power generation is at great risk, for now accelerated glacier melt means *more* water and hence *more* energy. Since the 1980s, accelerated glacier melt accounts for about 3–4% increase in energy generation.

We've talked about how advancing glaciers carve out paths and valleys in mountain systems. As they recede due to melting they leave behind large gaping spaces in the mountain. In some cases these spaces are naturally converted into glacier lakes. We read about the danger of precarious high-perched glacier lakes in Chapter 6 on glacier tsunamis. In other cases, man-made dams could be built to contain further glacial melt, ideal for generating further electricity. Engineers are already anticipating the likely scenario of advancing temperature rise and subsequent glacier melting, and they're thinking about where new glacier lakes will form, and where, through planning and construction, new energy generating hydropower plants fed from glacier melt can generate society's future energy needs.

Similar glacier-fed rivers generating power can be found in many high mountain environments, including in countries in the Himalayas, in California, in the Rocky Mountains, in France, Canada, China, India, as well as in the Andes in countries such as Peru, Bolivia, Argentina, and Chile, to name just a few examples. Currently about one fifth of the world's energy supply is provided by hydropower, much of that from mountain environments that are glaciated.

In Argentina, a massive project to dam the glacier meltwater beneath Argentina's National Glacier Park is in the design-phase. This project, called Nestor Kirchner

Dam, would flood the valley immediately below the glacier park, raising natural glacier-lake levels significantly, and generate much needed electricity for Argentina's energy-starved population (fig. 7.3).

Environmentalists who study Argentine glaciers (including myself) are concerned with the impacts that artificially raising glacier lake levels could have on massive glacier stability, not to mention the added risk of glacier tsunamis. This part of the Andes already has numerous glacier lakes below some of the its largest glaciers, such as the Perito Moreno, the Upsala, and the Spegazzini glaciers, among others.

These glaciers are already vulnerable to climate change. They cyclically advance and break large seracs (glacier pieces) into their lakes below. The fronts of these glaciers, which are largely submerged into the glacier lakes, are in equilibrium with their lakes and outside ambient temperatures. If the glacier lake levels were to rise (due to the damming of water for energy production), this would cover portions of the glaciers that are currently at equilibrium with the air, but not the water, generating a great disequilibrium in both temperature and structure at the glaciers' front terminus, potentially causing a collapse of the terminus and a sudden collapse (forward slide) of the entire glacier mass, as the terminus collapses and the glacier mass can no longer

Figure 7.3 The site of the planned Nestor Kirchner Dam downstream from glaciers and glacier lakes.
Downstream from National Glacier Park in Argentina, the projected Nestor Kirchner dam aims to take advantage of glacier melt water to generate electricity, but poses serious risks to the glaciosystems of several of Argentina's already vulnerable glaciers.
Source: Google Earth. GIS: 50 00 57.34 S, 72 12 47.11 W

sustain the enormous weight behind it. These glaciers are many miles long and the overall impact to the ecosystem could be disastrous.

Peru's Cordillera Blanca is the most extensive tropical, ice-covered mountain range in the world. Over 700 glaciers and over 400 glacier lakes provide meltwater to downstream communities and to several of Peru's hydropower plants, producing about 5% of the country's energy supply. Current glacier melt rates however are generating concern over how much Peru will be able to depend on future energy supply from mountain hydropower plants with ever-shrinking glaciers and snow cover. Between 1970 and the early 2000s, permanent ice cover had already shrunk by about a third.[19] Other risks (besides losing energy production capacity) to power generation in high mountain environments from glacier melt and vulnerability include the immediate risk to existing energy infrastructure just below melting glaciers posed by glacier tsunamis and other mountain surface failures (such as lahars) brought on by a warming climate.

Where large dammed lakes may already have been created to generate power, a sudden rush of water into such lakes from unstable melting glaciers could breach the dams, resulting in the even more deadly coupling of glacier lake water with dammed water to produce destructive mega glacier tsunami floods.

This has been a problem typical of hydropower plants in Peru, which have experienced precisely this occurrence in the past. Mark Carey, professor of history and environmental sciences at the University of Oregon, wrote a book about the risks of melting glaciers and devotes a chapter to the "Killing Ice of the Andes." Carey says that "Peruvians have suffered the wrath of melting glaciers like no other society on earth."[20]

The Peruvian government has teamed up with glaciologists and engineers to consider the consequences of glacier melt from high mountain environments for downstream communities. Benjamin Morales Arnao, a good friend and glaciologist from Peru, has devoted his entire life to this conundrum. He moved to Switzerland to study glacier tsunamis to better understand the dynamics of GLOFs in his homeland, Peru. Benjamin returned to Peru to help design systems to alleviate water accumulation in glacier lakes in high mountain environments to avoid these dangerous and sometimes catastrophic glacier tsunamis.

What is certain into the future is that glaciers in mountain environments will continue to melt at alarmingly high rates. Coupled with glacier retreat, this melting will in turn create more and more high mountain lakes and will change the local environments and ecosystems. These lakes in turn become significant sources of water supply as well as potential sources for hydroelectric power generation. In some cases, there may already be hydroelectric infrastructure in place that may be placed at risk from rapidly melting glaciers. These may suddenly be overwhelmed by excessive glacier melt water or sudden tsunami-like rushes from ice collapsing into glacier lakes. In other cases, they may be at risk of losing relevance due to decreased water flow in glacier-fed rivers where glaciers are dying. Whichever the risk, accelerated melting of glaciers has severe consequences.

Rivers, Animals, and Human Habitat

We've all probably seen by now the World Wildlife Fund image of a polar bear perched on a small piece of ice in the Arctic Ocean.[21] This single image of a bear clinging to melting ice (while not specifically a glacier, but rather an iceberg, which is likely to be a piece of glacier that has fallen into the sea) is extremely telling about how nature is adapting (or not) to melting glaciers. This image drives home a point that is as central to the impacts of ice melt (specifically of melting Arctic sea ice) on the feeding grounds for much polar wildlife as it is to human populations residing particularly in the northern extremes of the planet.

Ice sheets that extend over water or land provide fertile hunting grounds for many species. Polar bears are one of the most visible (and famous). These species have, over many hundreds and thousands of years, learned to hunt and reproduce in these delicate ecosystems and depend on the freezing and thawing cycles of the Arctic to ensure their survival.

The IPCC noted:

> Arctic landfast ice [in this case, ice on the sea that is at some point connected to land] is important to northern residents as a platform for travel, hunting, and access to offshore regions. Reports of thinning, less stable, and less predictable landfast ice have been documented by residents of coastal communities in Alaska, the Canadian Arctic, and Chukotka. The impact of changing prevailing wind forcing on local ice conditions has been specifically noted including impacts on the landfast ice edge and polynyas.
>
> There is evidence that the combination of loss of sea ice, freshening, and regional stratification has affected the timing, distribution and production of primary producers. Satellite data show that the decline in ice cover has resulted in an increase (of more than 30%) in annual net primary production in ice-free Arctic waters since 1998.[22]

Ice loss has also resulted in earlier phytoplankton blooms. Earlier spring ice retreat and later autumn ice formation are changing the natural seasonal cycles of the biology of the oceans with cascading effects on Arctic benthic (seabed) community biodiversity and production. In the Barents Sea, evidence suggests that factors directly related to climate change (sea-ice dynamics, ocean mixing, bottom-water temperature change, ocean acidification, river/glacial freshwater discharge) are impacting the benthic species composition.

Glacier and ice melt from ocean-covering ice sheets are causing vast ecosystem transformations. For the polar bear, this means that food once captured by wandering over hard ice is much more difficult to come by where ice has melted and only open oceans exists. The polar bear in such an environment is left unable to capture its prey in the ocean water, in which its ability to swim is far less than its prey's (seals). The inevitable and tragic consequence of melting ice for polar bears is starvation. (fig. 7.4)

Another phenomenon that shows the impacts of the changing climate on polar bears is the sudden cross breeding of polar bears with their southern cousin, the grizzly bear. The Guardian reported recently that sightings of a "pizzly bear" or a "grolar bear" have risen in northern areas of Alaska and Canada. This is a hybrid species of bear that results from the mating of polar bears moving south in search for food and grizzly bears moving north for the same reason.[23]

Figure 7.4 Polar bear starving due to melting ice in the Arctic region (the bear's hunting ground).
Source: Kerstin Langenberger

We know also that melting glaciers are drastically affecting the spawning grounds of migratory river salmon species in places like Alaska. Henry Fountain of the New York Times traveled to the US Pacific Northwest and published a story examining how glacier melt is affecting the natural world and specifically fish spawning grounds.[24] He delves into the intricate workings of glacier-fed streams in some of the most fish-rich areas of the world.

> Glacier-fed ecosystems are delicately balanced, populated by species that have adapted to the unique conditions of the streams. As glaciers shrink and meltwater eventually declines, changes in water temperature, nutrient content and other characteristics will disrupt those natural communities.
>
> Lots of these ecosystems have evolved with the glaciers for thousands of years or maybe longer [quoting Jon Riedel, a geologist with the National Park Service].
>
> Streams that are mostly fed by glacial meltwater often have unique species that have adapted to the cold conditions. Reducing or eventually eliminating the contribution of this meltwater will raise stream temperatures. Even a small temperature increase can have potentially negative effects. "Certain species like cold water," [quoting Alexander Milner, a professor of river ecosystems at the University of Birmingham, in England]
>
> The impact to larger species like salmon and other fish may be more complex. In this part of the world salmon are just really important for ecosystems and cultures [quoting Jonathan Moore, an ecologist at Simon Fraser University in Burnaby,

British Colombia]. Salmon and similar fish are born in the freshwater streams, then head to sea to grow, often for years, before returning to fresh water to spawn and die. Water temperatures must be just right for salmon to develop, so as glacial meltwater declines and stream temperatures rise, populations may be affected.[25]

The North Cascade Glacier Climate Project studies the impacts of glacier retreat, increased glacier melt, and runoff on rivers and subsequently on salmon species.[26] One of their studies looks closely at the greater Salish Sea ecosystem area in the Pacific Northwest region. They have recorded notable glacier reduction in the past three decades, as well as significant variability in stream flows, such as in the Fraser River.

The Whitechuck Glacier, for instance, supplies flow to the headwaters of the Whitechuck River. The glacier began a rapid retreat in 1930, following a growth phase during the recent Little Ice Age. By 2001, the northern branch of the glacier had completely disappeared. The contribution of water flow into the river by the glacier has been reduced some 65–80%, and since 1950, glacier runoff has diminished some 5.7 million cubic meters annually, or between 0.55 to 0.65 cubic meters per second. The result, say the researchers, is less water for salmon that feed downstream and less food for stream invertebrates on which salmon feed downstream in the Sauk and Skagit Rivers.[27] "Overall, salmon are being faced with increasing climate stress on the top of the long term habitat alteration at the beginning and end of their life cycle," say Oliver Grah and Jezra Beaulieu, of the program. Oliver Grah in a video about the impacts of accelerated glacier melt on northern salmon species, says:

> The Nooksack River is a very important river system for nine species of Pacific salmon, and of greatest importance to the Nooksack Indian Tribe are spring Chinook Salmon. The conditions of the watershed have become worse in regards to salmon survival. Today there is only about 8% of the number of salmon than there were when Europeans first pioneered in this area.
>
> There has been this huge decline in salmon stocks. Glacier melt supports sustained stream flow late in the summer, both the amount of stream flow and the temperature of the stream flow. The salmon need adequate stream flow and cold temperatures to hold over the driest part of the year and the warmest time of the year so they can spawn in the fall. What's happening with climate change is that the amount of runoff is decreasing during that time and water temperatures are getting warmer, creating a larger impediment to salmon survival. Nooksack tribal members rely on spring Chinook salmon for cultural, subsistence and commercial uses.[28]

Erik Schoen and a group of colleagues have looked at the changing climate in Alaska and the glacier runoff to the Kenai River. They examined the changes in climate, hydrology, land cover, populations, and fisheries over the past 30–70 years in this region. His findings indicate that while some salmon populations may be harmed, others may benefit. He stresses however that climate change is altering the flow and temperature of regimes of Alaskan salmon streams. Regional losses of glacial mass likely have both positive and negative implications for freshwater ecosystems and salmon populations. The net effects remain poorly understood. As glaciers lose mass, says Schoen and the co-authors of the study, short-term increases in meltwater inputs can reduce the production of zooplankton and Sockeye salmon in lakes, the diversity and production of microorganisms, algae, and benthic macro-invertebrates in rivers, and the quality of fish spawning habitat.[29] What is certain is that melting glaciers and climate change are bringing great change to the salmon population of the region.

Glacier melt also affects the flow of river streams, and here we may see drastic impacts to local flora and fauna and especially to water-dwelling species, such as fish, as well as other microorganisms the river ecosystem depends on to thrive. Glacier-fed rivers depend on an ecologically sustained and balanced *flow* of water as well as the sustainability of the *temperature* of that water year to year. Changes in river flow and temperature due to altering glacier melt patterns will drastically alter river and riparian ecosystems. Alexander Milner and his colleagues who are studying impacts of glacier melt on river systems reflect on this point :

> Changes in river hydrology and morphology caused by climate-induced glacier loss are projected to be the greatest of any hydrological system, with major implications for riverine and near-shore marine environments. Glacier shrinkage will alter hydrological regimes, sediment transport, and biogeochemical and contaminant fluxes from rivers to oceans. This will profoundly influence the natural environment, including many facets of biodiversity, and the ecosystem services that glacier-fed rivers provide to humans, particularly provision of water for agriculture, hydropower, and consumption.[30]

Salmon that spawn in rivers and then go out to sea, come back to spawning grounds year after year to lay eggs. The ecological balance necessary for these aquatic populations is key to maintaining delicate life cycles not only of the fish, but of all of the species including birds, insects, mammals, and other aquatic life that depend on the river's delicate ecological system.

If the flow of water is significantly altered, say for example as a result of increased water speed (due to an acceleration of glacier melt), or because of dwindling flow (due to a reduction of glacier mass), or if the river hydrology drastically changes temperature (due to higher or lower concentrations of cold water deriving from changes in glacier melt rates), these changes can greatly alter, or even destroy stream ecosystems.

The atmospheric scientist Angela Fritz offers some firsthand evidence of the rapid melting of the Greenland ice sheet and the impacts to river streams that she witnessed there. In 2012, we saw an especially intense warming of much of the Greenland Ice Sheet, where up to 97% of the ice sheet was warming unusually at one time.[31] The Watson River, one of the main rivers of the Greenland Ice Sheet that received the melting from the ice sheet, doubled its prior flow record! This intense glacier melt took out rivers and destroyed riverbeds and coastline. You can watch some terrifying videos of this unusual glacier-melt-caused river swell on YouTube.[32]

And while initial glacier melt acceleration will cause an *increase* in the glacier melt runoff into river streams, that initial increase will be offset quickly by a significant *reduction* in size of glacier mass. Milner and colleagues estimate that the overall loss of glaciers will decrease total runoff by 10–20%. More importantly, glacier-fed rivers (fig. 7.5) have a fairly constant flow pattern in dry and summer months, as they are fed largely by the meltwater of relatively mass-stable glaciers. If glacier mass is rapidly

Figure 7.5 A glacier-fed river stream on Mt. Cook in New Zealand will see drastic changes in river flow in a warming climate.
Source: Pixabay

reducing, this means rivers that are fed by glaciers will see a steady decline in water availability during warmer and drier months.[33]

A rapid increase in glacier melt can also be extremely problematic for navigable river streams, due to the inordinate amount of silt sent down the river and subsequent accumulation of mud on the riverbed. If combined with ground expansion resulting from glacier retreat, this can cause havoc for river-going vessels. Cornelia Dean describes, in a New York Times article, how accumulating silt in Alaska's rivers is drying up riverbeds during low tides, making navigation impossible.

> A few decades ago, large boats could sail regularly along Gastineau Channel between Downtown Juneau and Douglas Island, to Auke Bay, a port about 10 miles to the northwest. Today, much of the channel is exposed mudflat at low tide.
>
> "There is so much sediment coming in from the Mendenhall Glacier and the rivers it has basically silted in," said Bruce Molnia, a geologist at the United States Geological Survey who studies Alaskan glaciers.
>
> Already, people can wade across the channel at low tide or race across it, as they do in the Mendenhall Mud Run. At low tide, the navigation buoys rest on mud.[34]

The Mississippi Delta is another area of North America greatly affected by melting glaciers and silt sediment deposits. James Aber, who studies the effects of glacier melt and freezing on the Earth's surface, explains that during high glaciation (a very frozen Earth) with low resulting seas (because much of the ocean water was in the glaciers), the Mississippi River wedged out a deep valley along its flow. During past interglacial times during which glaciers rapidly melted, the Mississippi brought down a large amount of silt, filling the valley with sediment accumulation and raising land by about 9 meters (30 feet).[35]

In his recently published book, *End of Ice*, Dahr Jamail converses with famed glaciologist Daniel Fagre of Glacier National Park, focusing on the critical role glaciers and the cryosphere play in the survival of the biosphere in sensitive mountain ecosystems. I'd like to end this chapter with a few exquisite excerpts from chapter 2 of his fabulous book, in which Jamail and Fagre exchange thoughts on this delicate ecosystem.

[Jamail] Glaciers serve as a relief valve, or as a free reservoir system, which opens its spigots precisely when water is needed. They also, conveniently, provide very cold water far from the glacier, giving life to a variety of animals, including endangered insects and species like the bull trout—the polar bear of fish.

"So when the glaciers are gone, the safety net of cold water will vanish" Fagre explains. "So streams will warm up, or dry up, either of which are lethal to alpine aquatic biota."

[Jamail] A glacier also cools the air in a mountain basin, and cold water downstream has an impact on shrubs and other species. Without this cold water, species that haven't adapted to warmer temperatures will be negatively affected.

Heavy snowpack also keeps trees at bay. A diminished snowpack allows trees to enter the alpine areas, and, as Fagre explains, once trees make it into areas where they previously could not live, they are "ecologically released" and tend to grow fast. The trees, flourishing in the warming temperatures at high elevations, cause vegetation in the sub-alpine meadows to disappear, along with the animals, birds, bees, and butterflies they support. One major consequence, therefore, is the loss of alpine diversity. … The impact further downslope is that larger trees can grow in mid-elevation forest for longer periods of the year, using more of the moisture from the soil and depleting streams. This increases the risk of even larger fires. There is more fuel and less moisture in the soil.

Water- and heat-stressed trees are unable to respond to beetle invasions triggered by warmer temperatures. Fagre also tells me that it's now already so warm that one beetle species is experiencing two life cycles each year instead of just one. In addition, the warming during winter means fewer beetles are dying off.

All of these phenomena combine to affect one of the major ecological services that mountains provide—water storage and delivery. … "You can count on all alpine glaciers in the world to be gone by 2100," Fagre says.[36]

8

Invisible Glaciers

... Will They Save Us?

[follow #rockglacier on Instagram to see and receive daily images of the world's amazing but *invisible* rock glaciers]

As glaciers melt away forever (or at least until the next ice age arrives), *invisible* glaciers will remain for a considerable time longer. Let me explain.

In the late winter of 2019, I called up my good friend Jared Blumenfeld, who had directed San Francisco's Environment Department in 2007. He spearheaded the elimination of plastic bags from supermarkets, a bold move that would go viral, globally. In 2009, he made San Francisco the first city to mandate the composting of organic waste. He later became the head of the US Environmental Protection Agency for Region 9, covering most of the western United States and Hawaii. Jared is now California's environment secretary under the governorship of Gavin Newsom.

I asked Jared, on that call, "Hey, have you ever heard of a rock glacier"?

"No ... I've heard of glaciers, but I've never heard of a rock glacier!" responded this extremely seasoned environmentalist and environmental policy expert that probably knows more about the environment, about environmental policy, and about the vulnerability of environmental resources than most people on Earth.

Noting his curiosity, I continued. "Basically, rock glaciers are these huge bodies of ice underneath the Earth and if someone doesn't point them out to you, you can't see them," trying to convince California's highest environmental authority (and my friend) that I wasn't pulling his leg, that there really were *invisible glaciers*, and that he should go with me to see them as they weren't that far away from where he was living in Sacramento. A short drive to Yosemite National Park, in the heart of the Sierra Nevada Mountains, would prove it. My aim was to expose these wonderful icy resources to prominent environmental policymakers in the hope that they would help protect them as hydrological resources for future generations.

We ended up going to the Sierra Nevada that summer to visit these enigmatic "invisible" glaciers with Sierra Nevada rock glacier specialist Connie Millar, Senior Research Ecologist with the US Forest Service, David Herbst, a research biologist with the Sierra Nevada Research Laboratory, Sara Aminzadeh, a California water policy expert, and Adam Riffle who worked with me at the Center for Human Rights and Environment (CHRE) as coordinator of cryoactivism for the western United States. Jared ended up recording a wonderful podcast of the experience for his extremely interesting, provocative, and innovative environmental podcast series, *Podship Earth*, which I encourage you to listen to![1]

Rock glaciers are indeed invisible glaciers, well, invisible until, as I told Jared, someone points them out to you. You could be standing on a glacier more than a mile long (1.6 kilometers), half a mile wide (nearly 1 kilometer), and more than 300 feet deep (100 meters) and not see any ice anywhere!

How can that be?

The Rock Glacier Experiment in Your Refrigerator

To answer this question, let's start right in your refrigerator with an experiment that is great to help kids (and you too) understand the functioning of a critical piece of the cryosphere in high mountain environments.

Take a Tupperware container (preferably transparent) and fill it about half-way with small rocks or pebbles. Pour water over the rocks approximately level with the rocks—it doesn't have to be exact. Put the container in your freezer and let it thoroughly freeze. This replicates the freezing nighttime temperatures of high mountain environments after a good rain or snowfall. As water accumulates in these environments (through rain or snow, for example), it trickles down, percolates in between rocks, and freezes at night.

Once the water is frozen, remove the Tupperware from the freezer. You'll have a big glob of ice and rock. Let everything sit in the Tupperware at room temperature for a good 20 to 30 minutes or until about *half* of the ice has melted. Don't let it all melt or you will have to begin again. You'll notice that as it melts the meltwater drips down to the bottom of the Tupperware. We are copying the *thawing* warmer daytime temperature of the mountain.

Next place the container back in the freezer and let thoroughly freeze again. We are now replicating the next *nighttime* mountain freezing temperature. Once fully frozen, take the container out again. If your Tupperware is transparent you will notice a thin layer of solid ice at the bottom of the container and a new blob block of ice and rock at the top. Let it sit at room temperature again for another 20 minutes or so, never letting it fully melt. We are again replicating the *daytime* thawing that takes place in a high mountain environment. More water from the blob melts again and trickles to the bottom of the container. Once about half of the blob has melted, return the container to the freezer and let thoroughly freeze. Do this five or six times. Eventually most or all of the rocks will be on top and most or all of the ice will be on the bottom. We've just recreated the freeze-thaw cycle of rocks and water that happens in the mountains.

Voilà! You have created a rock glacier in your refrigerator!

The Freeze-Thaw Dynamics of Rocks and Ice in Nature

This process of ice penetrating to the bottom between rocks through the melting of ice in repetitive thaw cycles, and then refreezing over and over again, happens in nature all the time, especially in high mountain ecosystems where it can get very cold during the night and warm considerably, or at least, stay above freezing during the day. Through many years, decades, centuries, and even millennia, this "geo-cryogenic process" (in other words, a geological process interacting with water and ice) of cyclical freezing and thawing, just as in your Tupperware, generates massive bodies of ice nestled underneath thick rock layers.

Once this large mass of ice and rock begins to grow very large, sometimes as big as a normal glacier, the sheer weight of the rock and ice, and the lubrication from the water accumulating at the bottom of the rock glacier, combined with the incline of the mountain and a bit of gravity, all begin to create a slide down the mountain of the icy rock debris. This slide mimics the same displacement of a normal glacier, albeit much more slowly. While normal, visible, exposed white glaciers that we may be more familiar with might advance several feet or meters per day, covered rock glaciers usually only advance several feet or meters per year! Sometimes they only move inches.

As they slowly *ooze* down a mountainside, if you got a broad look from above (as from a drone flying high overhead or through satellite images) they look something like a big rocky tongue with a lobular form at the lowest extreme (fig. 8.1). Some confuse these slipping rock and ice formations with remnant lava flow. Others actually walking on the surface of these rock glaciers think that they are walking on a rockslide or on a remnant glacier moraine. Moraines are mounds of rock left to the side and at the front of a former glacier after its retreat up the mountain as it melts away. Many

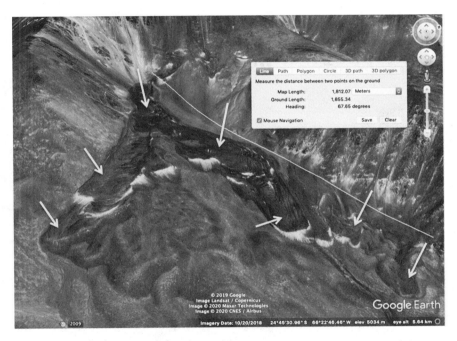

Figure 8.1 Rock glaciers in Salta, Argentina.
These rock glaciers appear as lobe-like rock debris flow. This is an extremely dry area of the country that depends on meltwater from rock glaciers and the broader periglacial environment to nurture the hydrology of downstream ecosystems. These visible rock glaciers are about 2 kilometers long (1.2 miles) and cover an area about 1 kilometer wide (0.6 miles). Few people in Salta Province are aware of these subsurface rock glaciers in the mountain ranges above their cities.
Source: Google Earth. GIS: 24 45 31.86 S, 66 22 46.42 W

avid and regular hikers walk frequently on rock glaciers and have no idea they are walking on moving ice!

The rocks that are pushed and moved to the surface of rock glaciers can be small or colossal, even as big as a small house. Walking on a rock glacier is no easy task, and the instability of the rock glacier itself—because it is actually moving—makes them especially prone to rockfall or shifting, so if you're ever walking on a rock glacier, be attentive to its instability, it might save your life! (fig. 8.2)

Rock glaciers exist in areas somewhere between the bottom end of the last visible glacier (on your way down a glaciated mountainside) and before the forest line (where plants begin to grow). There is usually (not always) a swath of land between these two points about 500 meters in width (1,600 feet) where no glaciers or plants are visible.[2] (fig. 8.3) This is because in that strip of land, water freezes and thaws cyclically, preventing both plant growth (the plant cells are destroyed by the freezing temperature) and development of normal uncovered visible glaciers (temperatures are too warm for exposed ice to survive).

This cyclically freezing and thawing terrain is called the "periglacial environment." The higher you go in this area, the colder and more frozen it is for most of the year, and conversely, the lower you go, the nearer you get to the forest line and to warmer and more actively thawing areas. These thawing areas provide meltwater to the ecosystem below permanently! It's important to understand that this area is constantly

Figure 8.2 Adam Riffle, The Center for Human Rights and Environment's (CHRE's) former Cryoactivism Coordinator, Western United States, barely visible slightly below the center of the image, walking on the surface of Barney Rock Glacier in the Sierra Nevada Mountains, California. Don't be fooled, there is ice underneath!
Source: JDTaillant

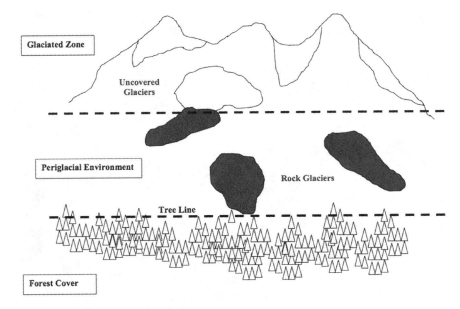

Figure 8.3 The periglacial environment.
Source: JD Taillant

going through temperature changes throughout the year and could be higher or lower depending on the season and the weather. The entire swath is considered here as per the yearly average temperature, which will determine if surface ice (the upper limit) and plant life (the lower limit) can survive.

Particularly interesting in this barren fringe are the visual results of the freezing and thawing cycles and the "cold weathering" (breaking of rocks due to freezing) or "frost heave" (upward push of rocks). The resulting effect is not random. Patterns form and large and significant shapes emerge.

California is full of rock glaciers and periglacial areas. I've inventoried, on Google Earth, California's rock glaciers and have counted approximately 800 rock glaciers and other periglacial features throughout the Sierra Nevada, where we visited with Jared during the end of summer 2019.[3] Surprisingly—for a state that suffers severe and prolonged droughts and where water conservation is a top political, economic, and social priority—no one, not even the highest environmental authority, my friend Jared, knows about rock glaciers! (Well, he does now!)

The situation is similar in other parts of the world where the periglacial environment thrives and plays a critical role in local water supply. The Central Andes, particularly in Argentina and Chile, for example, are rich in periglacial features. Alaska, which is very well known for its majestic white uncovered glaciers visited by thousands of people each year, unbeknownst to much of the population, and certainly to most visiting tourists, Alaska also has many very evident, magnificent, and yet mostly ignored, rock glaciers (fig. 8.4). And yet, despite their inconspicuousness, Rock glaciers and other ice-rich subsurface geological formations continuously contribute to

Figure 8.4 The Atlin Rock Glacier in Northern British Colombia.
This is particularly illustrative of the distinct glacier and periglacial environments,
with high mountain glaciers above (white visible ice) and a flowing rock glacier
below that reaches almost to the lower lake elevation. We see clearly how white
surface glaciers can only survive at high elevations. However, thanks to insulating
rock cover, rock glaciers can survive at much lower elevations. We can also
distinguish the likely periglacial environment as the barren brown areas below the
visible glaciers and above the tree line.
Source: Kirk Miller. Wikipedia Commons. GIS: 59 33 08.50 N, 133 51 07.72 W

local water supply, and sometimes very substantially. It's at least curious that practically no one has ever heard of rock glaciers!

Connie Millar, a researcher and scientist with the US Forestry Service who went
with us to the highest points of the Sierra Nevada, talked to my friend Jared about the
functionality of rock glaciers. You can hear the full conversation between Connie and
Jared on Jared's Podship Earth Podcast but here are a few notable excerpts:

When we talk about rock glaciers, the first thing people do is roll their eyes. Then they
think you're fooling them because [it] doesn't make sense. A lot of slopes that look just
like scree slopes or maybe regular moraines from old glaciers are actually what we call
rock glaciers. They are rocky debris that has ice underneath the rocky surfaces.

It is very hard to determine the ice content underneath because they are covered with
maybe 10 feet of rock mantle. But they do form glacial ice, they will move and creep

very slowly, at most a meter a year, which is maybe a tenth of what ice glaciers do. And because they are covered with this heavy rocky mantle and it's thick enough, it insulates them from the warm air outside.

If it were very thin [the rock mantle] it would conduct heat and maybe melt them faster. But because the rocky mantle is so thick and because there is air between these rocks, a unique microclimate sets up which actually supercools the interior, and serves like a freezer to keep the ice cold, even if it is above freezing out in the external temperature [where exposed ice would melt].

[Rock glaciers] are more and more important not only to get them on the radar as hydrological reserves but because as other water resources in mountains they are melting. They will really be the water towers of mountains for the future. And because they're so widely distributed they are important not only in the upstream for biota and the plants and animals up in the mountain but also for downstream resources like mountain meadows and of course for basin aquifers for human uses.[4]

I have scoured the Earth's mountain ranges on Google Earth to identify rock glaciers in various mountain ranges, in order to better understand how widespread rock glaciers and periglacial environments are (figs. 8.5 and 8.6). I have found rock glaciers in Greenland, Canada, Alaska, California, Oregon, Washington, Nevada, New Mexico, Colorado, Montana, Wyoming, Peru, Bolivia, Chile, Argentina, China,

Figure 8.5 This spectacular three-tongued rock glacier in Salta, Argentina is 500 meters wide (1,640 feet), 2 kilometers long (1.2 miles), and more than 40 meters thick (130 feet). Note the typical 30–40 degree inclined front slope.
Source: Google Earth. GIS: 25 00 37.58 S, 66 21 17.00 W

Figure 8.6 The Gibbs Rock Glacier near Dana Peak in California's Sierra Nevada flows to and feeds Kidney Lake.
Source: Connie Millar. GIS: 37 53 41.60 N, 119 12 18.33 W

Slovakia, Romania, Kosovo, Serbia, Bulgaria, Armenia, Azerbaijan, Montenegro, Afghanistan, Iran, Nepal, Norway, Russia, Sweden, Mongolia, and Turkey. I've even found rock glaciers *on Mars*!

Yes, Mars! In fact, scientists looking for water and ice on Mars have used knowledge of the geo-cryogenic process occurring in rock glaciers (the same one we experimented with in your refrigerator) to identify presumed geo-cryological formations on the martian surface. The photograph in figure 8.7 below shows some of these likely frozen features visible on Mars.[5] Compare figures 8.1 of the Argentine rock glacier and 8.7 of the martian rock glacier and you see how similar they are. The physics of rock glacier formation seems clearly similar throughout the universe!

Where and How Can We Find Periglacial Environments?

In Chapter 5, on the thawing Earth, I briefly covered the use of the permafrost mapping tool[6] developed by Stephan Gruber at the University of Zurich, which utilizes an online tool applicable in Google Earth to map permafrost areas around the globe with incredible ease.

This tool looks at air temperature data picked up by satellites and allows us to identify where the mean *surface* temperature is at or below 0°C on a yearly average

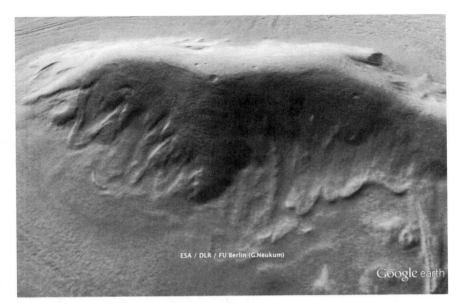

Figure 8.7 Rock glaciers on Mars.
These are possible rock glacier features visible in images taken by NASA. Compare
this image to figures 8.1 and 8.14, of rock glaciers on Earth to see the similarities.
Source: Google Earth

(it cannot measure temperature below the surface). This sure beats the old way
of doing things when you had to trek up to these inhospitable places, take your
shovel and put a thermometer in the ground, and take such measurements regularly throughout the year to determine if in fact the ground is ever frozen. Other
ways of finding frozen ground include identifying the presence of ice in the earth
or observing rock debris for signs of ordering, such as the very distinct flows ex-
hibited by rock glaciers, or the octagonal rock-sorting patterns sometimes created
by cyclically freezing and thawing grounds. You might think that alien visitors to
Earth decided to place rocks in special geometric formations to leave behind signs
of their presence, but in fact, freeze-thaw cycles are the actual culprits (apologies to
the UFO fanatics out there).

Rocks ordered in geometric shapes on the ground in very cold environments
that have cyclical freeze-thaw cycles are an indication of the frost heave and cyclical
freezing processes that are occurring at the site. The identification of rock glaciers and
other frozen earth regions is made very easy with Stephan Gruber's Global Permafrost
Zonation Index, that gives your computer this capacity without you having to trek up
the mountain. With this Google Earth driven permafrost model, I find rock glaciers
throughout the world right from my balmy Florida home!

Try it out! Download the file to you home computer and open the it in Google
Earth. Now find your favorite tall mountain and see what it shows. You may have to
click or unclick boxes in the left pane to determine all of the information you want to

Figure 8.8 Permafrost regions in the Unita Mountains, Salt Lake City.
The Global Permafrost Zoning Index is an easy tool to use for identifying likely
permafrost (frozen grounds).
Source: Google Earth

see (or not see), including various layers and legend boxes. Unclicking ruggedness in
the pane helps to focus only on frozen ground.

Figure 8.8 shows the Salt Lake City area with the Unita Mountains to the east. With
the permafrost mapping tool activated, you can see clearly the areas of the planet where
the temperature at the surface is at or near the freezing point year round. It's not a guarantee that you will find ice-rich permafrost at this site, or rock glaciers, but if it's there,
I guarantee you it will be in the areas mapped by the model (for me, this inverse relationship has never failed, that is, if I find a rock glacier, it's always at a spot that the
model had identified its possible presence). On your home computer, green and yellow
(lighter shades if the picture has been converted to black and white) are fringe areas of
uncertainty, while the colder areas along the ridge of the mountain in shades of red,
purple, and blue (darker shades) are much colder. Note the direct relationship of these
frozen grounds to Salt Lake City's massive water bodies (on the left edge of the picture).

Permafrost and Mass Wasting

Just as with the glaciosystem of exposed glaciers, if the surrounding characteristics of
the geology are just right, rocks and ice will gather more in one place than in another,

and the systemic effects of cyclical cold weathering dynamics will be clearly seen. It could be that a steep ravine, or a very slanted mountainside, produces abundant rock avalanches. If a heavy snowfall occurs on the surface, the avalanches will mix snow, ice, and rock, which will necessarily accumulate at the base of the slope, where the avalanche comes to a stop. There, a mixture of rock, snow, and ice starts to accumulate.

This mix becomes a dynamic *ice-saturated* mass, growing and expanding as new ice forms from snow and water that enters the mass and freezes. It begins to go through cyclical processes similar to that described in the rock glacier experiment earlier in this chapter.

The mass of ice and rock contracts when parts of it melt off and then expands when refreezing occurs and more material accumulates. These cycles reproduce themselves day after day, season after season, year after year, decade after decade, and even century after century, forming a thick ice-saturated rocky earth mass. This can be in the form of continuous ground that is constantly stretching and contracting, it can form a large body of rock and ice, or it can accumulate in a crevasse, at the base of a mountainside, or in between mountainsides.

This continuously shifting ice-saturated rocky surface layer can, when a critical weight is reached and if it rests on a slope, begin to slip downhill, thanks to the humidity in the ground and the effects of gravity. The entire ground surface, mixed with rock and ice may begin to slide downhill, just as a glacier creeps forward toward lower elevations. Scientists call this downslope movement of debris under the influence of gravity "mass wasting." *Mass wasting* can occur in both frozen and non-frozen ground, but periglacial areas are especially conducive to mass wasting because they contain rock fragments (created as water seeps into rock fissures, expands, and breaks the rocks during freezing), expansive and dynamic movement of rock and ice caused by the freezing and thawing cycles, and lubrication, provided by meltwater.[7]

Depending on the concentration and accumulation of this material, it may be present in thin layers, in concentrated piles, or in other forms, shapes, and sizes. The various forms are similar in that they contain ice and rock, they are going through freezing and thawing processes, they are constantly modifying their form and shape, they are storing water, and when the temperature rises, they are releasing water into the environment. Scientists give these periglacial forms different names, according to the various attributes that distinguish each, based mostly on location, ice content, and movement. From an environmental resource perspective, the importance lies less in these *distinctions of type* and more in the *hydrological value* and properties of each, which are in fact very similar. A few of the more common elements found in the periglacial environment are:

- Active rock glaciers;
- Inactive rock glaciers;
- Talus lobe rock glaciers;
- Protalus rock glaciers;
- Permafrost;[8]
- Solifluction (or ground creep);
- Relict rock glaciers.

All of these are of hydrological importance. The first four (with the suffix *rock gla-cier*) are basically different variations of the same phenomenon. They are bodies of dynamic ice and rock that rest on the surface, of diverse shape and size, and to varying degrees, they move vertically and/or laterally. *Active* rock glaciers are the most dynamic, moving because of the their significant ice content. *Talus* rock glaciers are simply rock glaciers located at the lateral bases of a mountain valley, but don't necessarily flow down the middle of the valley. *Inactive* rock glaciers still have ice in them but are "deflating" in size while ceasing to move and so will look smooth on the surface. They are harder to identify in satellite images to the untrained eye. Inactive rock glaciers are on the road to leaving the periglacial environment (or may have already left). The last of these, *relict* rock glaciers, are the remains of a rock glacier that has passed through the active to the inactive and to the final, end state where there is no longer a permanent ice core. These relict rock glaciers are sometimes referred to as *fossil* rock glaciers.

I have met scientists and many mining company consultants who want to mine glacier and rock glacier areas and who disregard the hydrological importance of relict rock glaciers. I have done so myself in previously published documents, until I began to go deeper into the world of rock glaciers and periglacial features, discovering their importance through site visits and conversations with scientists. Connie Millar (mentioned earlier) helped me understand the role of relict rock glaciers, and even glacier moraines, in providing hydrological services. More about this later.

Each of these types of rock glacier is generally found in permafrost areas. *Permafrost* (as we learned in Chapter 5) is a generic name given to permanently frozen grounds—*perma* (from permanent) and *frost*, suggesting a frozen state. Any humidity within the ground is in ice form. As science would have it, however, not all permafrost is *permanently* frozen, and this anomaly in the end, from an environmental standpoint, is what most interests us, since the unfreezing of frozen ground could release water into the environment, and the cyclical freezing and thawing can not only release water into the environment, but can also provide a recharging effect to sustain, distribute and regenerate water reserves.

As we noted in Chapter 5, scientists classify various types of permafrost.

Permafrost Categorization	
Continuous	frozen 90–100% of the time
Discontinuous	frozen 50–90% of the time
Sporadic	frozen 10–50% of the time
Isolated	frozen 0–10% of the time

Source: French, 2008. P. 94.

Curiously, even the continuous and *permanently* frozen grounds (the 90–100% of the time category) aren't *always* frozen!

The lower fringes of periglacial areas (which correspond to the lower limits of rock glaciers) that thaw and freeze cyclically are generally what we refer to as discontinuous permafrost or sporadic permafrost; that is, these grounds are frozen 10–90% of

Figure 8.9 Creeping grounds of the periglacial environment (permafrost) in the Central Andes, visible as curved, ripple-like waves or crevasses on the surface. While this is a photograph, these grounds are also distinguishable in high-resolution satellite images.

Source: Juan Pablo Milana. GIS: 29 01 30.04 S, 67 50 07.03 W

the time, which is quite a large range! Another important characteristic of permafrost is that the presence of frozen grounds is not a function of humidity. That is, completely dry grounds can also be frozen, simply because they are at a temperature below the freezing point. Another anomaly is that freezing may not occur at 0°C (or 32°F) since saline concentrations in the water may lower the freezing point to somewhere between 0°C and −2°C.

Solifluction or *gelifluction*[9] or *creeping ground* is a layer of permafrost that is actually *moving* on the surface (fig. 8.9). It moves because of a combination of dynamics between the cryogenic processes that create contraction and expansion of the surface layer, the incline (gravity), and the characteristics of the environment. This is ice-rich ground that literally creeps over the surface, creating visible folds on the surface of the Earth as the grounds move forward, which can be recognized, sometimes even in satellite imagery.

Types of Rock Glaciers

We've already mentioned rock glaciers extensively in this chapter. You may have seen a few by now on the Instagram feed #rockglaciers. They are my favorite of the periglacial features. They're also the coolest! (Sorry, the pun is intended.) Rock glaciers are among the most significant *and magnificent* elements of the periglacial environment. They contain colossal amounts of ice and, like exposed white visible surface glaciers, can help us define the limits of periglacial areas. While the lowest area of the mountain where regular glaciers survive denote the lowest point of the *glacial* environment (the 0°C line), the bottom (or lowest point) of rock glaciers denote the lowest point of the *periglacial* environment. I will briefly touch on a phenomenon in the section on inactive rock glaciers later in this chapter, through which that lower limit of rock glaciers may be lower than the tree line.

As explained earlier, rock glaciers are formed by cyclical freezing processes involving the mixing, freezing, and thawing of snow, ice, earth, and rock, emplaced in specific areas of the periglacial environment, which then begin to move downhill in recognizable patterns. They live and thrive in the periglacial environment and deteriorate when they slip too far downslope into warmer areas.

Because rock glaciers are in this *transitional* area between the glaciated areas of a mountain (where ice visibly survives on the surface), and the vegetation line (or the treeline), and because they are right at the place where everything freezes, they have lower fringes that are permanently melting away. Some parts of the rock glacier, however, at higher elevations or deep in its core ice, are permanently frozen. Those parts of the rock glacier at higher elevations, though, are usually just lower than the lowest parts of visible surface glaciers. That means that rock glaciers can survive at lower elevations than normal visible glaciers. This will be very important in future climate scenarios where warming mountain ecosystems can no longer sustain exposed glaciers. Now you may be starting to understand where I was headed with the title of this chapter.

Another characteristic of rock glaciers is that they have different layers of rock on the surface as well as different layers of ice in the interior. The ice nearer to the surface, because it is closer to the changing outside temperature, is experiencing cycles of freezing and thawing, while ice deeper inside the core of the rock glacier may be in a more permanently frozen state. The deeper you go into a rock glacier, the more solid the ice: the more protected it is from the outside ambient temperature, the less likely it is that temporary fluctuations of ambient temperature will melt the ice core. It is precisely this cyclical freezing and thawing process combined with the stable ice core that gave rise in the first place to the rock glacier.

As climate change warms ambient temperature over the long run, however, and as the freezing isotherm range creeps up the slope of a mountain (that is, as the freezing point on the mountain moves higher due to climate change), if the temperatures don't fall again, even the deep core ice of a rock glacier will begin to melt, slower than normal glaciers, but it will still melt.

This *surface versus core ice* difference is key in the evolution and ecological balance of rock glaciers. The surface layer is the *active layer*, since it is freezing, thawing, contracting, expanding, and moving, according to variations in ambient temperature. It's also the layer that releases water. The cooler core, because it is more protected from ambient temperature variations, is more rigidly frozen and thus, more stable. The ice core acts largely as a water reservoir, as well as the principle cooler for the rock glacier in general. In this regard, the rock glacier's ice mass, protected from the ambient temperature, is its own natural refrigerant!

Active/Inactive/Fossil Rock Glaciers

The rock glaciers just described, which are in constant activity and moving, are referred to as active rock glaciers. They are located in frozen terrain, which means they go through freezing and thawing cycles, they are constantly receiving new water and snow, they incorporate new rock, and they move, and therefore, they are considered "active."

As we move downslope, or as the freezing range moves upslope, active rock glaciers can be overcome by warmer temperatures and find that their new location is no longer in the permanently frozen zone, that is, they've moved out of the periglacial environment. When this occurs, they begin to melt, and they do not replenish the ice and water content they need to remain active (that is, moving). In terms of a rock glacier's "activity," this is a degenerative state. The rock glacier loses its dynamic *active* state, which is replaced by a sedentary *inactive* state. Sustained warm temperatures will eventually penetrate the surface and begin to reach the inner core of the rock glacier, slowly melting the ice away and deflating the rock glacier.

As the frost heave action ceases and the rocks stop moving, the crevasses, ridges, and curvatures of the surface will smooth out in time, and the vibrant rock glacier will become progressively calm, much like a deflating ball. Over time, which may be a few decades, or sometimes centuries, the ice content of the rock glacier will melt until all of the core ice is gone. The rock remains, assorted in a similar fashion to the original rock glacier, but with much smoother and rolling contours. It may be covered with vegetation (not in all cases) and will become what is called a *relict rock glacier* (or fossil rock glacier).

This "inactive" phase may be happening halfway down the Atlin Rock Glacier in figure 8.4. Note that about halfway down this rock glacier it crosses through the treeline, and the lower half is completely surrounded by trees and vegetation. While this may be difficult for the untrained eye to see, if you look carefully, in the upper half of the rock glacier, above the treeline, you can make out distinct surface activity, while below it the rock glacier's surface is smoother. While it is impossible to determine from examining a picture, this could mean that above the tree line we have an *active* rock glacier and below it an *inactive* section. The analysis is inconclusive, but some clues are present for us to speculate. You can see to additional rock glaciers, Sourdough Peak in Alaska (fig. 8.10, with similar dynamics to the Atlin rock glacier) and Mercedario Rock Glacier, and active rock glacier in Argentina (fig. 8.11).

In an inactive rock glacier, the dynamics that once made the active rock glacier thrive have ceased. It's important to understand however, that although inactive or relict rock glaciers have lost their mass, weight, and permanent ice, and lack the dynamics necessary to create mass movement, the body of rock that forms the relict rock glaciers can nonetheless still act as temporary or seasonal hydrological reserves that capture and freeze water during winter, and release it slowly over a warmer season. This, like a glacier or active rock glacier, is simply another way for Mother Nature to capture water in high mountain environments and regulate water flow to ecosystems below.

Figure 8.10 Massive rock glacier at Sourdough Peak, Alaska.
Source: NPS Natural Resources. GIS: 61 23 22.20 N, 142 44 27.17 W

Genesis of Rock Glaciers

Curiously, scientists don't all agree on the genesis of rock glaciers, particularly because sometimes they are connected to normal uncovered white glaciers or to debris-covered glaciers—which have some very different characteristics from typical rock glaciers. The presence of rock glaciers in the vicinity of, below, or adjacent to normal uncovered glaciers, as well as their physical connection to some debris-covered glaciers, and the difficulty of fully grasping how these ice bodies actually interrelate to one another, lends itself to unresolved debates about their origin. Are glaciers and rock glaciers two different ice-related resources intermingling simply because they are in contact? Does one (a glacier) become the other (a rock glacier) because of the warming climate? Or are they each unique and independent?

What everyone *can* agree on is that rock glaciers form where the surrounding environment is conducive to their formation, and that the basic ingredients are: snow, rock debris, water and freeze-thaw cycles, as on talus slopes, and a stable very cold (freezing) climate, as well as temperatures that do not permit the survival of exposed ice on the surface. Glaciers and rock glaciers are generally not located in the same environments. Rock glaciers are located where the average yearly temperature is above the freezing point but close to it, where normal exposed glaciers could not survive. There has to be snow with a predominance of cold (freezing) temperatures all year round. There has to be an extraordinary accumulation of snow and rock at a given site. The site should ideally be at the foot of a very slanted slope (without being too slanted), which allows for a continuous and cyclical processes of accumulation and mixing of rock and ice and water, melting, freezing, expansion, and contraction, and there should be continuous entry of new water, ice, and rock into the rock glacier.

Once the accumulation of rock and ice reaches a certain critical weight, gravity starts to play a determinant role in the dynamics of the overall mass. If placed on a slanted slope, the lubrication at the base (produced by meltwater) will create the ideal environment for a massive (but very slow moving) slide to occur. Thanks to the unevenness and coarseness of the surface, the mass will not completely quickly slide away, but instead it will creep slowly downhill. The downslope creep of a rock glacier is not as noticeable as that of a normal uncovered white glacier, which may advance many feet per day, and many hundreds of feet per year. Instead, the rock glacier moves much more slowly, sometimes only a few meters per year, if even that much.

The rolling, flowing, churning movement of the ice and rock on the surface of a rock glacier forms typical patterns in the way the rock is sorted. These are similar to those we described earlier for creeping permafrost grounds. Crevasses, wrinkles, folds, and concentric curves appear as the rock is deposited and sorted on the surface and around the edges. It's very difficult to distinguish these features when you're actually on the rock glacier, but from a birds-eye view, especially with good satellite imagery, they're clearly revealed. Normally rock glaciers display forward and vertical movement of the rock flow. Seen from above, an active rock glacier appears to flow down a mountainside as lava would. In fact, in satellite images, many rock glaciers are confused with past lava flow from dormant or extinct volcanoes, and vice versa.

One view is that rock glaciers *are not* related to glaciers at all, but that they form on their own and have independent dynamics and systems unrelated to normal uncovered glacier processes. In this view, uncovered white glaciers are simply the accumulation of snow, transformed to ice, that may flow downslope, while rock glaciers only derive from a mixture of rock, snow, and the cyclical freezing and thawing processes (the cryogenic process) of the periglacial environment. These would be, then, "cryogenic" rock glaciers.

If this interpretation were correct, it would be great for rock glaciers in a changing and warming climate, since we wouldn't need a glacier in order to have a rock glacier. Rock glaciers could simply be created by the interactions of rocks, water, and ice. Regular glaciers would need a much colder environment to sustain them, which in current climate scenarios presents serious challenges.

Others point to long valley glaciers, which begin as uncovered white glaciers, pass through a debris-covered phase when a thin rock mantle is added, and then become full rock glaciers with a much more significant presence of rock in the interior, nearer to their terminus. In this example, we see an evolutionary explanation of rock glaciers, and these would be referred to by some scientists as "glaciogenic" rock glaciers, that is, deriving from glaciers. Others discard this theory, suggesting that glaciogenic rock glaciers are in fact not rock glaciers but debris-covered glaciers and that the seeming transition of a glacier to a debris-covered glacier to rock glacier are in fact two separate systems.[10] In this last example, the transition is accomplished by the fact that the rock glacier feeds off of the rock being pushed by the upper glacier system, but that it is in the end a different and distinct system.

Whomever we prefer to listen to on rock glacier genesis, distinguishing a debris-covered glacier from a rock glacier may not always be that easy. But there may be some clues. In debris-covered glaciers, the rock cover is thinner than in rock glaciers. There may be just a few centimeters or meters of rock in debris-covered glaciers (see fig. 8.15), while many meters might cover a rock glacier (see examples in figs. 8.11, 8.12, 8.13, 8.14 and 8.16). Also in debris-covered glaciers, the thin mantle sometimes breaks open, revealing ice. This is rare in rock glaciers. In fact, I've never actually seen the inside of a rock glacier while I have seen many ice cores of debris-covered glaciers. Further, debris-covered glaciers tend to form small ponds or lagoons on the glacier surface (called thermokarsts) that seem not to occur on rock glaciers.

Perhaps the real answer is somewhere in between, and certain rock glaciers form simply because of the accumulation of rock and snow in certain areas (such as at the foot of steep slopes) and others do in fact derive from glaciers as these massive ice bodies deteriorate into lower and warmer environments.

In Argentina, when the National Glacier Inventory took place, the new glacier and periglacial environment protection law mandated that scientists register both of these ice bodies in an official inventory. Much debate occurred at the IANIGLA (Argentina's Glacier Institute) over how to distinguish rock glaciers from debris-covered glaciers.

Figure 8.11 An active rock glacier at the head of the Mercedario River (note the river's birth stream) in the Andes Mountains, Mendoza, Argentina.
Source: Juan Pablo Milana. GIS 31 58 30.36 S, 70 10 32.81 W

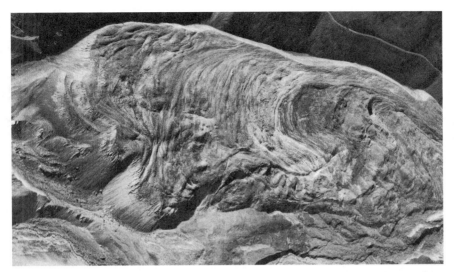

Figure 8.12 Aerial view of a massive rock glacier in Argentina, more than 40 meters (130 feet) thick and 2 kilometers long (1.2 miles).
We can see lines and furrows denoting movement.
Source: Juan Pablo Milana. GIS: 31 57 41.92 S, 70 03 12.02 W

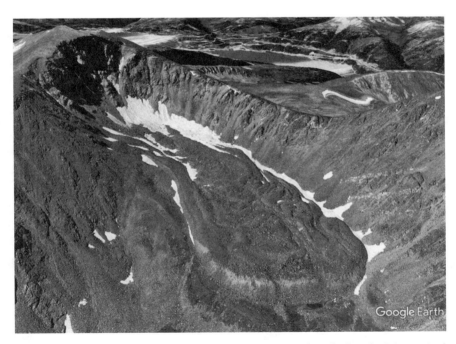

Figure 8.13 This rock glacier near Copper Mountain, Colorado (Rocky Mountains), is 700 meters long (2,300 feet), 300 meters wide (985 feet), and about 30 meters thick (more than 100 feet).
Source: Google Earth. GIS 39 25 28.90 N, 106 06 58.52 W

Figure 8.14 Four rock glacier systems visible on these slopes of the Kaçkars Mountains, Turkey, near the Black Sea.
Source: Google Earth. GIS: 40 49 59.61 N, 41 08 14.65 E

Figure 8.15 Debris-covered glacier in the Chola Pass to Gokyo, in the Everest-Himalayas area. Note the hikers for size reference.
Source: Jorge Garcia-Dihinx. GIS 28 00 22.92 N, 86 51 53.98 E

Figure 8.16 The Calingasta Rock Glacier in San Juan, Argentina, is massive at 3.7 kilometers long (2.3 miles) and 500 meters wide (1,640 feet). If it could be bottled, it could provide all of Argentina's population with over a year's worth of water!
Source: Google Earth. GIS 31 15 17.03 S, 70 10 24.51 W

All of a sudden, this dilemma actually became an important challenge in the implementation of the law. The scientists had a tough time deciding, but in the end, they chose not to. The simpler solution was to create a *new* category when a transition is present. "Debris-covered glacier with rock glacier," . . . and that's how they ended that discussion.

How Big Is a Rock Glacier?

Rock glaciers, like regular white uncovered glaciers, come in all sizes, ranging from a few dozen meters wide and several hundred meters long, to several 100-meter city blocks wide (330 feet) and several kilometers long (several miles). In terms of thickness, they can be anywhere from 10 to over 100 meters thick (33–330 feet). That's taller than a 30-story building! They can be of all sizes, but generally cover areas ranging anywhere from 1 hectare (2.5 acres) or 0.01 km² (about the size of a football field) to much larger rock glaciers that could cover more than 100 football fields![11]

Generally, rock glaciers that slither down contained slope-sided valleys, have the tendency to form into glacier-like tongue-shaped forms, or spatulate (lobate) forms, flowing down the valley floor (see figs. 8.5, 8.6, 8.10, 8.12 and 8.14 for examples). They have a typical slanted front cut sharply at a 30–40 degree angle.

How Much Water Does a Rock Glacier Hold?

First of all, we should say that very little research has been done on rock glaciers and even less on their hydrological contribution to ecosystems. Thankfully that is slowly changing, but we still don't have reliable and widely accepted data about the water content of rock glaciers. We know that they contribute water (or at the very least, retain moisture and ration its distribution over time) because their genesis derives from continuous melt-thaw cycles. We just don't know how much water they contribute because few people have studied them.

Globally, there are few rock glacier specialists. The few we have are generally *geologists* who have taken a liking to the cryosphere. Some geographers have also jumped into the realm of rock and ice. More recently, following conflicts regarding the impacts of mining on glaciers and periglacial environments, hydrologists have also begun studying ice and water discharge from glaciers *and* rock glaciers, as policymakers need to make educated estimates of the natural resource value of these periglacial features. Those that specialize in rock glaciers are sometimes called geocryologists.

Estimates of the water content of rock glaciers vary considerably. Some experts suggest a 50% ice, 50% rock and earth composition. Others suggest more ice if the rock glacier is active, with healthy active rock glaciers containing as much as 80% ice. Inactive rock glaciers may have considerably less than 50% ice at their core, while relict rock glaciers may have already extinguished their ice core completely. The ice content of the Himalayan *debris-covered glacier* depicted in figure 8.15 is much more significant than that of the most active rock glacier.

Let's make a quick and very conservative calculation of the water content of the fairly small rock glacier (Box 8.1)[12] we saw in the photo from the Rocky Mountains shown in figure 8.13.[13]

If we consider much larger rock glaciers, for instance the Calingasta rock glacier in San Juan Province, Argentina,[14] the hydrological relevance of the rock glacier is mind-boggling.

The Calingasta rock glacier is 3.7 kilometers long (2.3 miles) and about 500 meters wide (1,600 feet). It's about 40–80 meters thick (130–260 feet), depending on where you measure it. If we do the same math for the Calingasta rock glacier and also assume 50% ice content, we get:

1,850,000 m^2 of ice

1 m^3 of ice is 1,000 liters of water, but we'll only take 50% of that (to be conservative), so that's 1,850,000 m^3 × 500 liters, or 925,000,000 liters (that's nearly 1 billion liters) for every meter layer of ice. We will calculate conservatively, that the Calingasta is only 40 meters (130 feet) thick (although we know it is thicker), which makes for 37,000,000,000 liters, or 37 billion liters of water contained in the Calingasta rock glacier.

At two liters per person per day of drinking water, that's 82 million liters of water consumed daily by Argentina's total population of 41 million. So, 37 billion divided by 82 million gives every Argentine about 451 days of water, or over a year's worth of drinking water for the *entire* Argentine population. If Denver had a rock glacier like the Calingasta, they'd have daily water consumption for about 50 years.

Box 8.1 Water in a Small Rocky Mountain Rock Glacier @ 39°25'40.67" N 106°06'59.67" W

The surface area as seen from above can be estimated as a rectangle of about 200 × 600 meters.

The surface is = 200 m × 600 m = 120,000 m².

We consider that 1 m² of surface, viewed volumetrically is 1 m³.

There are approximately 1,000 liters per 1 m³

so, if we have 120,000 m² (surface), then we have 120,000 m³ in volume per meter of thickness

120,000 m³ × 1,000 liters per m³ = 120,000,000 liters per meter of glacier thickness

but we will be conservative and assume ONLY 50% water in each m³

(recall that many rock glaciers have up to 80% ice content)

so 50% of 120,000,000 liters per meter of thickness is = 60,000,000 liters per meter of thickness.

If we consider that this rock glacier is about 20 m thick, but that perhaps 10% of the top layer is pure rock cover, we can guess about 18 meters of rock glacier thickness

so, 60,000,000 liters × 18 meters = 1,080,000,000 liters (or 1.08 billion liters) of water by conservative estimates in this small rock glacier!

That's still an unfathomable amount. So, let's take it to a city-level analogy. If we could bottle that water for human consumption, let's say, what a person consumes daily (close to 2 liters per day), how much water would that be?

We'll take the city of Denver as an example. Denver has a population of 1.3 million people. At two liters of water consumed per day per person, the city population consumes 2.6 million liters of drinking water daily.

We had 1,080,000,000 liters available from the rock glacier. Divide the glacier's water by the number of liters consumed daily by the entire population, and you get

1,080,000,000 / 2,600,000 = 415 days of daily water consumption for the city population.

The small rocky mountain glacier could provide drinking water to the entire city of Denver, Colorado (1.3 million people) for well over a year!

Rock Glacier and Periglacial Water Regulation

In the examples just presented, we've unrealistically bottled the rock glacier water for consumption, so that the reader could get a sense of the amount of water contained in the ice mass. But in real life, glaciers and rock glaciers don't get bottled up and distributed, nor do they melt away overnight. In fact, as we explained in Chapter 2 on water supply, glaciers and periglacial areas work very efficiently at providing smaller amounts of water over long periods of time, after the seasonal snow has melted, in a slow but regulated manner to the environment below, making them a renewable water source (so long as the glaciosystems in which they are located remain healthy).

In a place like California, the disappearance of glaciers in the high Sierra Nevada is a tragedy for water supply. Thankfully, for now, California has an extensive (albeit little known and little understood) periglacial environment that continues to feed water to the downstream hydrology. In this sense, rock glaciers and the periglacial environment have already outlived glaciers and continue to provide water, despite climate change—for now.

Concluding Remarks on Rock Glaciers and Periglacial Environments

If before reading this book you had never heard of a rock glacier or the periglacial environment (which is likely for most), you now know of a natural resource that for most of the world's population is a complete mystery. I could stop here and claim a huge victory, I've gotten one more person to learn about rock glaciers and periglacial areas!

What I am most happy about is that you the reader can already identify rock glaciers in a satellite image and will now have a newly acquired appreciation for a fantastic adaptation Mother Nature devised to capture humidity in the winter months for use in warmer and drier seasons.

The swath of land between *glaciers above* and the *treeline below* is an extremely active hydrological fringe of our mountain environments, playing a critical role in the storage and provision of water. It is evident from the examples shown here, that rock glaciers, and periglacial environments more generally, are immense water towers, holding enormous amounts of water in relatively confined environments, ready to release it when it is most needed.

As we consider the impacts of climate change, and the likelihood that many or most glaciers will eventually succumb to a warming environment, we can note that rock glaciers and the periglacial environment will be in a better position to survive. It may also be the case that many glaciers that exist today, could morph into rock glaciers as their environments become too warm to sustain their glacier health. Because their surface rock cover isolates them from the warmer air, rock glaciers can withstand heat better than exposed glaciers. This means that the role rock glaciers and other periglacial features play in water provision will become more and more important if global warming intensifies. That doesn't mean that they are immune to climate change, it just means that they will survive longer than current exposed surface glaciers. How much longer, we don't know. That will depend on how fast the climate changes and the Earth warms.

9

A Race to Save Everything

In the introduction, I recounted the time French glaciologist Bernard Francou and I were trekking up to the Antisana Glacier in Ecuador and reflecting on the type of work we were doing to try to *save glaciers*, or at least to slow their demise. We came up with the idea of "cryoactivism," that is, *activism* to protect the *cryosphere* (the frozen environment). I ended up writing an article for an academic journal defining the term hoping that others might read about it and that it would catch on.[1]

Once you've seen a colossal glacier, you're stricken by its majestic beauty. Learning of their vulnerability, your first thought might be, how can these enormous inventions of Mother Nature be protected if our climate seems to be irreversibly changing? Is it a foregone conclusion that they will all simply disappear? Is it even possible to "save glaciers"?

The answer is yes, but we have to flesh out the implied challenges, put things into perspective, and think about these challenges in pieces.

Glaciers are formed and melt away naturally in periods of tens of thousands of years. It takes about 80,000 years to create an ice age and about 20,000 to melt one away. We are probably near the end of a 20,000-year inter-glacial period, and at some point, the Earth will shift back into glacier-creation mode. Some believe that it was already doing this, but that the industrial revolution and subsequent intense spewing of CO_2 into the atmosphere, and resulting global warming may have effectively interrupted the return of the next glacial age. Whether or not we're ending or beginning a glacial phase in geological history, we shouldn't presume that we can simply sit around and wait for the next ice age to come on its own because eventually Nature will kick into gear, and all will be fine. It won't, necessarily.

In my conversations with Gabrielle Dreyfus of the Institute for Governance and Sustainable Development (IGSD), the ice-cubologist, we discussed precisely this question. Are we about to head back into an ice age? Well, we know that the CO_2 concentrations that are generally present when we *have* slipped into one before (around 280 particles per million or ppm) have been suddenly greatly exceeded due to anthropogenic emissions. We're about at 410 ppm now. Is it possible that we have *missed* the onramp to the next ice age? If so, the question is, when will the next one come along?

What is certain is that in the last two or three hundred years, we have not been helping our planet maintain its ecological balance and if there is indeed a window entry point for an ice age, and if it has anything to do with CO_2 and artificially induced global warming (which it seems reasonable to presume), we should be worrying. If we look at the cycles the Earth has gone through before to move into and out of ice ages, it certainly looks like we should be shifting slowly, very slowly into the next one. But what we have been doing instead is bringing about accelerated de-glaciation. That is, we are increasingly causing glaciers to melt and making our planet warmer and warmer, while also destabilizing the natural order of things that need to be realigned if we are to continue a natural evolutionary ice-age cycle.

Through the highly polluting atmospheric emissions that have come from intense industrialization beginning in the 18th Century and through modern lifestyle choices that we have made and continue to make about energy generation and consumption, we are likely causing an even more extreme acceleration of glacier meltdown than would otherwise have happened naturally and that might have even begun to reverse in the last few centuries. It's impossible to know for sure. We may have been comfortably in a millenarian cycle of glacier and temperature stability before the next ice age launched. Instead, our planet is warming, and warming fast. But have we missed the ice age train?

David Archer, in his book *The Long Thaw*, suggests that we may have actually altered our planetary ecological and climate balance so much that we may *not* be able to naturally shift back into a slow and prolonged glacial freeze (as is normal) for many tens of thousands or even hundreds of thousands of years to come. If CO_2 consumption and emissions continue at the pace we are seeing, it is possible that the onset of the next glaciation will be greatly delayed. Here is an excerpt of his conclusions:

> During the glacier inception from the last interglacial period 120 thousand years ago, CO_2 remained high, at a typical interglacial level of 280 ppm, until after the ice sheets started to grow. If it wasn't a drop in CO_2, that caused the ice sheet to form, the other usual suspect would have to be a change in the Earth's orbit. The intensity of sunlight varies in different rhythms at different latitudes and around the seasons, but the amount of ice on Earth seems to listen particularly well to the intensity of sunlight in the northern hemisphere summer.

> The long lifetime of fossil fuel CO_2 in the atmosphere means that human activity will affect the trigger sunshine value for a long time into the future. The larger the CO_2 release, the greater the shift in the trigger. Natural evolution of climate, from the natural interglacial CO_2 concentration, was a near miss; it was touch and go whether an ice sheet would begin to form or not. With rising CO_2 the trigger moves farther and farther out of reach. It becomes less of a near miss.

> If mankind ultimately burns 2,000 Gton C (this is about the business as usual forecast for the coming century), then it looks as though climate will avoid glaciation in the 50 millennia as well, waiting until the next period of cool summers 130 millennia from now. If the entire coal reserves were used (that is, 5,000 Gton C), then glaciation could be delayed for some 500 millennia, half a million years. The Earth could remain in an interglacial state until the end of not our current period of circular orbit, but the next circular time, 400 millennia from now.

> On the surface of it, this would seem like a good thing. I would not put this forecast forward as an argument in favor of CO_2 emissions, however. The potential dangers of warming are immediate, while the potential next ice age in a natural world was not due for thousands of years. By releasing CO_2 humankind has the capacity to overpower the climate impact of Earth's orbit, taking the reins of the climate system that has operated on Earth for millions of years.[2]

What Archer is positing is that through excessive CO_2 emissions that are not natural to Earth (but rather *anthropogenic*—or human caused) we have somehow *missed the ice age train* for the next glaciation and interrupted the natural glaciation cycle that we are living in. The ice age train has left the station and now we simply must wait around for the next ice age train to come in 100,000 years or so. If we continue to emit even more CO_2, that train may take a different route and not show up for another 500,000 years. In the meanwhile, we could see the climate on our planet turn aggressively warm, with extremely high levels of CO_2, making conditions on Earth unlivable for humans and unsustainable for the ecosystems we are familiar with.

Archer also notes that while we may prefer to avoid a terrible ice age, even if we did hop on the ice age train now, the large cooling changes in climate would happen slowly and not be completed for many thousands of years. Remember, it takes about 80,000 years to create a full-on ice age. This would help prolong our mild and livable climate for many hundreds or thousands of years to come. That would be a far better scenario than a continuously and rapidly warming planet that makes life quickly unsustainable and in some places, unbearable.

If we miss the ice age train, the dangers of current climate change trends, and the adverse impacts caused by a continuously warming climate, could be devastating for human society. Sea level rise, ocean temperature increases, plant and species die off, irregular and unpredictable inclement weather, etc., would become the norm (and maybe already has), putting great strain on our global ecological balance. We are starting to feel those impacts now. In such a scenario (where the train won't be returning for another 100,000 years, or maybe even 500,000 years), the predicament facing us, the human race on Earth, is pretty bleak.

The central problem for us in this tragic scenario is our global climate and how it affects us. That's what's out of sync and threatens our long-term sustainability on this Earth. The Earth will survive after us, just as it survived after the dinosaurs when the planet then became unsustainable for life. In the same way, the Earth will survive humans. The question is really, how much longer do humans want to survive on Earth?

Meltdown is about melting glaciers, which cause (among other things):

- *more* global warming
- *more* sea level rise that destroys coastal communities and ecosystems
- *more* fresh water shortages at key moments when we need water the most
- *more* warming of the oceans
- *more* unstable and violent weather patterns
- *more* destruction of habitat and feeding grounds for flora and fauna
- *more* land shifts that can destroy current ecological stability
- and *more* dangerous environments that can lead to catastrophic incidents like glacier tsunamis (GLOFs).

Melting glaciers aren't the original *cause* of our climate problems, they are a *symptom* of a much greater problem (global warming). But they are also a force, which in turn feeds back into the problem and makes it even worse. Scientists call this a feedback loop intensifying a tendency. Our climate is changing drastically, perhaps beyond repair. Scientists call this moment when things become unreparable, the tipping point.

Once we have reached, and gone beyond the tipping point, we cannot return to the earlier state, we cannot repair the damaged climate.

In terms of *saving glaciers*, we certainly *can* save glaciers, if we fix (and also reverse) climate change. This could mean that we stop warming the planet and therefore stop melting glaciers. One question will remain, however: can we fix what we've already destroyed? That seems unlikely. Nonetheless we do have some options.

There are two ways that I see to approach the issue, and frankly both could be pursued together, they are not at all exclusive:

- A *temporary fix*, as a band aid to help us *adapt* to glacier loss, by preserving, re-cuperating, and maybe even *regenerating* glacier ice so that we can continue to benefit from the glacier services we now enjoy, such as water supply and cold en-vironments, services that we are losing or that in many cases we have already lost.
- A more *permanent fix*, which is a bigger challenge, namely to *mitigate* and re-vert and resolve the broader climatic change problem that we face, which put glaciers in peril in the first place. This involves reducing or eliminating climate change causing pollution and maybe also sequestering the emissions we've al-ready emitted.

Correcting the greater problem, the current global warming, through actions that re-duce pollution that causes global warming would be one way of saving glaciers. It would clearly be the most ideal way. It would require significant changes in lifestyles, actions by industry, laws to tackle climate change, and policies and programs devel-oped by public officials to reduce the emissions causing climate change. We can also consider engineering climate repair, such as devising technology that can rapidly cap-ture CO_2 that has already been emitted, or technology that can reduce warming by making the Earth more reflective (fixing the albedo changes caused by melting ice sheets). This *ecological engineering* may be at a scale that is beyond our capacity to de-ploy, but then again, when humanity started to build the technology that took humans to the moon, we didn't even have the materials needed to ensure that our rocket could get back into the atmosphere without burning up, and we figured that one out.

So what do we do? Three things.

Mitigation, Adaptation, *and . . . Hope?*

There are three approaches to fixing our melting glacier problem and working to ad-dress climate change. Two of them are *proactive* and have to do with how the climate change community approaches climate change challenges, through *adaptation* and through *mitigation*. The third is more fatalist and presents a riskier proposition: *hope*.

Let's take the last first: *hope*.

I include *hope* because I hear a lot of people profess an "it is what it is" attitude. They say we cannot do anything about climate change, that it is not caused by people, and that what is going to happen will happen. In such a scenario, the melting of glaciers is simply normal (though some refuse to believe it is happening) and as such, it could also cease suddenly by natural causes. Yes, this is a possible result of the Earth's natural time and evolution. In this scenario, Mother Nature is in charge, she does what she wants, and if she wants to fix it she will. But if she doesn't, we're toast.

Essentially, in this scenario we're hoping for the onset of a new ice age or at the very least a sudden re-stabilization of our global climate through the Earth's own natural climate adjustments. Sure, a *natural* return of our planet to a phase where we begin a new ice age could be one solution to melting glaciers. We don't know when, or if, this will occur. It could be still thousands of years away. It will eventually happen, as it has for many hundreds of thousands and even millions of years. It could start now. It could start in a few hundred years, or maybe a few tens of hundreds of thousands of years from now. I wouldn't bet my money, however, on this happening in our lifetime, or in that of our children or grandchildren, or for many generations yet to come.

We can also *hope* for another Little Ice Age, like the one that occurred between 1300 and around 1800 in much of the world. That would be a great occurrence. Those were a few centuries during which it suddenly started to snow, a lot, and the world got colder. I wonder if in fact that wasn't the onramp to the return of an ice age. Some suggest that it may have been just that, and that we may have stalled it by inventing the internal combustion engine and discovering how to burn fossil fuels to create energy. A Little Ice Age, if one were to return, would help recharge currently receding glaciers, particularly smaller mountain glaciers that we really depend on for water supply, and maybe put them back on a healthy path. But a Little Ice Age would probably not do much for the collapsing Antarctic glaciers. They're colossal and would take 100 millennia to rebuild. Such a hoped for haphazard chance that Mother Nature would send us a winning lottery ticket only big enough to patch our climate problems for a few centuries seems like too little too late for humanity and its long-term sustainability.

Frankly, I'd rather worry about the fact that we are destroying the interglacial period we are fortunate to be living in and do something within our reach to reverse course. We can't blame Mother Nature for our ill-chosen path and let her deal with our reluctance to leave it. She created a pretty nice environment for us and we're screwing it up. Thankfully, enough people *have* realized our error and many have decided to confront climate change (despite what the climate change skeptics say).

Those who take the "mea culpa" response, recognize anthropogenic climate change, and are willing to take action to stop it, are divided into two camps, *adaptation* and *mitigation*. These two technical words define actions around different approaches, one for us to accustom ourselves to climate change (so it doesn't harm us as much), which is *adaptation*, and another to do something to stop it and fix the problem, which is *mitigation*. In each case there definitely *are* things that we *can* do to help stop the melting of glaciers, and of course, to stop climate change.

Adaptation

The adaptation agenda is comprised of actions to address the impacts of the changing climate, to deal with bad weather, to deal with sea level rise, to deal with crop failure, and to address water supply issues. These are actions that permit us to move on with the changing climate, while reducing its negative impacts on us. The idea is that we *adapt* to climate change. Adaptation is also an agenda for focusing on the emergencies that are most pressing: such as famine, water shortages, environmental refugees, wildfires and flooding. All of these tragic impacts caused by climate change generate some very pressing emergencies, which we must deal with immediately or suffer brutal consequences. We must address the impacts that are already happening so that we can move on and live (in less harmful ways) with climate change.

So how do we *adapt* to climate change and its impact on *glaciers* and the loss of the *benefits* they now provide but no longer will, or at least not to the same extent, because of climate change?

First of all, we *can* work to protect the glaciers that we still have and ensure that they can survive as long as possible. We certainly shouldn't be doing anything to glaciers or their glaciosystems that make them any more vulnerable than they already are. Let's look at some examples of people that are actually working to *protect* glaciers, *conserving* their ice, trying to *slow* their melt and in some crazy cases, there are even people *creating* glaciers from scratch. These would all be examples of *adaptation*. And while some of these examples may seem far-fetched, many are actually working, and as we see what others have done with varying degrees of success, we begin to realize that there is a lot that we can do, to save glaciers, or as a last resort, to make new ones.

We can work to protect, conserve, save, and even help regenerate small glaciers. That is actually already being done in certain parts of the world. Some people are experimenting with glacier fabrication to provide water to downstream ecosystems where that water supply is diminishing due to climate change.

The Glacier Man

In India, Chewang Norphel, or the "Glacier Man" as some call him, is actually fabricating his own glaciers. Chewang, a retired engineer, is in his late seventies now, and was concerned with dwindling water supply in his agricultural lands. Understanding the concept of snow accumulation and ice formation in a glacier (the glaciosystem I described in Chapter 1), he devised a way to dam winter snowfall, to create temporary ice, and have an accumulation of ice at his disposal that would slowly melt and release its water into the downstream environment for crop use. Essentially, he created an artificial "glaciosystem" to make his own glaciers.

These "artificial glaciers" as he calls them, provide millions of liters of water to downstream communities that would otherwise not have had that water for much of the year. With his ingenious artificial glaciers, he has improved crop yield and even introduced new crop cycles during winter months, a time when agricultural activity was impossible before Chewang made his artificial glaciers. He points to the benefits of creating your own glaciers, including recharging groundwater, increasing cash crop farming, fuel, livestock fodder, and incomes, mitigating climate change, and improving soil quality. He is now looking to replicate his model in water-stricken countries like Kyrgyzstan and Kazakhstan.[3]

Figure 9.1 An artificial glacier stupa in India built by local mechanical engineer Sonam Wangchuk.
Source: Sonam Wangchuk

Sonam Wangchuk, a mechanical engineer in the Ladakh region of Northern India is another glacier man who decided to conduct his own experiment to *make* glaciers.[4] Ladakh is a very dry area of the planet that gets very little rainfall and depends on glacier melt for local agriculture, but that glacier melt is running dry lately. Sonam was inspired one day by ice preserved in the shade under a bridge and realized that it could survive longer than expected if he could build a glaciosystem. So he built a two-story stupa or *cone* of ice (fig. 9.1). With an intricate system of water running through tubes, and utilizing elevation and physics to push the water upwards, spraying 60 meters (200 feet) into the air, Wangchuk created a fountain effect with water dripping from the pipe in cold weather condensing to form a two-story ice cone, more than 6 meters (20 feet) tall and containing 150,000 liters of water (nearly 40,000 gallons). Because a cone has limited surface area relative to volume, the ice is less exposed to the sun, and Sonam can conserve ice in the form of a freezing waterfall. Importantly, while other exposed snow/ice melts away by March, Sonam's ice stupa keeps the water through mid-May, precisely when he needs it for irrigation.

Chile's Glacier Maker

Cedomir Marangunic, a Croatian-Chilean geo-cryologist and glaciologist working in the Andes Mountains, has also spent a considerable amount of time thinking about how to design glaciers and even create the more enigmatic subsurface *rock glaciers* (see Chapter 8 for more on rock glaciers). His work on creating glaciers was central to the debate about mining impacts to glaciers in Chile and Argentina. Concerned about the destruction of glaciers by mining operations, Cedo (pronounced *chedo*), as his friends call him, began to think about how to conserve ice in these high mountain environments already at risk due to climate change as well as intruding industrial activity (mining).

One option could be to move glacier ice from one area to another. In other words, if a certain area were not conducive to ice sustainability (in this case because mining companies were dynamiting and removing glacier ice to get at minerals), large ice blocks could be transported to artificial glaciosystems (areas that are conducive to ice survival). This may seem like a radical idea, but in fact, it is a millenarian practice of communities in Pakistan, for example. Pakistani tribes "grafted" pieces of glacier ice, as you would graft a branch of a tree to create a new tree, by taking chunks of ice in winter to caves at higher grounds where it was cooler in summer months, and put the ice into contact with other ice and elements (such as water, saw dust, wheat husk, charcoal and salt) which are said to help the ice grow in size. This was a way to not only conserve ice to be able to use it year round but also to "grow" glaciers. These locals were in fact recreating one of the most important functions of glaciers, which is the regulation of water basins by producing slower melt flow into the downstream environment.[5]

But let's get back to our crazy, mad, experimenting glacier scientist, Cedo in Chile, who was challenging himself to create glacier reserves in the Central Andes, where they are extremely vulnerable. Cedo experimented with lots of ideas, including placing chicken wire or a wooden wall structure in high snowfall and wind-intense areas so that when snow fell in the wintertime, they would cause a flowing wind loop pattern, over the artificial wall, and a subsequent snow drop and accumulation of snow just over the wall, adjacent to the structure. The idea was to continuously drop snow onto a single location, more so than would normally fall there, to create an intense snow accumulation area of what would become perennial ice that could survive longer than the snow fallen on the ground nearby.

The chicken wire experiment was similar in nature to the wooden structures depicted in figure 9.2. In the same way that more snow accumulates in certain parts of your back yard as the wind carries it in certain directions (and not in others), these experiments attempted to mimic the particularities of natural glaciosystems and foster the re-accumulation of snow and ice where it is otherwise dwindling.

Finally, Cedo also experimented with covering snow and ice with rock debris, to recreate a rock glacier (or debris-covered glacier) environment, another way to attempt to conserve ice while the climate is warming. Hired by a mining company to attempt to build a functioning rock glacier, he moved 30,000 tons of ice from one mountain location to a spot that would likely be conducive to rock glacier formation and conservation. He placed a thin film of rock cover on the new glacier that would help conserve the ice, and then waited for the warmer weather to measure results. The results of the artificial rock glacier were significant. Whereas the natural glacier where the ice had come from was receding at about 15 centimeters (6 inches) per year, the artificial, debris-covered glacier receded only about 3 centimeters (1.2 inches), proof that glacier engineering can effectively "save the glaciers."

In yet another crazy experiment many years ago, during an unusually severe drought, when downstream communities needed *more* water from glaciers above, Cedo experimented by flying a small airplane over a glacier to spray-paint it black to

Figure 9.2 A wooden structure placed strategically over a snowfield during the summer that utilizes wind flow patterns during snowfall to foster the accumulation of ice just over the wall. A researcher in the picture evaluates ice accumulation.
Source: Geoestudios

temporarily increase glacier melt and provide local streams with more water that year. Like I said, pretty crazy stuff, but interesting!

A Glacier and Sawdust Experiment in the Peruvian Andes

My good friend Benjamin Morales Arnao, of Huaraz, Peru (the town that was destroyed in 1941 by a glacier tsunami), is a globally recognized glaciologist who has spent the better part of his life confronting the risks of unstable glaciers in high mountain environments below dangerously precarious glaciers and glacier lakes. Benjamin is the director of Peru's new INAIGEM (The National Glacier and Mountain Ecosystem Research Institute of Peru) which several cryoactivists, including Benjamin, helped create. A victim himself of glacier instability, he has designed systems to drain many of Peru's extremely dangerous glacier lakes that, due to collapsing glaciers above, have caused deadly glacier tsunamis.

Benjamin is also concerned about accelerating glacier melt in his native country and the impacts this will have on water supply for Andean communities and ecosystems. He has experimented with placing sawdust on seasonal snow cover that is slowly

Figure 9.3 Sawdust experiment on glaciers.
An experiment in Peru utilizing sawdust to cover snowfall surfaces showed a remarkable benefit in just 3 months, conserving up to 3 meters (10 feet) of snow/ice as compared to uncovered areas. Note the woman standing to the right. The uncovered area of the glacier is at her feet, while the protective cover of sawdust conserved a substantial amount of ice.
Source: Benjamin Morales Arnao

turning into ice. This snow generally completely melts away by the end of springtime. In his experiment, Benjamin used many hundreds of bags of sawdust to cover and isolate the ice from the burning sun and was able to conserve over 3 meters (10 feet) of snow, while the rest of the ice around the control area melted away (fig. 9.3).

Utilizing sawdust as a conservation agent is novel for glaciers, but it's not a new idea for conserving ice. Our grandparents (at least mine), before the widespread arrival of electric refrigerators, utilized sawdust to conserve ice. When the iceman came around to deliver large ice blocks a few times a month (just like Kristoff in the Disney movie *Frozen!*), sawdust was the element of choice (hemp bags were also used) to cover the ice and help conserve it so that homes could have constant refrigeration. A large ice block could survive more than a week if kept in a cool location (usually the cellar) and in a large wooden chest with the insides covered in aluminum or metal with good insulation in between. Those were the old refrigerators before our time. Benjamin in Peru probably had his own ice chest in the cellar when growing up, and from that memory derived this concept in the high mountains of Peru to experiment with successful glacier conservation, another example showing that "saving the glaciers" is definitely possible, we just need to think wisely, *and big*.

Painting Mountains

Also in Peru, Eduardo Gold, a self-trained glaciologist, thought about utilizing the albedo effect that we talked about in Chapter 4. Eduardo received a grant from the World Bank to protect cool areas of the Chalon Sombrero peak at 4,756 meters above sea level (15,600 feet). His project utilized a homemade, environmentally friendly concoction to paint rocks white. The opposite of Cedo in Chile who painted the mountain black to melt glaciers and extract more meltwater, Eduardo was painting the mountain white to increase reflectivity, and by doing so, he was able to show that white rocks indeed keep the mountain cooler than darker rocks.

Blankets on Ski Slopes

In the European Alps, as we have mentioned already, glacier retreat is posing a serious concern for ski resort operators, since the annual winter snow fall and snow cover is generally decreasing. There will come a time when ski resorts will simply be out of snow for skiing. This will mean a great reduction in the economic productivity of these areas, including of restaurants, hotels, shops, transportation, etc., as fewer and fewer patrons come to the region to ski. One solution we all know about is the use of artificial snow makers that are placed strategically on ski slopes to generate snow where too little snow falls. However the costs of snow fabrication can be steep, which is why at Germany's Zugpritze and France's Val Thorens ski resort operators have come up with another idea. While for most people blankets help keep you warm, for glaciers, blankets, if they are white, can have a great cooling effect. Ski resort owners in this part of the world have been rolling out white tarps to cover snow-covered mountains to help protect and conserve snow for continued use and its longer survival.[6]

Tiny Glass Balls for the Arctic

Marisa Endicott offers a fantastic story of human ingenuity employed to save Arctic ice.[7] A group of scientists are experimenting with some unlikely inventions to stop the melting of the sea ice. It's costly, it's crazy, but it just may work they say. This group of researchers found that millions of tiny spheres (they are more like very small grains of sand) spread in a layer across swaths of Arctic ice reflect sunlight and help keep the ice frozen. The idea belongs to an inventor and engineer named Leslie Field, founder of Ice911 (now called the Arctic Ice Project), a nonprofit founded to promote arctic ice restoration. Field, like many of us, grew frustrated listening to more and more news about the melting glaciers and sea ice. In 2006, she saw Al Gore's climate documentary *An Inconvenient Truth* (many of us watched this) and felt she just had to do something. She calls her idea "embarrassingly simple." Her answer was a bunch of hollow glass spheres made of silica, that reflect light and help keep the sea ice cold.

All of these experiments are examples of ways to *adapt* to climate change and explore ways to "save the glaciers." Obviously, with each of these examples, *size and magnitude* is a significant factor. They are also cases that address glacier melt in mountain environments (except for the Arctic sea ice experiment). None of these cases help us reproduce or save the colossal glaciers of Alaska, or Patagonia, and much less the deteriorating glaciers of Antarctica.

For the most part, reproducing these experiments at grand scale seems to be an impossible Herculean task, but the examples *do* show us that there are ways to reproduce glaciosystems and the glacier-creating dynamics of Mother Nature in places where even the smallest glaciers play a fundamental role in ecosystems and human survival.

These examples may end up being the only choice for some mountain communities who would otherwise run out of water completely when glaciers recede beyond critical points.

Part of the challenge in saving glaciers is the need for education, realization, attitude, and action. We need to *learn* more about how our planet functions. It's pretty crazy that, given the enormous importance of glaciers to our planetary and local ecosystems, we know so little about them.

I've mentioned before that only 2% of the world's water is freshwater, and 75% of that water is in glaciers. And yet, until very recently, not a single country in the world had a single law aimed at protecting glaciers. In fact, practically not a single country even *mentioned* the word glacier in their laws. Given how conscious we've become over the past several decades about the environment and the importance of conserving natural resources, I find it mind-boggling that we hadn't (until very recently) taken steps to legally protect most of our fresh water. That seems extremely odd, doesn't it?

It is important to *learn* about glaciers and how they interrelate with our lives and with the Earth's ecosystems. They evidently play a much more important role than we have attributed to them in the past. We need to learn about why they are important, why and how they are vulnerable, and what we can do to save them.

Once we've learned about their importance, it is critical to take a proactive stance and act to protect them, just as we are now acting as a society to address climate change. I recently received a beautiful book by M Jackson called *The Secret Lives of Glaciers*. Jackson is a geographer, geologist, and glaciologist living in Eugene, Oregon, and writes about the relationship between glaciers ... *and people*. As she says in her recent TED Talk, science has generally *not* focused on that relationship. Glaciers are usually a topic of science and about science, discussed between scientists but not for people and society. The relationship between glaciers and people is rarely explored, but may be at the heart of ways to address their vulnerability.[8]

Establishing an effective glacier protection framework means educating ourselves and our children about the importance and vulnerability of glaciers. It also means developing ideas, initiatives, and programs to protect them. The experiments described here show we *can* create environments that are conducive to glacier creation and glacier conservation. Glaciers can be protected so as to prolong their existence, their health, and their sustainability. If we realize that we are losing our glaciers, fully value their importance to our global climate, and learn more about how they provide fundamental security to our lives, we might be driven to act to more effectively to address climate change once and for all.

Replacing Glaciers

In Chapter 2 on the rising seas, I mentioned the work of Thai landscape architect, Kotchakorn Voraakhom, who designed a city park in the heart of Bangkok, Thailand. Kotchakorn was seeking to replace the role of the seasonal rains in her city because the urban jungle created by humans no longer assisted with water absorption/retention. Instead of being absorbed and stored, water in the urban environment repeatedly floods the streets and other infrastructure. To address this, Kotchakorn designed a massive park in the heart of the city and on an incline, to capture water, let it flow downhill, and store it so that it would be available for the dry season to water the vegetation. Sound familiar? She begins her explanation of this ambitious urban project by saying that the water pattern problems derive from the north. Clearly, these are problems with glacier melt flows from the Himalaya that have impacted Asian hydrology.[9]

Glacier Laws

In my previous book, *Glaciers: The Politics of Ice*, I devoted a number of chapters (the odd chapters) to telling the story of Argentina's Glacier Protection Law and how we got that law passed in 2008, only to have it vetoed immediately because of influential mining companies that wanted to dynamite glaciers to get at gold. The law came back in 2010 and remains in place to this day. In fact we celebrated its 10-year anniversary just a few weeks before I sent this manuscript off to the publisher. What is still incredible to me, however, is that it remains the *only* law in the world that was created specifically to protect glaciers. Since getting the law passed we have tried advancing similar efforts in California, Chile, Peru, and Kyrgyzstan, but none have come to fruition.

Laws are important because they help us set limits on human behavior that can harm the greater public good. Not too many people think about laws and their origins, but simply accept them as rules that they must abide by.

Laws are usually grounded in the evolution of the customary behavior of societies. Over time, a certain societal behavior is deemed detrimental to the greater good, and so a law appears to curb that behavior, for the betterment of society. An example is cell phones and texting and driving. Twenty years ago, no one had a cell phone. There was no law regarding texting and driving, or even phone use and driving. Then cell phones arrived, and people started using their phones while driving. Then texting became a thing, and all of a sudden people were texting and driving, causing accidents that killed pedestrians and other drivers. Once this problem reached a level where it became a widespread societal problem, the first laws appeared to prohibit the use of cell phones and particularly texting and driving, all for the good of society.

This evolution from the appearance of a problem, to the perception of a problem, to recognizing the impacts of a problem, to regulations to address the problem, followed by actions to resolve it, is how laws develop and how a society regulates itself in order to promote safer, more harmonious and more sustainable cohabitation.

Glaciers, and climate change more broadly, provide a similar example. We have begun recently (in the past several decades) to identify climate change (and glacier melt) and its relevance and importance to our way of life. We have also identified the sources of the problem (industry, cars, utilizing excessive fossil fuels, etc.) and now we are beginning to regulate our society in order to contain the problem and ensure a more harmonious future for ourselves.

Given that melting glaciers are one of the most visible symptoms of climate change, and that glaciers contain most (75%) of the planet's freshwater, it seems wrong that we don't have laws on the books to address and reduce glacier vulnerability.

In my previous book, and to some extent also here in *Meltdown*, I have laid out some of the reasons why we have no laws to protect glaciers anywhere (except for Argentina). In a nutshell, it has to do with society's lack of awareness of the importance of glaciers to our ecosystems and the fact that glaciers are so far away from us, in places that we never visit. It is also due to the fact that we simply don't know exactly how we are impacting glaciers and what to do about it. That is changing thankfully. *Meltdown* hopefully contributes to raising this much-needed awareness.

First and foremost, we need glacier laws. A state like California, for example, that depends entirely on its glacierized Sierra Nevada Mountains to provide water to its very large population and to its enormous agricultural sector, should be a leader in glacier protection laws. It is not. In fact, Californians mostly ignore the fact that there are nearly 1,000 rock glaciers in the Sierra Nevada Mountains. California's exposed white surface glaciers are already nearly extinct, and yet that has not moved the state to mobilize legislation to conserve ice. Rock glaciers and permanently frozen grounds rich in hydrology will be critical feeders into their river systems and yet, they are largely unrecognized by the state's very extensive environmental legal framework. I am not saying that a law to protect permafrost will from one day to the next instantly preserve these cryospheric resources in the Sierra Nevada. A law has to do with much more than merely having a formal legal protection mechanism. A law, the development of the law, its approval, and then its implementation is a learning process through which a society becomes aware of and embraces a social objective.

Many countries, many states, and many provinces of countries around the world, depend on glacier melt to survive and yet none have glacier laws, and only a handful have initiatives and programs in place to protect their glaciers.

Basically, a glacier protection law (and one that also includes periglacial environment/permafrost protection) should:

1. Affirm the importance of glaciers for all of the reasons we have learned about in this book: water storage, water provision, water basin regulation, human agricultural-industrial use, sea level stability, reflectivity effect, ocean temperature and composition stability, disaster risk management, methane gas containment, scientific value, tourism, ecosystems sustainability, etc., etc.
2. Affirm the importance of glaciers to the public interest
3. Affirm the importance of glaciers for both the local and the global environment as well as climate sustainability
4. Protect the glaciosystems that support the integrity of glaciers and allow them to thrive
5. Promote the study and research of glaciers and their glaciosystems
6. Promote glacier education for greater society
7. Identify the presence of glaciers in a given territory (promote glacier inventories)
8. Establish the value of glaciers and their need of protection
9. Establish prohibitions on activities that could damage or destroy glaciers
10. Establish governmental authority, monitoring programs, and enforcement mechanisms to protect glaciers

The functions of a good strong glacier protection law are multifold.

On the one hand, a good glacier protection law promotes awareness of the importance and significance, both local and global, of glacier resources. This will help society to better understand their place and role in the ecosystem and how the ecosystem functions. It also sets society and individuals on a proper course for orienting human behavior to protect and conserve glaciers and to promote glacier protection.

On the other hand, a good glacier protection law also helps ensure that existing glacier resources (that are likely very vulnerable) are fully protected and that nothing society does (such as industrial activity) places those resources at risk. We have seen the dreadful consequences of uncontrolled mining in glacierized regions of the Central Andes (mostly in Chile and Argentina). If you are interested in learning more about this, my previous book, *Glaciers: The Politics of Ice*, is devoted to that topic.[10] The point of this is to ensure that activities that take place near glaciers are not harming them. A strong glacier protection law will also likely encourage policymakers to promote global policies to address the issues that affect glacier health locally, like good international climate change agreements and their subsequent policies and programs.

A strong glacier protection law also provides key signals for specific societal actors—science, researchers, academics, and industry—to plan their activities properly so as to ensure glacier protection at a local and global level. Mining companies will not undertake mining projects in glacier terrain if a law prohibits it. This has been proven in Argentina, where planned controversial mining projects in glacier areas, valued at more than $25 billion USD, have been completely stalled or canceled thanks to a very clear glacier protection law that prohibits mining in glacier and periglacial environments. Meanwhile, global policies and agreements will prioritize glacier protection and promote measures to curb the human activities that destroy glaciers.

A good glacier law, and the mandate and obligation to study and research glaciers, also gives academics a backdrop on which to plan out academic research, studies, experiments, and other activity that could be geared to glacier protection, conservation, experimental creation, etc. Just as our friend Cedo in Chile is carrying out various initiatives to promote glacier conservation and even glacier creation, with a strong glacier protection law in place that fosters this sort of initiative, we are bound to see similar initiatives move forward in other areas.

We need more studies, for instance, on the important contributions of periglacial features like rock glaciers to ecosystems. Strong glacier protection laws with academic relevance could promote such studies, which would in turn lead to better national, regional and international policies to protect, conserve, better utilize, and promote more efficient use of periglacial hydrology.

Glacier laws can also be contagious. Since we helped get Argentina's Glacier Protection Law passed in 2010, several countries, including Chile, Peru, Kyrgyzstan, and others, have been considering passing their own glacier protection laws. Actors in some of these countries have reached out to seek assistance and guidance. Peru created its national glacier institute to study and protect glaciers. Legislative initiatives are underway in these countries to achieve glacier protection legislation. I should say however, as in Argentina when we first got the glacier law passed in 2008, the mining lobby in Chile and in Kyrgyzstan have successfully blocked the passage of their first glacier laws. Luckily in Argentina, we were able to get around a presidential veto that was specifically issued to ensure mining companies would not be hindered by a glacier protection law, and the law came back to stay in 2010.

In sum, if you are in a position to submit a bill to your local or federal legislative body to protect glaciers, go for it! So far, the Argentine National Glacier Act[11] and subsequent provincial (subnational level) versions of this act are the only glacier laws existing today anywhere in the world. They can serve as models for you! We need more glacier laws!

Mitigation: The 10-Year Sprint

Mitigation is the real issue, the big white elephant in the room that we are failing to address. Climate change is real and it's getting out of control. We can sit and debate whether we think people caused it or whether it is a natural phenomenon. The science is definitive on this point already, yet denial persists. Regardless, the fact is that at this point we need to take action to reduce the emissions of climate pollutants that are affecting our atmosphere and warming our planet. It is possible to reduce emissions and slow warming. We're not sure if we can reverse our deteriorated climate condition or if we will be able to restore the global environment and the global climate to pre-industrial conditions, or if we will be able to restore melted, deteriorated, or vanished glaciers, but science does tell us that there are many things we *can* do to slow and to stop climate change. In climate policy speak, this is what we call *mitigation*.

In terms of our glaciers it's not about making new glaciers, or slowing their melt, or creatively taking action to store ice in places where we can use it later. It's not about putting tarps on ski slopes so that we can ski during winters when we don't have enough snow fall. These are all examples of adaptation.

Climate mitigation is action that we can take as a local, national, and global society to reduce climate-polluting emissions (greenhouse gases) to actually stop the warming of the planet. Most people now realize that the CO_2 concentrations in our atmosphere are too high. We have also learned over the past decade that we need to make a strong effort to wean ourselves off of fossil fuels. It's no longer a matter of *if* we do this, but rather a matter of *when*, and the sooner the better. We need to rethink our society, our energy grid, our forms of production, and our industry, and reinvent them in ways that use less fossil fuels and more renewable clean energy. If our current energy mix is 80/20 (fossil fuels/renewables) we need to strive to make it 20/80 in the next decade or two. Time is short. By 2050, and preferably sooner, we need to have already created not only a carbon neutral, but a fossil fuel free, zero-carbon emission economy. It is possible.

Climate mitigation is a series of actions that local, state, and national governments can and must take, they are actions everyday people and businesses can and must take, and they are actions that industry can and must take, to change our climate course. For the past several years I have been working with numerous colleagues promoting such a course that will actually do more to fix climate change than simply efforts to remove CO_2 from the atmosphere. Much more.

We've all heard of the Paris Agreement,[12] reached under a recurring yearly global climate meeting referred to as the United Nations Framework Convention on Climate Change (UNFCCC).[13] It's called the Paris Agreement because the text was negotiated and agreed to by 196 countries in Paris, France, in late 2015. It was formally opened for country signatures on April 22, 2016, Earth Day. It is 25 pages long, with big type. Maybe you should read it. Most people haven't.

The most important thing that the Paris Agreement says is that we need to make sure global warming does not surpass an extra 2°C compared to pre-industrial levels, and that preferably we should try to keep it to 1.5°C because this would really reduce climate risks. Here's the actual text, from Article 1 paragraph (a):

> This Agreement, in enhancing the implementation of the Convention, including its objective, aims to strengthen the global response to the threat of climate change, in the context of sustainable development and efforts to eradicate poverty, including by:
>
> (a) Holding the increase in the global average temperature to well below 2°C above pre-industrial levels and pursuing efforts to limit the temperature increase to 1.5°C above pre-industrial levels, recognizing that this would significantly reduce the risks and impacts of climate change.

It also says a few more things, some of which I summarize here:

- we need to reach global peak emissions soon (Art. 4)
- we need to make rapid reductions after the peak (Art. 4)
- we need to follow the best available science (Art. 4)
- countries need to tell everyone what they will reduce and be ambitious (Art. 4)
- countries should preserve carbon sinks (Art. 5)
- countries should conserve and promote forestation (Art. 5)
- countries should promote climate adaptation (climate resiliency) (Art. 7)
- countries should ensure financing needs for climate actions (Art. 9)
- countries should work to promote technological solutions (Art. 10)
- countries should work to build the capacity of actions to address climate (Art. 11)
- countries should work to promote education and participation on climate (Art. 12)

These are very reasonable *asks* for all countries and *all* of us to work together to address climate change. This list is a basic action plan to tackle climate change, one that everyone should be working towards. However, there are forces that work against these objectives, including inertia, old habits, stupidity, tolerance of bad ideas, resistance to change, fear, laziness, unwillingness to do the right thing, greed, profitable (but contaminating) business arrangements that we don't want to give up, and of course, lobbyists who convince our political leaders to do the wrong thing, like the false idea that promoting the use of natural gas is a good idea because it's clean or that it's a transition fuel—neither are true. The reality is that if we continue on the path that we are on, our climate will worsen, our atmosphere will heat up, our seas will rise, our storms will become more violent, our forest fires will be more severe, our droughts will last longer, our cities will get hotter, and flooding will be more and more deadly.

Most of our conversations and actions for the past several years have been around the need to reduce CO_2 emissions in the air. This is indeed true. As a society we need to quickly stop using fossil fuels, because excessive CO_2 emissions derived from the burning of fossil fuels are overrunning the air. They're being trapped in our atmosphere and choking us and our environment. Essentially we have to get rid of the internal combustion engine that runs on gasoline, which would greatly reduce and eventually eliminate the excessive CO_2 emissions that have initiated global warming.

But even if we did that overnight (which is impossible), it will take years, centuries actually, to reduce the levels of CO_2 sufficiently to see global climate benefits. Although we might cut emissions drastically in the short term, only about 50% of CO_2 is indeed removed from the atmosphere. Between 20 and 40% remains for thousands of years. Additionally, much of the warming caused by CO_2 emissions has already been captured in the oceans, where it is held and will be released over hundreds of years. Given what we have seen with melting glaciers and accelerated global warming, clearly, waiting around centuries for CO_2 emissions reduction benefits, would simply not be enough.

We need a Plan B.

Think of CO_2 removal from our atmosphere as a *marathon* that we all need to run. It will take a while, many decades, even centuries, but eventually we *will* reach the finish line. We will completely eliminate the internal combustion engine (soon) and move entirely into zero-carbon energy. It's already starting to happen in many countries. California recently banned the production of vehicles using internal combustion engines by 2035 and trucks by 2045. The federal government in the United States also has set forth executive policy aiming to electrify transportation. We may not see the full transition into electric vehicles in our lifetime, but future generations will see this drastic change and we all need to start taking the steps to make it happen as quickly as possible.

In the meantime, however, we need to run another race, a 10-year sprint, during which we *will* see significant and immediate benefits. This 10-year sprint is absolutely critical to make the short-term climate gains we need to avoid falling off of the climate cliff. If we do not run this 10-year sprint, game over. It will be too late.

Don't get me wrong, we still need to run the CO_2 reduction marathon, *in parallel*, but the 10-year sprint is critical, it's a life or death situation, if we want to avoid catastrophic tipping points beyond which there will be no return.

So what is this 10-year sprint? It's immediately reducing what are called, *super pollutants*, or as scientists call them, *short-lived climate pollutants*. If we reduce these super pollutants in the next year, or few years, we will see big gains. If we reduce them aggressively over the next decade, beginning now, we will see enormous gains. Reducing super pollutants will improve our air quality, will be good for our climate, and will buy us time so that we do not reach tragic climate tipping points. Oh yes, and it's good for glaciers!

Let me explain.

These super pollutants are many times worse than CO_2 for the climate. The good thing is that getting rid of them is fairly easy. It can be done quickly and also has immediate benefits that we can see right away.

I'll give you one example everyone witnessed during the onset of the COVID-19 pandemic. Within weeks after most of the world stopped everything, we saw phenomenal improvements to air quality. Suddenly, people hundreds of miles away from the Himalayas could see the snow-capped mountains for the first time ever. That's because many industries and vehicles that emit *black carbon* were simply turned off. Black carbon (or soot) is one of these super pollutants that soils our air (it's also bad for our health), and as we explained in Chapter 4 on albedo, it's terrible for glacier health.

What are the super pollutants that we must focus on (the so called, *short-lived climate pollutants*)? They are:

- methane
- black carbon
- hydrofluorocarbons (HFCs) (which are basically dirty refrigerants)
- tropospheric ozone

Eliminating just these four super pollutants could hold back global warming by a bit more than 0.6°C by 2050. That sounds like a small amount, but in fact, it's more than a third of the 1.5°C that the Paris Agreement has called for. Comparatively speaking, if we held CO_2 emissions to 440 parts per million (ppm) by 2050 and to the slightly lower level of 420 ppm by 2100 (we are at 410 now), we would only avoid 0.1°C of heating by 2050. Reducing super pollutants gets us six times that amount by 2050.

The great thing about focusing on reducing these super pollutants is that there are immediate strategies (that industries are already willing to go along with) within our reach, that we can put into place now, which will have enormous benefits for society, for human health, for the climate, and ... yes, for glaciers. For the businesses and industries that must also collaborate, there are also many win-win strategies, through which phasing out or capturing and reusing super pollutants actually saves money.

Let's go back to my good friends at the Institute for Governance and Sustainable Development (IGSD), run by Durwood Zaelke, who is perhaps one of the most knowledgeable persons I know on solving our climate problem, and particularly on super pollutants.

Durwood has been fighting a global battle to bring scientific experts on climate change together with politicians and those working on environmental policy to find ways to tackle *the most urgent climate problems first*, and to find *the best climate strategies* to do *the most good first*, now, in the next weeks, months, years, and decades, before our climate catastrophe gets away from us and becomes unresolvable.

IGSD published a special report on super pollutants,[14] revealing the science of the short-lived climate pollutants, the benefits of phasing them out, and the strategies available for immediately beginning to fight global warming—effectively in time to avoid catastrophe. Durwood has been leading a team of advocates to work this strategy through the world's leading global climate agencies, like the UNFCCC, the Montreal Protocol, the UN's Climate and Clean Air Coalition (CCAC), and through actions by national and state governments.

Super Pollutant Science, Benefits, and Phaseout Strategies

CO_2 emissions are responsible for about two thirds of global warming.[15] In other words, CO_2 emissions are responsible for about two thirds of the increase in planetary temperature. And while fast and aggressive CO_2 mitigation is essential to combat climate change, it is not enough. Fast and aggressive mitigation is also needed for other non-CO_2 emissions that are responsible for 40–45% of that net increase in planetary heat. These emissions include: black carbon, tropospheric ozone, methane, and hydrofluorocarbons (HFCs). Because these pollutants survive for relatively short times in the atmosphere (from days to a few years or decades—unlike CO_2 which remains for centuries), they are considered "short-lived climate pollutants." The super pollutants, especially black carbon and tropospheric ozone, can also be very harmful to people and environments in the areas where they are emitted.

Methane is a powerful greenhouse gas that is many more times harmful than CO_2 for the climate. A full 60% of methane emissions are due to human activity. Black carbon (a component of soot) emitted for example by the dirty combustion of engines, the burning of wood, or from flaring in oil and gas production, is terrible for human health while also warming the atmosphere because its dark particles absorb light and melt glaciers when deposited on glacier surfaces, as we have seen in Chapter 4. Hydrofluorocarbon (or HFCs—commonly found in refrigerators or AC systems) are human-made chemicals that can be many thousands of times as harmful to the climate as CO_2. Tropospheric ozone (a key component of smog) is a major air and climate pollutant, bad for people and very bad for the climate.

So let's consider the benefits that would follow if we act quickly to reduce these super pollutants (the list is impressive):

- Cut 90% of the warming by super pollutants projected within a decade
- Avoid 0.6°C of global warming by 2050
- Avoid 1°C of warming in the Arctic by 2050
- Avoid 0.84°C of warming in the Arctic by 2070
- Cut the current rate of global warming by 50%
- Cut the current rate of warming in the Arctic by 66%
- Cut the warming over the Himalayas by 50%
- Avoid temperature rise of 1.5°C by the year 2100
- Avoid permafrost thaw in the Arctic and in other regions
- Help stabilize regional climate systems
- Reduce heat waves
- Reduce droughts
- Reduce forest fires
- Reduce floods
- Reduce hurricanes in middle latitudes
- Slow the shift of the monsoons
- Slow the expansion of global desertification
- Slow the increase in cyclones in the tropics
- Slow the melting of Arctic sea ice
- Slow the melting of glaciers

- Cut the rate of sea level rise by 25% and cumulative sea level rise by more than 20%
- Slow the pace of other climate impacts
- Provide critical time for us to adapt to unavoidable climate impacts

Those were climate benefits of reducing super pollutants; here are some people benefits:

- Save millions of lives a year by preventing premature deaths from air pollution
- Significantly reduce climate related illnesses, such as asthma
- Improve food security by increasing agricultural productivity
- Expand access to sustainable energy for billions of people
- Protect infrastructure and create good green jobs
- Provide low-lying states at risk from sea-level rise more time to adapt

Not addressing super pollutants and not phasing them out immediately is not an option if we are to keep to the maximum warming targets the world established in the Paris Agreement. If, for example, we delay reducing super pollutants until 2030 it will become almost impossible to keep warming to 2°C by the end of the century. A combined strategy, however, of aggressive CO_2 and immediate super pollutant mitigation can avoid 0.6°C of warming by 2050 and 2.6°C by 2100.

When it comes to methane gas and black carbon, our lifestyle choices are critical if we are to reduce emissions of these pollutants and slow climate change now. First of all, it is absolutely essential to reduce our dependency on fossil fuels and change the technology that we purchase so that we can reduce these emissions. You will hear ad nauseam commercials (essentially lobbying) from the oil companies about how *natural gas* has made us energy independent and that natural gas is a transition fuel to clean energy. BS! Let me say that again, BS! Natural gas (methane) is *not* a transition fuel. It's worse than CO_2; according to the EPA, if we look at a century of impacts, methane is 25 times worse than CO_2 (that is, methane traps 25 times more heat than CO_2).[16] But if we look at the short term (over a 20 year period, when methane is causing most of its damage) it's a whopping 86 times *more potent* as a greenhouse gas than CO_2.[17]

So when you hear those ads from the petroleum association or from oil and gas companies, talking about "clean" natural gas, just know that this is pure fabrication and paid for by the oil and gas industry. Methane is not a clean fuel. No fossil fuels are clean in any way shape or form. We need to get rid of them, and while it won't happen overnight, it has to happen soon. Energy systems that run on natural gas have high leak rates, as do the various extraction and processing phases in the production of oil and gas. Oil and gas companies don't like to admit it but the methane leak rates for their operations are very high, and with accidents like we saw at Aliso Canyon in California in 2015–2016, where 190,000 metric tons of natural gas leaked, uninterrupted, into the atmosphere for 5 months, the impacts to the climate are horrendous.[18] The carbon footprint of the Aliso Canyon blowout is thought to be greater than the BP Deepwater Horizon spill in the Gulf of Mexico.[19] Another source of significant methane leaks are the hundreds of thousands of abandoned oil and gas wells (that's in the United

States alone, globally, the number is in the millions) that if improperly sealed, may also be leaking methane gas, uncontrollably. These emissions are destroying our climate many times faster than CO_2 emissions, and we must address them.

We must immediately stop methane emissions because they are *the* immediate short-term threat to reaching tipping points, right alongside black carbon emissions, which are intensely contributing to changing the Earth's albedo by causing surface darkening with resulting surface heat absorption which in turn leads to accelerated glacier melt.

Here are some of the actions that can and *must* be promoted to reduce super pollutant emissions, focusing on the three pollutants over which we have direct influence (methane, back carbon, and HFCs):

Methane

- Identify methane emission sources, wherever they are, and work to reduce them
- Reduce and eliminate oil and gas production, particularly natural gas
- Debunk the myth that natural gas is a transition fuel
- Cease utilizing natural gas and biomass fuel for cooking or heating
- Reduce food waste
- Capture methane emissions from landfills and wastewater
- Detect and repair leaks in all stages of natural gas systems and oil/gas production
- Eliminate venting in oil/gas production, and recuperate gas leaks for reuse
- Promote low-methane emissions technology or recuperation in cattle and dairy production
- Promote low-methane emissions in agricultural waste
- Promote solar water heating
- Plug abandoned oil/gas wells
- Increase albedo efficiency of building surfaces to improve cooling/heating efficiency
- Increase the general energy efficiency of buildings to keep heat/cold in or out as needed
- Avoid or ban natural gas installations in all new buildings, and change out natural gas systems in old buildings where possible

Black Carbon

- Identify black carbon emission sources and work to reduce them
- Cease wood burning for cooking and heating
- Cease the use of gasoline powered cars/vehicles
- Eliminate internal combustion engines where possible
- Reduce risks of forest fires
- Eliminate diesel engine use or install air filters in them
- Cease/ban open burning of biomass
- Reduce/eliminate emissions from off-road vehicles, boats, ships, etc.
- Promote solar water heating
- Eliminate wick burning lamps
- Improve vehicle/ship fuel efficiency

HFCs (dirty refrigerants with high global warming potential)

- Identify HFC emission sources and work to reduce them
- Ban, eliminate, reduce high global warming potential (GWP) HFC refrigerants
- Monitor, repair leaks of refrigerants in AC and refrigeration systems
- Replace high GWP with low or zero GWP refrigerants (natural refrigerants)
- Improve thermal efficiency of home/business/industry so as to reduce the need for cooling
- Increase albedo of building and city surfaces to avoid cooling needs

Reducing super pollutants, or for that matter, CO_2 emissions has to do with lifestyle changes and commitments. It has to do with taking decisions to change things up, to be more aware of our footprint on the planet and our effects on the climate, and to make changes where we can, to do our individual part, as well as convincing others to do the right thing for the climate and for our collective sustainability.

In some cases, small choices can make a difference. Our decisions on what we eat, where we make purchases, and how sustainable our purchases are, can make an important impact if we all are making similar decisions, and we can spread the word, convince our family members, and live by rules to make our lifestyle choices more sustainable. Sometimes simple decisions taken collectively and at large scale, such as how the entire planet responded to COVID-19 by wearing a mask, can help stop planetary catastrophes. By the time this manuscript went to print, we were already embarked on getting the entire global population vaccinated against COVID-19. Let's think of ways to do the same for climate (see Chapter 10 for a COVID/Climate analysis).

We need to begin by taking better notice of the planet we live on and come to terms about what we are doing to sustain it, to improve it, or to cause its demise. Many people don't stop to ask this question. We all should, all the time.

Sometimes very large personal decisions can have long-lasting benefits (or impacts) and can be very significant in tackling climate change. For example, when we make very important decisions or investments that are long lasting and very significant to our lifestyles, such as when we purchase a vehicle, or a second vehicle, or decide to install solar panels on our roofs, or when we make important changes to our energy efficiency (getting more thermally efficient windows and doors), or when we decide to plant trees, or put a new layer of tar or paint on our roof, or paint our homes and places of work. These are moments when a good decision can be a very significant one, with long lasting positive climate benefits. We need to take our emissions seriously, considering what we consume, where it came from, if it was shipped from overseas or grown locally. We need to think about the car we drive or if we can take a bus, bike, or walk instead. We need to consider how we cook and what fuel we use to cook. We also need to think about how and where we play. What about our homes? Are we running our AC too cold, or do we leave it on when it's not needed? Do we need to run the heater or would putting on a sweater work just as well?

Where do *you* begin?

The first step is to address the simple things. You'll find your own path from there. Examining your monthly budget and recurring expenses is a good first step. You can probably save quite a bit of money while also helping the climate.

- Energy efficiency: Make an effort to heat and cool with less by better managing your energy use. Don't waste it. Don't leave windows open if you're using AC or open them when it's possible and at the right time of the day to get natural air-cooling. Shade windows from sunlight during summer. Don't abuse AC or heating. If you can afford it, invest in an efficiency retrofit in your home. All of these actions will reduce your utility bill and help the climate!
- If you can install solar energy, particularly if you can take advantage of on-bill options where you may not even have to put down any money to get solar panels to power your home, go for it!
- When you're buying an AC or refrigeration appliance (refrigerator or air conditioner), buy something with a good energy efficiency rating and/or a low-GWP gas. Do you know what HFC gas your AC uses? Go check its global warming potential now as see if you can lower it! Be informed! Don't let the system dictate your level of pollution, chose it yourself, and dare to eliminate it. Maybe you can chose a natural refrigerant gas if it's available. Do some research before buying something solely on visual appeal or for the brand name, or merely for the price. Spend a little more if necessary, the planet needs it. It may not be much more than you spend on coffee during the year.
- Walk, bike, run more, instead of driving. Try to get moving without fossil fuel whenever you can, your health will be better for it, and so will the planet.
- Get an electric vehicle instead of a gasoline-powered one, or if you need two cars in the family, chose strategically and use the cleaner one for your local miles and try to reduce the use of the one you power with gas. Nowadays, some electric vehicles also have hybrid systems where you can do extra miles on gasoline but most local miles on electricity. I bought a used Nissan Leaf with low mileage for under $10,000 USD, fully electric and basically dropped my gasoline consumption (of my other car) to a tenth of what I was spending, with the added benefit that the electric vehicle has no oil changes and no servicing whatsoever. It has been a huge savings and it's zero emissions. People are often dissuaded from purchasing electric because they fear they won't be able to do long trips, and yet they may not do long trips except once a year. So rent a car instead that day! It'll be a lot cheaper than what you spend on oil changes and maintenance for the year, I guarantee it.
- Find a local farmers market for veggies and other things that you regularly buy at supermarket chains that bring produce from across the country or internationally. You can buy local eggs, honey, bread, fish, and many other products that reduce emissions from the transport of foods and help the local economy as well as the climate.
- Reduce waste: compost, buy things with less packaging, combine purchases when you can. Make single buying trips for multiple items, or order many items on a single shipping order. Avoid frivolous and unnecessary purchases. Avoid purchasing single-use items that are non-reusable or made from non-recyclable materials.
- Don't buy water in small bottles, filter water from your tap instead and use a thermos.

- Recycle, buy, *and sell* used when you can, buy things with less packaging, buy in bulk, particularly cleaning products, this will save you money and reduce consumption.
- If you can, donate money, or better yet, time, to a local non-profit working on sustainability issues, climate change, or providing low-income housing assistance for energy efficiency or solar panel installation. This will not only teach you ways to do these things but will also place you in contact with climate-friendly people whose enthusiasm will be contagious and help you become more climate-friendly. It's also a great way to give back to your community.
- Take overland vacations instead of flying, and visit places where you can learn about the environment and our climate. If you can, visit glaciers!
- Oh yes, I almost forgot. Vote wisely!

A lifestyle change, or at least an honest reflection as to whether or not we can do better, is in order. Decisions we make, such as those I just listed, can lead to significant climate benefits or climate impacts, and the collective sum of these decisions across society can be very contagious, and can help us reach or fail to reach our climate challenge. Every little bit counts.

Conclusions on Cryoactivism and What You Can Do to Save the Glaciers

It should be clear from all that has been covered in this chapter that there are definitely ways to save glaciers. We can all be "cryoactivists." The first step is to think about our climate emergency and how we contribute to either worsening or improving our climate. If you're on the wrong side of that answer, maybe it's time to change.

Collectively, we can make a difference, but it starts with a single person, and that person is each and every one of us. And if you simply can't come up with strategies to change the planet, volunteer for or give a donation to an organization that can, maybe one that works on glacier protection, or one working on the conservation of mountain ecosystems. Volunteer for an organization that installs solar panels for low-income homes or improves their energy efficiency, and that way you learn how to do the right thing, and then you can go back to your home and do the same! There are probably a number of organizations in your city that currently work on climate resiliency or strategies. That could be a first step. Get involved with your climate. Think globally, act locally. It's a cliché, but it works.

And regarding glaciers, take a vacation, go visit one, you won't regret it. Head up to Alaska, to the European Alps, to Canada, to Glacier National Park, to the Sierra Nevada, the Rocky Mountains, to the Central Andes, or to wherever you can find these beautiful frozen creations of Mother Nature. Go see a glacier, and when you're there, think about the vulnerability of these magnificent natural beauties. I assure you, if you haven't woken up yet to our climate emergency, or are just a bit lazy about doing so, a trip to a glacier is just the inspiration you will need to turn a new leaf.

I'd like to end *Meltdown* with excerpts from an inspiring speech given by the governor of California from the site of the state's most devastating climate wildfires, which destroyed millions of acres of land in 2020. Governor Gavin Newsom, tired of the denials of so many politicians who refuse to act on climate change, doubled down, saying that all that has been done to date is woefully inadequate and that we must do more, do it better, and do it faster. Here are his remarks, and I should say, a few days after making this speech, Governor Newsom issued an Executive Order banning the use of the internal combustion engine in cars and trucks built in California.

Our goals are inadequate to the reality we are experiencing. It is a perfect storm. The debate is over around climate change. Just come to the State of California. Observe it with your own eyes. It's not an intellectual debate. It's not even debatable any longer. We are experiencing extreme droughts, extreme atmospheric rivers, and extreme heat.

In the last few weeks alone we've experienced the hottest August in California history. We had 14,000 dry-lightning strikes over a three-day period. We're experiencing world-record temperatures, 130 degrees, arguably the hottest recorded temperature in the history of mankind in the State of California, just a few weeks ago. It was 121 degrees in Los Angeles County, Burbank Airport 114 degrees. It was 103 degrees in one part of the State of California at 3 o'clock in the morning. ... Ashes are falling hundreds of miles away from these fires. Fires that we are experiencing in the north of California, all the way to the border of Mexico, twenty-eight active large-scale fires that we are currently battling in the State of California.

The economic impact ... just ask the folks in Butte County, $2.2 billion just to clean up the debris from the fire. I want to emphasize the economic consequence of our neglect. The cheapest way to deal with this is to invest in the future, in a low-carbon, green growth future, to decarbonize our economy. To change the way we produce and consume energy. The biggest cost is in our neglect. The biggest cost is *not* accelerating and fast-tracking low-carbon strategies.

And California is doing this, 5:1, we have more green jobs than we have fossil fuel jobs. We're proving this paradigm. You can grow your economy, 3.8% average GDP growth in the last five years in the State of California, as we move to accelerate the decarbonization of the economy.

But again, it's not enough.

It's not enough to do it alone. California does not live on an island. While it's the largest state in our union it is not even large enough to have the consequences of making a greater impact in terms of reducing greenhouse gas emissions. And that's why we need to get other states on board.

These efforts will save taxpayers money and will lower the cost of consuming energy, allowing us to build more climate resiliency into our environment and to leave something a little bit better than this for our kids and grandkids.

[Governor Newsom then turns around and points to the devastation caused by the fires burning immediately behind him.]

Some people want to roll back vehicle emissions standards [he says incredulously] so that you can spend more money at the pumps and produce more greenhouse gas emissions to create more of what you see around me. That's beyond the pale of comprehension. We're fighting against that, and we will prevail as long as more people come to this cause. Forgive me if I'm being a little bit long-winded but I am a little bit exhausted that we have to continue to debate this issue.

This is a climate-damn emergency! This is real. And it is happening. This is the perfect storm. It is happening in unprecedented ways year in and year out. You can exhaust yourself with your ideological BS, by saying that 100 years ago you have done this or that. [He is referring to then President Trump's argument that California's problem with wildfires was because the state had not properly raked the leaves of its forests.]

The reality here are the mega-fires that we are experiencing that come from the mega-droughts that we have experienced. One hundred fifty million dead trees, in our forests, in the Southern Sierras, and those mega-droughts impacting the mega-fires. There is something going on, not just bad practices related to forestry.

When you have a heat belt over the entire west coast of the United States, when you have record-breaking temperatures, record droughts, then you have something else at play. That's exactly what the scientists have been predicting for half a century. It is here, now.

California, folks, is America fast-forward. What we're experiencing right here is coming to communities all across the United States of America unless we get our act together on climate change. Unless we disabuse ourselves of all of the BS that is being spewed by a very small group of people that have an ideological reason to advance the cause of a 19th century framework and solution. We're not going back to the 19th century. We're not apologists to that status quo. We believe in the fresh air progress vs. the stale air normalcy.

That's California. We're going to lead here in the future. We're going to accelerate our low-carbon green growth strategies. We're going to create more economic opportunities in this space, more resiliency, a sustainable mindset. We're going to advance this cause in partnership with hundreds and hundreds of sub-national and national leaders of states around the rest of the world. We want to build on this leadership.

We've got to fast-track our goals, if we're going to be judged well in the future.

Everything we've done is inadequate.

Continue to do what you've done, you'll get what you've got. I'm explaining this to my four kids, to my little four-year old, who moved from talking about a novel corona virus, to asking me why he can't play outside [because of the fires]. That's not the world I want to leave to my kids. It's not the world *you* want to leave to your kids. This is not the world that *anyone* should be experiencing. And we don't have to.

It is our decisions and not our conditions that will determine our fate and future. I'm very, very proud of California's leadership, in the absence of national leadership [referring to the Trump administration's efforts to roll back climate and environmental regulations]. I recognize our responsibility to accelerate those efforts.[20]

10

Why for COVID but Not for Climate?

This chapter was originally published on April 24, 2020, as an opinion piece by the Center for Human Rights and Environment: https://center-hre.org/why-for-covid-and-not-for-climate-an-inter-generational-perspective/.

By Jorge Daniel Taillant (age 52) and Amelia Murphy (age 20), executive director and researcher, respectively, Center for Human Rights and Environment (CHRE).

CHRE's executive director and its new research intern got together to reflect on their intergenerational perspectives on the dynamics of climate change in the context of the COVID-19 global pandemic. Here are the results.

A few months ago, as I (JDT, age 52) was discussing what was needed to address our spiraling-out-of-control climate change crisis with my kids, we talked about the urgent and global need for industries to stop and reconfigure production methods in an effort to reduce pollution. We spoke of politicians needing to earmark billions, and even trillions, of dollars of climate finance to introduce structural change. We spoke of media unifying an effective and impacting message of the need to respond to the climate crisis, and that this message would be the sole communicational priority. We talked about people immediately changing their consumption and lifestyle habits in a unified global response to climate change where everyone came together and took the far-reaching and necessary actions that were needed to save our planetary ecosystem or otherwise face oblivion. Sadly, we ended that family conversation, as usual, on seemingly idyllic hopes, realizing (or thinking rather) that this sort of global and unified action of all of society to reverse climate change was simply impossible.

And now we have it.

The response is here. In a matter of months, weeks, days, hours, industries halted production, people stopped what they were doing, habits changed, politicians came up with the money, trillions even, the media unified a message, and the world has, largely, united. We are suddenly doing what we need to do.

In a matter of days, human habits have drastically changed. We've stopped cars and grounded airplanes and ships, bringing vehicular emissions to nearly zero. We stopped spewing smoke into the air from industry and ceased dumping waste into our rivers. Practically overnight, the global hustle and bustle of urban activity suddenly stopped. And the environment likes it. Coyotes are roaming the streets of San Francisco, fish are reappearing in suddenly cleaner bays and rivers throughout the world, peacocks are roaming Spanish villas, and mountaintops are again visible on the horizon lines in otherwise highly polluted cities. We've stopped polluting and Nature is happily responding.

But alas, the world's sudden environmental rebound is not due to our tackling climate change, but rather, due to our unified efforts at stopping the COVID-19 pandemic.

What we can affirm from this occasion is that intense and sudden social emergencies (like hurricanes, earthquakes, floods, ... *and pandemics it seems*) spawn rapid reactive responses. In life-or-death emergencies, people move, they act and react, they are willing and able to change long-standing embedded habits, even if merely for a few days to respond to a life-threatening emergency.

So why for COVID but not for CLIMATE?

So, if we are doing it for COVID, why don't we do it for climate change, which ultimately will impact many more millions if not billions of people than the current global health pandemic? According to the World Health Organization, climate change has killed more people than COVID in the same amount of time [this was at the time of the original publication date]. In fact, the most conservative estimates from researchers state that air pollution reductions due to the sudden halt of industry has saved at least 20 times the number of lives in China as deaths directly stemming from the virus, and yet no one is stopping economies because of air pollution and climate change deaths.

What's more, both crises disproportionately affect similarly vulnerable groups. We know that communities of lower socioeconomic standing that have less access to basic public services like clean water and sanitation, that have lower local air quality and poorer living conditions and infrastructure, that have inadequate health care, are generally the *least* resilient and *most* vulnerable to the effects of climate change. In the United States, Black and other communities of color are disproportionately exposed to air pollution due to antiquated redlining policies and inadequate infrastructure investment, and are thus facing increased risk of mortality due to COVID.[1]

We see now that in a global health crisis such as COVID-19, these same populations are also more vulnerable due to their living conditions and to the basic infrastructure inequities they experience in daily life, particularly in regard to poor living conditions, limited or no sanitation, and poor air quality. These communities suffer disproportionately, and in many similar ways, to both the COVID and the climate crisis.

So, why are we so rapidly scrambling today and immediately to stop COVID-19 while not doing the same to detain the climate crisis that we know we have been facing already for many decades?

The short answer is: fear.

Fear, particularly immediate fear of death, can drive collective action like few other emotions. It's an innate and instinctive trait. Animals, and humans too, want to survive above all else. It's pure instinct. Gunshots heard in a crowd make masses suddenly scramble for cover. A building fire or an earthquake that risks building collapse will drive us into the clearing in seconds. A hurricane or an arriving tidal wave will make us run for the hills. And a sudden pandemic that is highly contagious and that has a high mortality rate for us or for our family will also make us do things we never thought possible. And our reactions to these imminent risks are fast and efficient.

So, what's the difference? Climate change is far worse for us as a planetary society than this pandemic, and yet, we are direly slow to react to the escalating climate crisis, or at least, we don't seem to think or feel that it is so urgent that we should stop everything to deal with it (even though we should do just that). We are not even close to achieving the social mobilization, the habit changes, the industrial mitigation and structural change, the policy changes, the reduction of emissions, or to obtaining the climate finance we need to reverse climate change trends and avoid a climate catastrophe.

In terms of finance to adapt our economies, we need about 2 trillion dollars per year at a global scale through the year 2050 to keep the world to a 1.5 degree increase in temperature, which would help us avoid irreversible tipping points for the climate. The United States alone allocated that very amount to the COVID response through recent stimulus packages approved by Congress, and that decision happened in about 2 weeks! The financing for COVID has since tripled in the following 2 or 3 weeks, but it seems we can't get that for the Earth in a whole year?

What's wrong with us? As a society, we don't fear climate change, or at least we don't fear it enough to quickly prioritize actions to do something about it. The second factor driving the response (or lack of response), after fear, is time.

Let's use a few analogies to help understand the intricate relationship between fear and time in the COVID and climate crisis respectively.

We can think of climate change like a bullet about to leave a gun aimed for our heads. If it were a real gun and a real bullet, we'd react immediately and try to run for cover. But the climate change bullet is a bullet that won't immediately kill us, and we know it. It will take time for it to reach us, maybe a decade, or several decades, or even a century. We might not even be around when it hits, and instead of killing us, it will kill our children or our grandchildren, or someone else's children, which makes the climate change bullet less frightening to us. Others have used the frog in a pot with hot water analogy to understand climate change responses. A frog thrown into a pot with boiling water will immediately try to jump out because it realizes that life depends on it now (that's the COVID crisis), but if the frog is placed in the pot with the water cold and then the fire is turned on, the frog will stay in the pot, enjoying the warming water, until eventually it dies (that's like our response to the climate crisis).

So, while avoiding immediate death is instinctual, and we all react to it quickly, time gives the benefit of thought and the ability of procrastination, and in this case, reaction to fear depends more on the instantaneous emotion present (or not present) than about instinct. As a result, in the climate crisis, our fear factor is affected by the time factor and our subsequent instinctual urgency to react even though we realize we may eventually die, is greatly diminished. We are the frog in the pot of cold but rapidly simmering water.

While we may agree that climate change is serious or even deadly, we also know that climate change is something in constant evolution, it's progressive, with a much longer time horizon than a speeding bullet about to leave the gun pointed at us. Slowly boiling water buys us time and the luxury of putting off difficult decisions until a later date (the policies, the finance, the habit changes, etc.).

The COVID bullet, meanwhile, has left the chamber, with the likelihood that it hits us high. The water is boiling now, and society is demanding immediate action from elected officials. Politicians feel urgency and pressure to act out of both fear for themselves and their families, as well as from the social and political pressure that is exerted by their constituencies. Combined with the fact that political actions and reactions happen within short-term electoral and administrative cycles (2–4 years), you get the response we are seeing from political leaders: quick decisions, directives, programs, action, and financing.

On the other hand, the climate bullet might kill you (or your descendants) in say 50, or 100, or even 1,000 years; the water in the pot seems nice and cool for the moment, and while it's going to kill you some day, maybe, you're willing to enjoy the waters for a while—no need to jump out just yet. Maybe it's a bit warmer and drier in your city, but for many it's not a life or death matter. Climate impacts, or at least their immediate consequences, for the most part, are not happening within electoral cycles, and hence, the option for political leaders to procrastinate or let some other politician, in the future, take the necessary and difficult action, that will require controversial large spending, is much more amenable and tolerable for society.

People more generally are also not feeling the "immediate" climate burn. Yes, there are some who are beginning to get it. If you live in Maldives and your home and entire country will soon be underwater maybe in the next few years or decades, you are already reacting, and looking for a place to live. If you live in the Florida Keys or even in Miami where heavy rains flood your neighborhoods and may not leave for a while (or ever) you're all of a sudden looking at Zillow to see if your home is declining in value, and you're maybe looking for an exit opportunity in the property market. If the heat in your home is unbearable, your AC bill through the roof, and your skies tinted orange and grey from wildfire smoke, maybe you're considering how to adapt your home to this new volatile environment, or maybe you're already planning to move. You may also be waiting and expecting your local government to help you out and calling on your politicians to start planning and taking action now.

Most people don't have the climate change gun to their head and don't see that a climate change bullet is coming, or if they do, they don't think it will hit for now and so, they procrastinate. People respond to emergencies when the emergencies are visibly upon them—when death is near. You don't hunker down for a hurricane until the hurricane is visibly hours away. You don't run from your building until the walls start to shake from the earthquake. You don't leave your home because of a flood until the water starts rising in your neighborhood and maybe even then until it's actually inside of your living room. We procrastinate.

The sort of social, political, governmental, and industrial response that we need to address climate change is of the same massive proportion that we are seeing as a response to COVID-19, in fact, it is more urgent and more existential, and yet we don't get it. The window for effective mitigation to slow feedbacks and avoid tipping points is shrinking to perhaps 10 years or less,[2] including the window to prevent us crashing through the 1.5°C guardrail that the UNFCCC has told us we cannot surpass if we hope to contain irreversible climate collapse.[3] If we don't act soon, seas will rise irreparably, wildfires will engulf much of the United States, pandemics will become the norm. In the United States, 10 years buys us about two presidential terms. Who we elect in the next 10 years will carry the legacy of permanent climate change.

Can we learn something from the COVID-19 response that all of us are now living which can help us re-embark on a path to address climate change effectively once we move on from the pandemic?

Climate activists have been trying to create narratives about the climate crisis to mobilize people to act. They've tried to crank up the "fear" gauge to get us to react like a frog thrown into boiling water—with little results. Our society has been trying to drum up environmental awareness since we coined terms like "sustainable development" in the 1980s. Narratives on climate change awareness are much more recent (since Al Gore started touring the world with his PowerPoint presentation about climate change trends, in the mid 2000s), but in the end, we have made little progress in getting the world to change course quickly and effectively.

We have mostly gotten past the barrier of climate change denial (although some that deny science still exist and hold legislative power). And we have reached some global consensus on what needs to be done to reverse trends (e.g., commitments to reduce emissions as stated and outlined in the Paris Agreement, switching to renewable energies, reducing dependency on fossil fuels, changing habits, etc.).

The problem is that it's been much more difficult moving from theory to practice. Changing collective habits to avoid climate tipping points is a hard task. The clock is ticking and hope that we will be on time before it's too late is dwindling. And while much of society may be on-board with the need to change (thanks largely to ubiquitous social media telling us so), we have seen roadblock after roadblock from the public and private sectors to get things moving quickly.

In contrast, finance and political action to address COVID-19 is moving and it's moving fast. We have seen a response to COVID that environmentalists and for that matter, most of society, thought was impossible. In a matter of days, we've completely changed systems and habits previously thought unchangeable. And while it would be harsh and inappropriate to speak of this tragedy as an opportunity, what the response to COVID has done, possibly most importantly, is shown us and proven that we *are* capable of massive social, political, and economic change, in the face of an existential crisis, something that most of an older generation of people and environmentalists advocating for addressing climate change thought impossible.

For the younger of the two co-authors of this article (AM, age 20), this crisis may simply be a blip in my timeline, in an era where the volatility of socioeconomic crisis may become the norm and where living with a pandemic or other climate emergencies like wildfires or massive sea-level rise and flooding may be a recurring fact of life.

What we both agree on as we set out to write this article is that our generations see this crisis quite differently. The older generation (and now I am back, JDT age 52), largely to blame for much of current climate change crisis, is scrambling to address the problem we've caused. The bullet has fired, and it is moving fast toward us. We've seen the deterioration and recognize that we are on the edge of collapse. COVID is the materialization and confirmation of our global fragility. We have been dreading our planetary collapse and now, here it is, in the form of a global health pandemic.

For the younger generation (back to AM age 20), we see this crisis quite differently. While staying home on a computer seems novel to the Boomers, it's not so different for us in Generation Z. We point optimistically to the rebound of nature, to the coyotes in San Francisco, and to the blue-green waters of Venice as a sign that we can dream of a cleaner world, cleaner than the one we are inheriting. We can suddenly see the magnificent Himalayas from some of the world's most polluted cities. Change is possible.

The unique position of a younger mind is the capacity to worry less about the tragedy of the past and think more about the opportunities of the future.

While the views of what got us into this crisis and our vision of hope on how we might get out of it may be fundamentally different, the underlying response we need is in fact not generational, it is value-based. And the values that determine how we respond, are generally shared among people, whoever they are.

What we've learned from the response to the COVID crisis, that is extremely important for us to apply to our climate crisis, is that as a society, we care. And that when we have to respond to save our lives, we do. It's not just fear that drives us, it's fear of losing what we most care about and love. Whether we are 82, 62, 42, or 22, or even 12, we care about our families and our well-being. We have all seen most of what we consider "normal" in our lives stalled, stopped, drastically withered away, and/or placed at great risk (school, work, friends, family members, and nearly all other aspects of life). The COVID crisis has taken hold of everything, showing us, teaching us perhaps, just how fragile we are.

We've seen the damage our global vulnerability can cause for us and for the way of life we once thought unchangeable. We have seen our parents and teachers, our doctors, our children, and delivery people, our supermarket workers, our police, our paramedics, face great danger and insecurity, and yet work to protect us. We've learned that without strong, visionary leadership, millions are left unemployed, without income, or stuck at home working or learning online, or worse, suffering chronic impacts of an invisible virus, with little-to-no support.

And so, as a group and as a society, because we have to, we stay home, we change our habits, we shut down our economy, all in order to protect our families, our neighbors, and ourselves. As we devise new policies and actions to move out of COVID and back into our "normal lives," we must keep in mind that we are immersed in a reigning climate crisis that, thus far, we have been unwilling to change for. But thankfully the COVID crisis has shown us that we *can* change.

Our task is not to crank up the fear gauge and make people terrified of the certain-destructive future of climate change, but to all stop for a moment and evaluate what we care about most deeply. The COVID-19 pandemic has given us a window to look through, to show us our faults, to show us our errors, and our fragility, but also to show us our strength as a family, a global family that can *and must* have hope that we *can* change.

As policymakers move forward, the goal should not be to simply rebound to our pre-COVID norms, but to progress toward a better normal. And while better is subjective, we can all agree that it means improved health and safety for ourselves, for our loved ones, and for the most vulnerable populations. Air quality is a key variable. The worse your air is, the higher the likelihood that your population suffers from chronic respiratory illness and the higher your vulnerability as a community to pandemics like COVID. It is with this intersection that COVID and climate have a common vulnerability zone. Public officials and policymakers must consider how to prepare for the next wave of disease, through clean air and pollution policies and programs that are good for pandemic prevention and preparation, but also good for climate.

Economically, we must reopen economies and be sensitive to job creation, but not simply any jobs. By transitioning economic resources from high carbon intensive resources to more climate-friendly ones, we can create jobs in recycling, reforestation, nature conservation, water management and waste reduction, and environmental risk management, and promote renewable energies instead of ones dependent on fossil fuels, all while making our economies more resilient and more climate friendly. In 2008, spending on low-income households and green infrastructure had the highest yields of the American Recovery and Reinvestment Act.[4] In the post-Depression era of the 1930s, millions of jobs in the United States were created simply to plant trees! It was a fast and easy way to create jobs in a post-recession/depression market, while also doing something significant for our environment.

Most of us aren't policymakers or public officials. We're mothers, fathers, sisters, brothers, business owners, essential workers, teachers, and more. What can we do in our daily lives to help achieve the change we need to address climate change? Firstly, we can use our time now to pause and reflect; reflect on whether or not we want to return to the pre-COVID norms, on how our footprints have changed. We can also think about what we want from our political leaders and make sure that we demand that from them (manifested through elections but also in being more active and engaged in pre-election policy discussions).

While in lockdown, we saw our planet and environment improve all over the world. Let's try to capture that! Can you waste less? Eat out less? Use less plastic? Drive less? What can *we* do, and more importantly, what can *you* do after this is all over to lessen our impacts on the environment? While we're all at home, working, studying, and shopping remotely, we can consider how to maintain our tele-habits and reduced emissions after COVID. Maybe you can continue eating at home more often, source your food locally, speak to your boss about continuing to work remotely (even just sometimes), or take a class online instead of on campus. If you're remodeling or updating your home to accommodate a new home-bound and electricity-dependent lifestyle, can you upgrade to energy efficient appliances or install solar? As businesses, can you reduce waste or work to reduce emissions in your day-to-day operations by having employees continue to work remotely? Can you rethink your 5- or 10-year plans to shift away from fossil fuels and into renewables?

At the end of the day, what is asked of us all is to refuse to sacrifice our futures and future generations for the temporary comforts of the old normal. Instead of going back to the way things used to be, can't we take this opportunity to strive for a better normal? By mitigating irreparable climate change, and financing adaptation to our new normal, we can create a society that embodies our hope for a healthy, sustainable, resilient future.

What's most important about our experience of the last few months is that we have learned we can change, we have the power to design a new and better normal, and the best thing is that in the end, it's your choice! JDT and AM

Notes

Foreword

1. For Biden's Climate Action Plan see: https://joebiden.com/climate-plan/.
2. For Biden's Executive Order on Tackling the Climate Crisis at Home and Abroad see: https://www.whitehouse.gov/briefing-room/presidential-actions/2021/01/27/executive-order-on-tackling-the-climate-crisis-at-home-and-abroad/
3. See: https://www.gov.ca.gov/2020/09/23/governor-newsom-announces-california-will-phase-out-gasoline-powered-cars-drastically-reduce-demand-for-fossil-fuel-in-californias-fight-against-climate-change/.

    For the text of the Executive Order: https://www.gov.ca.gov/wp-content/uploads/2020/09/9.23.20-EO-N-79-20-text.pdf.
4. For Greta Thunberg's moving speech on climate change see: https://www.youtube.com/watch?v=TMrtLsQbaok.
5. Associated Press. LA Times. August 20, 2020. https://www.latimes.com/world-nation/story/2020-08-20/greenland-record-melt-lost-billions-tons-ice-2019.
6. Claypool, Max, and Miller, Brandon. CNN. August 14, 2020. https://apple.news/AxEoUivIYSpuJQN6EPrQ7GA.
7. Mooney, Chris. Unprecedented Data Confirms that Antarctica's Most Dangerous Glacier Is Melting from Below. Washington Post. January 30, 2020. https://www.washingtonpost.com/climate-environment/2020/01/30/unprecedented-data-confirm-that-antarcticas-most-dangerous-glacier-is-melting-below/.
8. Taylor, Derrick Bryson. Antarctica Sets Record High Temperature: 64.9 Degrees. New York Times. February 8, 2020. https://www.nytimes.com/2020/02/08/climate/antarctica-record-temperature.html.

Introduction

1. For a definition of cryoactivism see: https://forum.lasaweb.org/files/vol47-issue4/Debates7.pdf.
2. See Benn, Douglass, and Evans, David. Glaciers and Glaciation. 1998. P. 4.
3. Taillant, Jorge Daniel. The Human Right . . . to Glaciers? Journal of Environmental Law and Litigation. Vol. 28. 2012.
4. See: https://glacierhub.org/2019/04/04/unearthing-rock-glaciers/.
5. For a podcast on a reconnaissance visit to California's Rock Glaciers with the head of CAL EPA and several rock glacier specialists see: https://www.podshipearth.com/rockglacier.
6. This may seem extreme, but it is real. A handful of laws around the world that I have been able to track down do mention the word glacier, but they are not glacier protection laws. Some laws protect national parks, for example Yosemite or Glacier National Park, or mountain environments, such as the French law of mountains, where glaciers might be located,

but those laws don't really focus on glacier protection. In fact very few laws even mention glaciers. The fact remains that laws (except for Argentina's new glacier protection laws) do not specifically address the specific dynamics of glaciers, and the needs for glacier protection. They do not define glaciers, nor are they designed to protect their glaciosystems or the specific functions of glaciers for our ecosystems and for the glacier's downstream basins.

7. See: https://en.wikipedia.org/wiki/K2.

8. See: https://www.youtube.com/watch?v=C1sS1OehnGw.

9. See *Chasing Ice* by James Balog the mesmerizing and fantastic documentary that registered one of the largest ice calvings ever caught on video see: https://chasingice.com.

10. Twila Moon, a glaciologist at the National Snow and Ice Data Center at the University of Colorado-Boulder. In: Jenny Howard. Alaskan glaciers melting 100 times faster than previously thought. Environment News, National Geographic. July 25, 2019. See: https://www.nationalgeographic.com/environment/article/alaskan-glaciers-melting-faster-than-previously-thought

11. Jamail, Dahr. The End of Ice: Bearing Witness and Finding Meaning in the Path of Climate Disruption. The New Press. 2019. P. 47.

12. See: https://www.usgs.gov/news/glaciers-rapidly-shrinking-and-disappearing-50-years-glacier-change-montana; see also: https://www.sciencebase.gov/catalog/item/58af7022e4b01ccd54f9f542.

13. Quoted in: Glick, Daniel. The Big Thaw. National Geographic. See: https://www.nationalgeographic.com/environment/global-warming/big-thaw/.

14. Quoted in: Jamail, Dahr. The End of Ice: Bearing Witness and Finding Meaning in the Path of Climate Disruption. The New Press. 2019. P. 39.

15. Miller, Jeremy. The Dying Glaciers of California. Earth Island Journal. Vol. 28. No. 2. Summer 2013. Pp. 48–53. See: https://www.earthisland.org/journal/index.php/magazine/entry/the_dying_glaciers_of_california/.

16. See: https://www.cbc.ca/news/technology/how-western-canada-glaciers-will-melt-away-1.3022242.

17. See: https://lcluc.umd.edu/hotspot/glacial-retreat-himalayas.

18. Glick, Daniel. The Big Thaw. National Geographic. See: https://www.nationalgeographic.com/environment/global-warming/big-thaw/.

19. See: IPCC Report on Oceans And Cryosphere. Chapter 2 on High Mountain Areas. By Hock, Regine and Rasul, Golam. Intergovernmental Panel on Climate Change. 2019. P. 1-33

20. IPCC Report on Oceans And Cryosphere. Chapter 2 on High Mountain Areas. By Hock, Regine and Rasul, Golam. Intergovernmental Panel on Climate Change. 2019. P. 1-42.

21. See: https://chasingice.com.

22. See: https://www.theguardian.com/us-news/2015/mar/08/florida-banned-terms-climate-change-global-warming.

23. See: https://www.miamiherald.com/news/local/community/florida-keys/article236261848.html.

24. See: https://keysweekly.com/42/flood-free/.

25. See: https://www.nytimes.com/2019/12/04/climate/florida-keys-climate-change.html

26. Benn, Douglass, and Evans, David. Glaciers and Glaciation. Arnold. 1998. P. 4.

27. See: IPCC Report on Oceans And Cryosphere. Chapter 2 on High Mountain Areas. By Hock, Regine and Rasul, Golam. Intergovernmental Panel on Climate Change. 2019. Pp. 94-95.

28. See: https://news.ucar.edu/132773/2020-was-record-breaking-year-ocean-heat

29. See: In "When the Glaciers Disappear, Those Species will Go Extinct", by Henry Fountain, New York Times, April 1, 2019. https://www.nytimes.com/interactive/2019/04/16/climate/glaciers-melting-alaska-washington.html

30. See: http://www.antarcticglaciers.org/glacier-processes/glacial-lakes/glacial-lake-outburst-floods/.

31. See: IPCC Report on Oceans And Cryosphere. Chapter 2 on High Mountain Areas. By Hock, Regine and Rasul, Golam. Intergovernmental Panel on Climate Change. 2019. P. 1-33.

32. See: https://chasingice.com.

33. I wrote a blog about the tensions between academia and policy, in the realm of glaciology. You can read more about this at: https://usapecs.wixsite.com/usapecs/post/cryoactivism

Chapter 1

1. Benn, Douglas, and D. Evans, Glaciers and Glaciation. Arnold. 1998. P. 4, 39.

2. See: Google Search for "glacier." Note the definition may change over time.

3. See: https://en.wikipedia.org/wiki/Glacier.

4. See: https://www.usgs.gov/faqs/what-a-glacier?qt-news_science_products=0#qt-news_science_products.

5. See: http://center-hre.org/wp-content/uploads/Argentine-National-Glacier-Act-Traducción-de-CEDHA-no-oficial.pdf.

6. See: http://center-hre.org/wp-content/uploads/2012/07/Definicion-de-Glaciosistema-version-1-febrero-2012-english.pdf.

7. For a fascinating, brief documentary by Astrum on the discovery of ice and glaciers on the surface of the Planet Pluto see: https://www.youtube.com/watch?v=6l4kr36TzQ4&feature=youtu.be.

8. Gosnell, Mariana. Ice: The Nature, the History, and the Uses of An Astonishing Substance. Alfred A. Knopf. New York. 2005. P. 88.

9. White, Christopher. The Melting World: A Journey across America's Vanishing Glaciers. St. Martin's Press. New York. 2013. P. 140.

10. For short videos that explain how ice ages work see.:

    (1) https://www.youtube.com/watch?v=dJ5GYQrkvxI
    (2) https://www.youtube.com/watch?v=iA788usYNWA.

11. See: https://www.livescience.com/58407-how-often-do-ice-ages-happen.html.

12. The new divisions of the Holocene are Meghalayan, Northgrippian, and Greenlandian; for more see: https://www.sciencealert.com/international-chronostratigraphic-chart-holocene-added-ages-official.

13. For more on the Anthropocene see: http://quaternary.stratigraphy.org/working-groups/anthropocene/.

14. This section is largely adapted from NASA's Global Climate Change page at: https://climate.nasa.gov/news/2948/milankovitch-orbital-cycles-and-their-role-in-earths-climate/.

15. For a fantastic video explanation of the Milankovich Cycles see: https://www.youtube.com/watch?v=iA788usYNWA; see also, NASA animated descriptions at: https://climate.nasa.gov/news/2948/milankovitch-orbital-cycles-and-their-role-in-earths-climate/.

    For a very clear animated explanation of the Milankovich Cycles and Glaciation see also: https://www.youtube.com/watch?v=ztninkgZ0ws&authuser=0.

    See also: https://www.skepticalscience.com/Milankovitch.html.

16. See: Benn, Douglass and David Evans, *Glaciers and Glaciation* (London: Arnold, 1998), 6–7.

17. See: https://nssdc.gsfc.nasa.gov/planetary/ice/ice_mercury.html.

18. See: https://phys.org/news/2016-12-ice-ages-linked-earth-orbitbut.html.

19. See: Steffen, Will, et al. Trajectories of the Earth's System in the Anthropocene. Proceedings of the National Academy of Sciences of the United States of America 115, 33. Pp. 8252–59. August 14, 2018. https://doi.org/10.1073/pnas.1810141115.

20. See: https://www.beg.utexas.edu/sites/default/files/media/0000/0804/Greenhouse%20-%20Icehouse%20Earth.pdf.

    See:

    1) https://en.wikipedia.org/wiki/Greenhouse_and_icehouse_Earth;

    2) https://www.astrobio.net/climate/greenhouse-earth/.

21. See: https://www.sciencedaily.com/releases/2016/01/160113160709.htm.

22. For information on the Little Ice Age see: https://en.wikipedia.org/wiki/Little_Ice_Age.

23. McGuire, Bill. Will Global Warming Trigger a New Ice Age. 2003 at: https://yaleglobal.yale.edu/content/will-global-warming-trigger-new-ice-age; see also: Atlas Pro. Could Global Warming Start a New Ice Age? at: https://www.youtube.com/watch?v=yyAuWeoTm2s; see also: https://www.whoi.edu/know-your-ocean/ocean-topics/climate-ocean/abrupt-climate-change/are-we-on-the-brink-of-a-new-little-ice-age/.

24. For aerial video footage of the glaciers on Heard Island see: https://www.youtube.com/watch?v=aVLFY71G1ZA.

25. See: https://www.bloomberg.com/news/features/2019-06-06/towing-an-iceberg-one-captain-s-plan-to-bring-drinking-water-to-4-million-people.

Chapter 2

1. See: https://www.miamiherald.com/news/local/community/florida-keys/article236261848.html.

2. Goodhue, David and Alex Harris, "Florida Keys Neighborhood's Been Flooded for over 40 Days," Miami Herald October 16, 2019. See: https://www.miamiherald.com/news/local/community/florida-keys/article236261848.html

3. Quoted in: Glick, Daniel. The Big Thaw. National Geographic. See: https://www.nationalgeographic.com/environment/global-warming/big-thaw/.

4. See: https://www.climatecentral.org/news/report-flooded-future-global-vulnerability-to-sea-level-rise-worse-than-previously-understood.

5. Flooded Future: Global Vulnerability to Sea Level Rise Worse than Previously Understood. Climate Central, 2019. P. 3. See: https://www.climatecentral.org/pdfs/2019CoastalDEMReport.pdf

6. IPCC Report on Oceans And Cryosphere. Summary for Policy Makers. Intergovernmental Panel on Climate Change. IPCC, 2019. Section B.3. See: https://www.ipcc.ch/srocc/chapter/summary-for-policymakers/

   Also: IPCC Report on Oceans And Cryosphere. Chapter 4: Sea Level Rise and Implications for Low-Lying Islands, Coasts and Communities. IPCC, 2019. P.3-23 and P. 3-31. See: https://www.ipcc.ch/site/assets/uploads/sites/3/2019/11/08_SROCC_Ch04_FINAL.pdf

7. IPCC Report on Oceans And Cryosphere. Chapter 4: Sea Level Rise and Implications for Low-Lying Islands, Coasts and Communities. IPCC, 2019. P. 3-23. See: https://www.ipcc.ch/site/assets/uploads/sites/3/2019/11/08_SROCC_Ch04_FINAL.pdf

8. My editor corrected my original statement indicating that there were no such stories, pointing out that in the 1980s there was a series called: The Clan of the Cave Bear, covering prehistoric times, by Jean M. Auel, speculating about what life and interactions were like between Neanderthal and modern Cro-Magnon humans. It's now next on my reading list!

9. Hoffman, Paul, and Schrag, Daniel. The Snowball Earth. 1999. P. 9.

10. Ibid. P. 1.

11. Hage, Melissa. Quoted in Ross, Rachel, "What Are the Different Types of Ice Formations Found on Earth?" Livescience. January 8, 2019. See: https://www.livescience.com/64444-ice-formations.html

12. Ibid.

13. Ross, Rachel. What are the Different Types of Ice Formations Found on Earth? Life Sciences. January 8, 2019.

14. Schriber, Michael. "Snowball Earth" Might Have Been Slushy. Research Features. NASA. August 2015.

15. Hoffman, Paul, and Daniel Schrag, The Snowball Earth. 1999. P. 5.

16. Ibid. Pp. 9–10.

17. Begert, Blanca. Back to the Future and Beyond with Climate Scientist Linda Sohl. Yale Environmental Review. October 22, 2019.

18. See: https://www.youtube.com/watch?v=9tkDK2mZlOo

19. See: https://apple.news/ArE8UcnVVPRyJ2njCGdU0zg.

20. See: Benn, Douglas, and D. Evans, Glaciers and Glaciation. Arnold. 1998. P. 4.

21. See: https://apple.news/A_OBWrI8vP2W-N8a8qThn0A.

22. See: Wilkinson, Jerry. Keys Geology. See: http://www.keyshistory.org/keysgeology.html.

23. Aber, James. Glacial Isostasy and Eustasy. www.academic.emporia.edu: ES 331/767. Lecture 9.

24. Ibid.

25. Quoted in: Daniel Glick. The Big Thaw. National Geographic. See: https://www.nationalgeographic.com/environment/global-warming/big-thaw/.

26. Muhs, Daniel, et.al. "Sea-Level History of the Past Two Interglacial Periods: New Evidence from U-Series Dating of Reef Corals from South Florida," Quaternary Science Reviews Vol. 30. 2011. Pp. 570–90.

27. IPCC Report on Oceans And Cryosphere. Chapter 4: Sea Level Rise and Implications for Low-Lying Islands, Coasts and Communities. IPCC, 2019. P.3-23. See: https://www.ipcc.ch/site/assets/uploads/sites/3/2019/11/08_SROCC_Ch04_FINAL.pdf

28. See: https://www.sciencealert.com/the-melting-arctic-has-revealed-five-new-islands-we-never-knew-were-there.

29. See: https://apple.news/Ag3qL6eVdRGy2lgbJ4GuSxA.

30. Flooded Future: Global Vulnerability to Sea Level Rise Worse than Previously Understood. Climate Central, 2019. P.6. See: https://www.climatecentral.org/pdfs/2019CoastalDEMReport.pdf

31. Flooded Future: Global Vulnerability to Sea Level Rise Worse than Previously Understood. Climate Central, 2019. Pp.8–9. See: https://www.climatecentral.org/pdfs/2019CoastalDEMReport.pdf

32. DeJong, Benjamin et al. Pleistocene Relative Sea Levels in the Chesapeake Bay Region and Their Implications for the Next Century. GSA Today. Vol. 25, No. 8. August 2015. P. 4.

33. Ibid. P. 9.

34. See: https://www.latimes.com/local/lanow/la-me-california-coast-storm-damage-20190313-story.html.

35. See: https://mail.google.com/mail/?tab=mm1&authuser=0.

36. Mooney, Chris. Unprecedented Data Confirms that Antarctica's Most Dangerous Glacier is Melting from Below. Washington Post. January 30, 2020.

37. See an Alaskan glacier melt in this video: https://www.nationalgeographic.com/environment/2019/07/alaskan-glaciers-melting-faster-than-previously-thought/.

38. Horton, Benjamin P. Is Sea Level Rising? Ted Talk. November 11, 2013. https://www.youtube.com/watch?v=rRrRgqKJEFw.

39. Zemp, Michael, et al. Global Glacier Mass Changes and Their Contributions to Sea-Level Rise from 1961 to 2016. Geophysical Research Abstracts. Vol. 21, EGU 2019-4975, 2019. EGU General Assembly. 2019. https://meetingorganizer.copernicus.org/EGU2019/EGU2019-4975.pdf.

40. Zemp, Michael et.al. Global Glacier Mass Changes and Their Contributions to Sea-Level Rise from 1961 to 2016. Nature. Vol. 568. 2019. P. 382 https://www.nature.com/articles/s41586-019-1071-0.

41. Ibid. P. 383.

42. IPCC Report on Oceans And Cryosphere. Chapter 4: Sea Level Rise and Implications for Low-Lying Islands, Coasts and Communities. IPCC, 2019. P.3-28 See: https://www.ipcc.ch/site/assets/uploads/sites/3/2019/11/08_SROCC_Ch04_FINAL.pdf

43. See: https://apple.news/A_OBWrI8vP2W-N8a8qThn0A.

44. See: https://apple.news/A_OBWrI8vP2W-N8a8qThn0A.

45. Sukee Bennett. Nova Text. April 22, 2020 at: https://www.pbs.org/wgbh/nova/article/warm-water-found-beneath-thwaites-glacier-antarctica/.

46. Rosier, Sebastian et.al. "The Tipping Points and early Warning Indicators for Pine Island Glacier, West Antarctica," The Cryosphere Vol. 15. 2021: Pp. 1501–16. See: https://doi.org/10.5194/tc-15-1501-2021

47. Quoted in: Los Angeles Times. By Associated Press. August 20, 2020. See: https://www.latimes.com/world-nation/story/2020-08-20/greenland-record-melt-lost-billions-tons-ice-2019.

48. Ibid.

49. Quoted in: https://apple.news/AxEoUivIYSpuJQN6EPrQ7GA.

50. See: https://www.theguardian.com/us-news/2015/mar/08/florida-banned-terms-climate-change-global-warming.

51. See: https://www.podshipearth.com/kingtides.

52. Flooded Future: Global Vulnerability to Sea Level Rise Worse than Previously Understood Climate Central, 2019, p. 10. See: https://www.climatecentral.org/pdfs/2019CoastalDEMReport.pdf

53. See: https://www.nature.com/articles/s41558-020-0874-1.epdf.

54. See: Befus, K. M. et al. Increasing Threat of Coastal Groundwater Hazards from Sea-Level Rise in California. Nature Climate Change. Vol. 10. October 2020. P. 948.

55. See: Barnard, P. et al. Dynamic Flood Modeling Essential to Assess the Coastal Impacts of Climate Change. Scientific Reports. Vol. 9. 2019. P. 4309.

56. See: Befus, K. M. et al. Increasing Threat of Coastal Groundwater Hazards from Sea-Level Rise in California. Nature Climate Change. Vol. 10. October 2020; quoted in: https://www.latimes.com/california/story/2020-08-17/sea-level-rise-flooding-inland-california.

57. Quoted in: https://www.latimes.com/california/story/2020-08-17/sea-level-rise-flooding-inland-california.

58. See: https://www.latimes.com/local/lanow/la-me-california-coast-storm-damage-20190313-story.html; for research by Barnard et al. see: https://www.usgs.gov/center-news/rising-seas-and-storms-could-seriously-damage-california-s-coast-within-30-years?qt-news_science_products=1#qt-news_science_products.

59. See: https://www.nature.com/articles/s41598-019-40742-z.

60. See: https://www.boston.gov/sites/default/files/embed/2/20161207_climate_ready_boston_digital2.pdf.

61. See: https://www.sciencedaily.com/releases/2018/06/180601134756.htm.

62. See: https://www.washingtonpost.com/climate-solutions/2020/02/19/boston-prepares-rising-seas-climate-change/?arc404=true.

63. See: https://www.ucsusa.org/resources/opa-locka-and-hialeah-florida-grappling-decades-storm-impacts-2015.

64. See: https://www.ucsusa.org/resources/opa-locka-and-hialeah-florida-grappling-decades-storm-impacts-2015.

65. Ibid.

66. See: https://apple.news/AFPZOvy3VSTqy0RMtJ622pw.

67. Quoted in: https://apple.news/AFPZOvy3VSTqy0RMtJ622pw.

68. See: https://www.sfwmd.gov/sites/default/files/documents/climate_change_and_water_management_in_sflorida_12nov2009.pdf.

69. See: https://apple.news/APBDcfD8JRXqELl4pVpLHWA.

70. See: https://apple.news/AhD7FdpYETC6VXi1ATgmgFw.

71. See: https://www.npr.org/2020/11/13/934257652/kotchakorn-voraakhom-how-can-we-better-design-cities-to-fight-floods.

72. See: https://www.ucsusa.org/sites/default/files/attach/2018/06/underwater-analysis-full-report.pdf.

73. Ibid. P. 2

74. McNamara et.al 2015, quoted in Underwater: Rising Seas, Chronic Floods, and Its Implications for US Coastal Real Estate. Union of Concerned Scientists 2018. p. 2.

75. See: http://dx.doi.org/10.1038/nclimate1979.

Chapter 3

1. See: https://www.statista.com/statistics/720418/average-monthly-cost-of-water-in-the-us/.

2. See: https://www.ksbw.com/article/monterey-peninsula-water-most-expensive-in-us/10269382#.

3. Permafrost is permanently frozen ground that may contain ice. We'll delve much more into this in Chapter 8.

4. Jamail, Dahr. The End of Ice: Bearing Witness and Finding Meaning in the Path of Climate Disruption. The New Press. 2019. P. 45.

5. Rock glaciers are considered part of permafrost environments.

6. See: https://www.usgs.gov/special-topic/water-science-school/science/how-much-water-there-earth?qt-science_center_objects=0#qt-science_center_objects.

7. Ice and water volume differ by about 8%. Water is denser than ice and therefore a cubic meter (m^3) of water contains a bit more water than a cubic meter of ice (not much more). But, for the sake of keeping the math simple, the calculation presumes a one-to-one ratio of ice to water, which actually results in a slight overestimation of water content in ice (by about 8%). To offset this, I've presumed that the glacier is actually thinner than it likely is; hence, in the end, these calculations are probably still very conservative and underestimate the true amount of water in these glaciers.

8. That's 8.9 million × 100 = 890 million. In a year (365 days), the population consumes 324,850,000,000 (or 324.9 billion) liters of water. Divide 1.2 trillion by 324.9 billion and you get 3.7 years of water supply!

9. Quoted in: Lovett, Richard A. Melting Glaciers Mean Double Trouble for Water Supplies. National Geographic. December 21, 2011.

10. Davies, Bethan. AntarcticGlaciers.org. Last updated: 6/7/20. Glaciers as a water resource. See: http://www.antarcticglaciers.org/glaciers-and-climate/glacier-recession/glaciers-as-a-water-resource/.

11. See: Baraer, Michael. Quoted in Lovett, Richard A. Melting Glaciers Mean Double Trouble for Water Supplies. National Geographic. December 21, 2011.

12. Inter-Governmental Panel on Climate Change (IPCC). See: https://www.ipcc.ch/srocc/about/faq/faq-chapter-2/.

13. See: https://www.youtube.com/watch?v=R7-4QWtMLm4

14. Cited in: Davies, Bethan. AntarcticGlaciers.org. Last updated: 6/7/20. Glaciers as a water resource. See: http://www.antarcticglaciers.org/glaciers-and-climate/glacier-recession/glaciers-as-a-water-resource/.

15. See: http://www.grid.unep.ch/activities/global_change/central_asia_glacier.php.

16. Xu, Baiqing, et al. Black Soot and the Survival of Tibetan Glaciers. Proceedings of the National Academy of Sciences. 2009. Vol. 106, No. 52. Pp. 22114–22118.

17. Wester, P., et al. The Hindu Kush Himalaya Assessment: Mountains, Climate Change, Sustainability and People. HIMAP and ICIMOD. 2019.

18. Cited in FAQ21: Summary for Policy Makers of the Special Report on the Ocean and Cryosphere in a Changing Climate. IPCC. 2019. See: https://www.ipcc.ch/srocc/about/faq/faq-chapter-2/.

19. Immerzeel, W.W., et al. Importance and Vulnerability of the World's Water Towers. Nature. Vol. 577. January 16, 2020. P. 364.

20. The Ocean and Cryosphere in a Changing Climate. UN IPCC. 2019. Pp. 1-52–1-53.
21. Cited in: Miller, Jeremy. The Dying Glaciers of California. Earth Island Journal. Vol. 28, No. 2. Summer 2013. https://www.earthisland.org/journal/index.php/magazine/entry/the_dying_glaciers_of_california/.
22. Milner, Alexander. Glacier Shrinkage Driving Global Changes in Downstream Systems. Proceedings of the National Academy of Sciences. September 12, 2017. Vol. 114. No. 37. P. 9770-9778.
23. Cited in: Miller, Jeremy. The Dying Glaciers of California. Earth Island Journal. Vol. 28, No. 2. Summer 2013. https://www.earthisland.org/journal/index.php/magazine/entry/the_dying_glaciers_of_california/.
24. Jamail, Dahr. The End of Ice: Bearing Witness and Finding Meaning in the Path of Climate Disruption. The New Press. 2019. P. 44.
25. See: https://www.nytimes.com/interactive/2019/04/17/climate/melting-glaciers-globally.html.
26. See: https://climate.nasa.gov/news/2989/ice-melt-linked-to-accelerated-regional-freshwater-depletion/.

Chapter 4

1. In this book, ice sheets (massive ice extensions) will be considered glaciers. This is different from sea ice, which develops over the ocean and can be seasonal.
2. Hoffman, Paul, and Schrag, Daniel. Snowball Earth. Harvard University. 1999.
3. See: https://www.theguardian.com/books/2019/jul/03/on-reflection-albedo-effect-word-of-the-week.
4. In my book Glaciers: The Politics of Ice, I tell the story of Eduardo Gold in Peru, a self-trained glaciologist, experimenting with whitewashing mountainsides to better reflect light. He won a World Bank prize in 2010 for his attempt at glacier creation by painting a mountain on the Chalon Sombrero peak at 4,756 meters above sea level (15,600 feet) near Ayacucho. He uses environmentally friendly ingredients: lime, industrial egg white, and water. The project is based on the idea that by changing the albedo (the Earth's reflective capacity) we can cool the ground, create a cooler surface temperature, and thus make it more ice friendly. He claims that surface temperature is many degrees cooler where he has painted the rocks white. Slowly, ice is accumulating at these sites.
5. See: https://smartsurfacescoalition.org.
6. This is based on a federal EPA study utilizing a city of 1 million people as a reference. See: https://www.epa.gov/heat-islands.
7. See: The Heat Island Group. Berkeley Labs at: https://heatisland.lbl.gov/coolscience/urban-heat-islands.
8. For a comprehensive description of roof albedo and actions that can be taken to cool roofs and reduce heat intensity, see: A Practical Guide to Cool Roofs and Cool Pavements. Global Cool Cities Alliance (GCCA), at: https://www.coolrooftoolkit.org/wp-content/pdfs/CoolRoofToolkit_Full.pdf.
9. National Snow and Ice Data Center (NSIDC). Thermodynamics: Albedo. April 3, 2020 See: https://nsidc.org/cryosphere/seaice/processes/albedo.html

10. Pirazzini, Roberta. Surface Albedo Measurements over Antarctica Sites in Summer. Journal of Geophysical Research, Vol. 109, D20018. Doi: 10.1029/2004JD004617, 2004. https://agupubs.onlinelibrary.wiley.com/doi/pdf/10.1029/2004JD004617.

11. Qian, Y. et.al. Sensitivity Studies On The Impacts of Tibetan Plateau Snowpack Pollution On The Asian Hydrological Cycle And Monsoon Climate. In Atmospheric Chemistry and Physics. Volume 11, No. 5. Pp. 1929–48. 2011. See: https://acp.copernicus.org/articles/11/1929/2011/

12. Xu, Baiqing. Black Soot and the Survival of Tibetan Glaciers. Proceedings of the National Academy of Sciences. 2009.

13. United Nations Environmental Program (UNEP). Recent Trends in Melting Glaciers, Tropospheric Temperatures Over the Himalayas and Summer Monsoon Rainfall over India. UNEP 2009. P.3 See: https://na.unep.net/siouxfalls/publications/Himalayas.pdf

14. The Little Ice Age is a period of especially cold weather that lasted several hundred years during the 1400s to the 1800s when, around the world, there was plentiful snow. It was especially significant for mountain glaciers in places like the European Alps, the Central Andes, and the California Sierra Nevada and Rocky Mountains, when these mountain glaciers grew significantly.

15. Painter, Thomas H., Mark G. Flanner, Georg Kaser, Ben Marzeion, Richard A. VanCuren, and Waleed, Abdalati. End of Little Ice Age in the Alps Forced by Industrial Black Carbon. Proceedings of the National Academy of Sciences of the United States of America. Volume 110, No. 38. September 17, 2013. Pp. 15216–21 See: https://www.pnas.org/content/110/38/15216.

16. Ibid. P. 15218

17. Thomas, J. L., C. M. Polashenski, A. J. Soja, L. Marelle, K. A. Casey, H. D. Choi, et al. Quantifying Black Carbon Deposition over the Greenland Ice Sheet from Forest Fires in Canada. Geophysical Research Letters 44, 15. August 16, 2017. PP. 7965–74. See: https://agupubs.onlinelibrary.wiley.com/doi/full/10.1002/2017GL073701

18. Quoted in: Los Angeles Times. By Associated Press. August 20, 2020. See: https://www.latimes.com/world-nation/story/2020-08-20/greenland-record-melt-lost-billions-tons-ice-2019.

19. Sasgen, Ingo, et al. Return to Rapid Ice Loss in Greenland and Record Loss in 2019 Detected by the GRACE-FO Satellites. Communications Earth and Environment 1. Article number: 8. August 2020. P. 2 See: https://www.nature.com/articles/s43247-020-0010-1

20. See: https://www.theguardian.com/world/2020/jan/02/new-zealand-glaciers-turn-brown-from-australian-bushfires-smoke-ash-and-dust.

21. Quoted in the Guardian, see: https://www.theguardian.com/world/2020/jan/02/new-zealand-glaciers-turn-brown-from-australian-bushfires-smoke-ash-and-dust.

22. See: https://theconversation.com/amazon-fires-are-causing-glaciers-in-the-andes-to-melt-even-faster-128023.

23. de Magalhaes, Newton, et al. Amazonian Biomass Burning Enhances Tropical Andean Glaciers Melting. Scientific Reports. 9. Article Number 16914. 2019.

24. See: https://blogs.ei.columbia.edu/2020/10/21/forest-fires-impact-glaciers/

25. Quoted in Phys.org; see: https://phys.org/news/2020-10-qa-year-forest-impact-glaciers.html.

26. See for example, five volcano eruptions caught on video: https://www.youtube.com/watch?v=VBTAcACmcgo; Calbuco Volcano (Chile): https://www.youtube.com/ watch?v=juwtnTB1RCU; Mt. St. Helens (USA): https://www.youtube.com/watch?v= AYla6q3is6w; Fuego Volcano (Guatemala): https://www.youtube.com/watch?v= 7MG9Z9sVROQ; Mt. Ruapehu (New Zealand): https://www.youtube.com/watch?v=h8W_sGYAQlc; Krakatoa (Indonesia): https://www.youtube.com/watch?v=NLhjNzQHphQ; Pinatubo Volcano (Philippines): https://www.youtube.com/watch?v=fSSvl_UcNB4; see also: https://www.youtube.com/watch?v=rjVZdlMlaiY.

27. This section was developed in conversations and consultations with the climate policy organization Institute for Governance and Sustainable Development (IGSD). All information in this section is attributable to IGSD unless otherwise noted. A special thanks goes to Durwood Zaelke, Gabrielle Dreyfus, and Kristen Leigh Campbell for their invaluable work.

28. For more information on Short Lived Climate Pollutants and their contribution to climate change see: http://www.igsd.org/documents/PrimeronShort-LivedClimate PollutantsNovemberElectronicversion.pdf.

29. Zaelke, Durwood, et al. Primer on Polar Warming and Implications for Climate Change. IGSD. May 2019. http://www.igsd.org/wp-content/uploads/2019/05/Primer-on-Polar-Warming.pdf.

30. Institute for Governance and Sustainable Development (IGSD). Primer on Polar Warming and Implications for Global Climate Change. May 2019. See: http://www.igsd.org/wp-content/uploads/2019/05/Primer-on-Polar-Warming.pdf.

31. National Snow and Ice Data Center. In Arctic Sea Ice News and Analysis. Rapid Ice Loss in Early April Leads to New Record Low. NSIDC. May 2, 2019 See: http://nsidc.org/arcticseaicenews/2019/05/rapid-ice-loss-in-early-april-leads-to-new-record-low/.

32. See: https://www.vox.com/22295520/climate-change-shipping-russia-china-arctic

33. Op.cit, NSIDC. 2019

34. Sasgen, Ingo, et al. Return to Rapid Ice Loss in Greenland and Record Loss in 2019 Detected by the GRACE-FO Satellites. Communications Earth and Environment. Vol. 1. Article number 8. 2020. https://www.nature.com/articles/s43247-020-0010-1.

35. See: https://www.noaa.gov/stories/what-are-atmospheric-rivers.

36. See: https://climate.nasa.gov/news/2740/climate-change-may-lead-to-bigger-atmospheric-rivers/.

37. Hobbs, W. R. Quoted in: Zhou, Chunxia, et al. The Characteristics of Surface Albedo Change Trends over the Antarctic Sea Ice Region during Recent Decades. Remote Sensing 11, 821. DOI: 10.3390/rs11070821. April 2019. P.2.

38. See: https://www.amnh.org/learn-teach/curriculum-collections/antarctica/extreme-temperatures/temperature-albedo; the article does not provide Stephanie's last name, sorry!

39. Parkinson, Claire L. A 40-y Record Reveals Gradual Antarctic Sea Ice Increases Followed by Decreases at Rates Far Exceeding the Rates Seen in the Arctic. Proceedings of the National Academy of Sciences. July 16, 2019. Vol. 116, No. 29. Pp. 14414–14423. See: https://www.pnas.org/content/116/29/14414.

40. See: https://www.nationalgeographic.com/science/2020/02/antarctica-pine-island-glacier/.

41. Quoted in the Guardian: https://www.theguardian.com/world/2019/jul/01/precipitous-fall-in-antarctic-sea-ice-revealed.
42. See: https://www.theguardian.com/world/2020/feb/07/antarctica-logs-hottest-temperature-on-record-with-a-reading-of-183c.
43. See: https://scitechdaily.com/evidence-of-antarctic-glaciers-tipping-point-confirmed-for-first-time-risk-of-rapid-and-irreversible-retreat/

Chapter 5

1. Depending on how you measure it (in the short or long term), methane is anywhere from 25 to about 80 times more potent than CO_2 in terms of its global warming potential. This is incredibly important since despite what the oil and gas industry have told us, natural gas is *not* a transition fuel toward a cleaner climate. If leaked into the atmosphere, it is actually *worse* than CO_2 in terms of global warming.
2. Davis, Neil. Permafrost: A Guide to Frozen Ground in Transition. University of Alaska Press. 2001.
3. Govorushko, S. M. Cryogenic Processes and Their Impact on Infrastructures. In Pokrovsky. Permafrost: Distribution, Composition and Impacts on Infrastructure and Ecosystems. Nova Publishers. 2014. P. 2.
4. Institute for Governance and Sustainable Development (IGSD). Primer on Polar Warming and Implications for Global Climate Change. May 2019. See: http://www.igsd.org/wp-content/uploads/2019/05/Primer-on-Polar-Warming.pdf. P.25
5. This section is adapted from Taillant, Jorge Daniel. Glaciers: The Politics of Ice. Oxford University Press. 2015. P. 216.
6. One example is Conoco Philips who conducted a joint venture project with Jocmeg in Prudhoe Bay, Alaska, attempting to extract gas hydrates from an ice field. The project's name is IGNIK SIKUMI #1, which means in the local tribal language, Fire in Ice. See: https://youtu.be/zfgnaeBPZKY.
7. You can download the Global Permafrost Zoning Index Map at: http://www.geo.uzh.ch/microsite/cryodata/pf_global/.
8. See: Kasischke, Eric S., et al. The Arctic-Boreal Vulnerability Experiment: A Concise Plan for a NASA-Sponsored Field Campaign. October 2010. P. 12.
9. Scientific American. As the Earth Warms, the Diseases that May Lie within Permafrost Become a Bigger Worry. November 1, 2016.
10. Pokrovsky, Oleg S. Permafrost: Distribution, Composition and Impacts on Infrastructure and Ecosystems. 2014. P. viii.
11. For a video of this phenomenon, see: https://www.youtube.com/watch?v=YegdEOSQotE
12. See: https://www.science.org.au/curious/video/exploding-methane-gas-bubbles.
13. Jamail, Dahr. The End of Ice: Bearing Witness and Finding Meaning in the Path of Climate Disruption. The New Press. 2019. P. 191
14. The Ocean and Cryosphere in a Changing Climate: Summary for Policy Makers. Intergovernmental Panel on Climate Change (IPCC), 2019. Pp. SPM-4, SPM-20
15. See: https://www.sciencedirect.com/science/article/abs/pii/S1871174X16300488.
16. See: https://www.independent.co.uk/environment/earth-permian-mass-extinction-apocalypse-warning-climate-change-frozen-methane-a7648006.html.

17. For more on the Holocene Extinction, see: https://en.wikipedia.org/wiki/Holocene_extinction.

18. See: http://www.igsd.org/wp-content/uploads/2019/05/Primer-on-Polar-Warming.pdf.

19. Jamail, Dahr. Op.cit. pp. 194–96.

20. Ibid. pp. 198–99.

21. See: https://www.washingtonpost.com/news/capital-weather-gang/wp/2013/07/25/methane-mischief-misleading-commentary-published-in-nature/.

22. See: https://www.aces.su.se/research/projects/the-isss-2020-arctic-ocean-expedition/.

23. See: https://www.theguardian.com/science/2020/oct/27/sleeping-giant-arctic-methane-deposits-starting-to-release-scientists-find.

24. Institute for Governance and Sustainable Development (IGSD). Primer on Polar Warming and Implications for Global Climate Change. May 2019. P. 28

25. A thermokast is a term for a type of land surface that occurs when ice melts in permafrost, creating small sinkhole-like pits and valleys as the ground settles unevenly. See: https://www.cnn.com/2019/06/11/americas/thermokarst-arctic-climate-change-intl-hnk

26. Govorushko, S. M. Cryogenic Processes and Their Impact on Infrastructures. In Pokrovsky. Permafrost: Distribution, Composition and Impacts on Infrastructure and Ecosystems. Nova Publishers. 2014. P. 20.

27. Quoted in: Simon, Matt. Permafrost Thawing so Fast, It's Gouging Holes in the Arctic. Wired. February 2020.

28. See: https://pdnpulse.pdnonline.com/2018/08/a-sense-of-real-fear-climate-change-photog-katie-orlinsky-on-documenting-arctic-melt.html

29. Troianovsky, Anton, and Mooney, Chris. Radical Warming in Siberia Leaves Millions on Unstable Ground. Washington Post. October 3, 2019.

30. Ibid.

31. Ibid.

32. See: As Earth Warms, the Diseases that May Lie Within Permafrost Become a Bigger Worry. Scientific American. November 1, 2016. https://www.scientificamerican.com/article/as-earth-warms-the-diseases-that-may-lie-within-permafrost-become-a-bigger-worry/.

33. Ibid.

34. Fox-Skelly, Jasmin. There are Diseases Hidden in Ice, and They Are Waking Up. BBC. May 4, 2017. See: http://www.bbc.com/earth/story/20170504-there-are-diseases-hidden-in-ice-and-they-are-waking-up.

35. Pikuta, Elena V. et al. *Canobacterium pleistocenium* sp. nov., a Novel Psychrotolerant, Facultative Anaerobe Isolated from Permafrost of the Fox Tunnel in Alaska. International Journal of Systematic and Evolutionary Microbiology. Vol. 55, No. 1. 2005.

36. Ackerman, Daniel. Ancient Life Awakens Amid Thawing Ice Caps and Permafrost. Washington Post. July 7, 2019.

37. Ibid.

38. See: https://www.newscientist.com/article/2191292-weve-dug-up-tiny-animals-from-beneath-a-frozen-antarctic-lake/.

39. Ackerman, Daniel. Ancient life Awakens Amid Thawing Ice Caps and Permafrost. Washington Post. July 7, 2019.

40. Fox-Skelly, Jasmin. There Are Diseases Hidden in Ice, and They Are Waking Up. BBC. May 4, 2017.

41. Doucleff, Michaeleen. Are There Zombie Viruses in the Thawing Permafrost. NPR. January 24, 2018. See: https://www.npr.org/sections/goatsandsoda/2018/01/24/575974220/are-there-zombie-viruses-in-the-thawing-permafrost.

42. See: http://www.igsd.org/wp-content/uploads/2019/05/Primer-on-Polar-Warming.pdf. P. 27.

Chapter 6

1. See: https://www.youtube.com/watch?v=8ScHkS_cqk4
Yet another GLOF took place near the time of final editing of this chapter, in Uttarakhand state in northern India. You can see that video at: https://www.cnn.com/2021/02/07/india/india-glacier-flash-flood-intl/index.html

2. See the incredible documentary of the eruption of Mount St. Helens: https://www.youtube.com/watch?v=fArB5Jz2wos&t=971s.

3. See: https://en.wikipedia.org/wiki/Mount_St._Helens.

4. See: Hambrey, Michael, and Alean, Jurg. Glaciers. Cambridge University Press. 1992. P. 124; see also: https://en.wikipedia.org/wiki/1980_eruption_of_Mount_St._Helens.

5. Jamail, Dahr. The End of Ice: Bearing Witness and Finding Meaning in the Path of Climate Disruption. The New Press. 2019. P. 35.

6. Hambrey, Michael, and Jurg Alean. Glaciers. Cambridge University Press. 1992. P. 119.

7. Ibid.

8. Kellerer, Andreas, Heinz Slupetzky, and Michael Avian. Ice-Avalanche Impact Landforms. The Event in 2003 at the Glacier Nördliches Bockkarkees, Hohe Tauern Range, Austria. Goegrafiska Annaler. Series A, Physical Geography 94, 1. 2012. P. 97, and Hambrey, Michael, and Jurg Alean. Glaciers. Cambridge University Press. 1992. Pp. 186-189.

9. Carey, Mark. In the Shadow of Melting Glaciers. Oxford University Press. 2010. P. 5.

10. Ibid. P. 7.

11. Maza, Jorge. Crecida de Diseño Generada por la Rotura del Endicamiento Glaciar Grande del Nevado del Plomo en el Río Plomo. Instituto Nacional del Agua. IT 175-CRA. 2016. Pp. 9-19.

12. Hambrey, Michael, and Jurg Alean. Glaciers. Cambridge University Press. 1992. Pp. 180-181.

13. Ibid. p. 181

14. Dickerman, Kenneth, and Whitlow Delano, James. How Climate Change is Affecting the Italian Alps. Washington Post. August 23, 2019.

15. Horstmann, Britta. Glacial Lake Outburst Floods in Nepal and Switzerland: New Threats Due to Climate Change. Germanwatch. 2004. P. 3.

16. See: ICIMOD. Inventory of Glacier Lakes and Identification of Potentially Dangerous Glacial Lakes in Koshi, Gandaki, and Karnali River Basins of Nepal, the Tibet Autonomous Region of China, and India. 2020. https://www.np.undp.org/content/dam/nepal/docs/reports/environment%20and%20energy/Inventory_Glacial_Lakes_2020_Full2.pdf.

17. Stokes, C. R., et al. Recent Glacier Retreat in the Caucasus Mountains, Russia, and Associated Increase in Supraglacial Debris Cover and Supra-/Proglacial Lake Development. Annals of Geology 46. 2007.

18. See: https://www.nps.gov/articles/taanfjord.htm.

See: https://www.washingtonpost.com/energy-environment/2018/09/06/one-biggest-tsunamis-ever-recorded-was-set-off-three-years-ago-by-melting-glacier/.

19. See: https://link.springer.com/article/10.1007/s10346-019-01225-4?shared-article-renderer.

20. See: https://www.nature.com/news/huge-landslide-triggered-rare-greenland-mega-tsunami-1.22374.

21. Hambrey, Michael, and Jurg Alean. Op.cit. pp. 71–72

22. See: https://www.nytimes.com/interactive/2021/04/13/climate/muldrow-glacier-alaska-mount-denali.html

Chapter 7

1. Summary for Policymakers. In: IPCC Special Report on the Ocean and Cryosphere in a Changing Climate. 2019. P. 3-11. You can download chapters or all of the report here: https://www.ipcc.ch/srocc/.

2. The Ocean and Cryosphere in a Changing Climate: Summary for Policy Makers. Intergovernmental Panel on Climate Change (IPCC), 2019. P. 3-12.

3. See: Zaelke, Durwood, et al. Primer on Polar Warming and Implications for Climate Change. IGSD. May 2019. P. 16. See: http://www.igsd.org/wp-content/uploads/2019/05/Primer-on-Polar-Warming.pdf.

4. On THC see: https://oceanservice.noaa.gov/education/tutorial_currents/05conveyor1.html.

5. On AMOC see: https://en.wikipedia.org/wiki/Atlantic_meridional_overturning_circulation.

6. The Ocean and Cryosphere in a Changing Climate: Summary for Policy Makers. Intergovernmental Panel on Climate Change (IPCC), 2019. P. 3-17

7. Green, Clare, et al. Simulating the Impact Of Freshwater Inputs and Deep-Draft Icebergs Formed during a MIS 6 Barents Ice Sheet Collapse. Paleoceanography and Paleoclimatology. May 2011. P. 1. See also: https://phys.org/news/2011-05-quantifying-glaciers-effect-ocean-currents.html.

8. See: https://climate.nasa.gov/news/2950/arctic-ice-melt-is-changing-ocean-currents/.

9. Quoted in: https://phys.org/news/2011-05-quantifying-glaciers-effect-ocean-currents.html.

10. The Ocean and Cryosphere in a Changing Climate: Summary for Policy Makers. Intergovernmental Panel on Climate Change (IPCC), 2019. P. 3-12. P. 3-19

11. Berwyn, Bob. Polar Vortex: How the Jet Stream and Climate Change Bring on Cold Snaps. Inside Climate News. February 2008.

12. Quoted in: McGowan, Jake. Arctic Ice Effects Global Temperatures, Jet Streams. Daily Targum. March 31, 2019.

13. Ibid.

14. Radford, Tim. Arctic Sea Ice Loss Affects the Jet Stream. Climate News Network. 2019.

15. Climate One Come Together. Can Greenland's Melting Affect the Weather. 2013. See: https://www.youtube.com/watch?v=d9Urq1mZ5KA.

16. Chen, Jenny. Melting Glaciers are Wreaking Havoc on the Earth's Crust. Smithsonian Magazine. September 1, 2016.

17. Dean, Cornelia. As Alaska's Glaciers Melt, It's Land That's Expanding. New York Times. May 17, 2009.

18. Fountain, Henry, and Solomon, Ben. Where Glaciers Melt Away, Switzerland Sees Opportunity. New York Times Special Report. February 14, 2019. See: https://www.nytimes.com/interactive/2019/04/17/climate/switzerland-glaciers-climate-change.html.

19. See: https://en.wikipedia.org/wiki/Cordillera_Blanca.

20. See Carey, Mark. In the Shadow of Melting Glaciers. Oxford University Press. 2010. P. 7.

21. See: https://image.slidesharecdn.com/yarmeymetadatabrokeringrdap2014-140326174649-phpapp02/95/rdap14-data-discovery-and-access-through-metadata-brokering-3-638.jpg?cb=1395856069

22. The Ocean and Cryosphere in a Changing Climate: Summary for Policy Makers. Intergovernmental Panel on Climate Change (IPCC), 2019. Chapter 3.

23. See: Pizzly or Grolar Bear: Grizzly-Polar Hybrid Is a New Result of Climate Change. Guardian. May 18, 2016.

24. Fountain, Henry. When Glaciers Disappear, Those Species Will Go Extinct. New York Times. April 17, 2019.

25. Ibid.

26. See: https://glaciers.nichols.edu/salmon/.

27. Ibid.

28. See: https://vimeo.com/129600600.

29. Schoen, Erik R., et al. Future of Pacific Salmon in the Face of Environmental Change. Lessons from One of the World's Remaining Productive Salmon Regions. Fisheries 42. 2017. See: https://www.tandfonline.com/doi/full/10.1080/03632415.2017.1374251.

30. Milner, Alexander M., et al. Glacier Shrinkage Driving Global Changes in Downstream Systems. Proceedings of the National Academy of Sciences. September 5, 2017. P.9770..

31. Climate One Come Together. Can Greenland's Melting Affect the Weather. 2013. See: https://www.youtube.com/watch?v=d9Urq1mZ5KA.

32. For the videos of the Watson River swelling see: https://www.youtube.com/watch?v=RauzduvIYog; https://www.youtube.com/watch?v=X8iUXX-JT90; https://www.youtube.com/watch?v=7SuJ1sFn_B0.

33. Milner, Alexander M., et al. Glacier Shrinkage Driving Global Changes in Downstream Systems. Proceedings of the National Academy of Sciences. September 5, 2017.

34. Quoted in: Dean, Cornelia. As Alaska's Glaciers Melt, It's Land That's Expanding. New York Times. May 17, 2009. See: https://www.nytimes.com/2009/05/18/science/earth/18juneau.html

35. Aber, James. Glacial Isostasy and Eustasy. ES 331/767. Lecture 9. (Quoting Blum et.al 2008) See: http://www.geospectra.net/academic/glacial/lec09/lec9.htm

36. Jamail, Dahr. The End of Ice: Bearing Witness and Finding Meaning in the Path of Climate Disruption. The New Press. 2019. Pp.41–43.

Chapter 8

1. You can listen to Jared Blumenfeld's podcast about the visit to the Sierra Nevada's rock glaciers at: https://www.podshipearth.com/rockglacier.

2. Five hundred meters is used merely as an example as reference for the reader, it could be more or less. There is no precise measurement.

3. For CHRE's report on rock glaciers and periglacial environments in California's Sierra Nevada (and the rock glacier inventory), see: https://center-hre.org/wp-content/uploads/Rock-Glaciers-and-Periglacial-Environments-in-Californias-Sierra-Nevada-December-17-2019.pdf.

4. Blumenfeld, Jared. Rock Glacier. Podcast in the Podship Earth Series. Episode 67. 2019. https://www.podshipearth.com.

5. Whalley, W. Brian, and Fethi Azizi. Rock Glaciers and Protalus Land Forms: Analogous Forms and Ice Sources on Earth and Mars. Journal of Geophysical Research 108, E4. P. 8032. 2003.

6. See: https://www.arcgis.com/home/item.html?id=cfca8703c575497bab271627a7f2160b.

7. French, Hugh M. The Periglacial Environment. Third Edition. Wiley. 2008. P. 224

8. Rock glaciers can be considered "permafrost" (they are actually one element of permafrost environments) and can also be referred to as "frozen grounds."

9. *Gelifluction* is a type of *solifluciton*, more specifically implying the presence of ice.

10. See: Barsch, Dietrich. Rock-Glaciers: Indicators for the Present and Former Geoecology in High Mountain Environments. 1996. P. 31.

11. Ibid. P. 22.

12. See: 39°25'40.67" N 106°06'59.67" W.

13. A proper calculation of the hydrological content of a rock glacier is a complicated process. The calculation offered here is simply a conservative approximation utilizing visual estimates for the sake of example. The numbers chosen are purposefully lower than the likely true measurements so as to calculate a *fair* lower limit of the actual number.

14. See: 31°15'57.33" S 70°10'40.10" W.

Chapter 9

1. Taillant, Jorge Daniel and Peter Collins, Cryoactivism. LASAFORUM. Fall 2016. Volume XLVII. Issue 4. 2016. Pp. 34–38. See: https://forum.lasaweb.org/files/vol47-issue4/Debates7.pdf.

2. Archer, David. The Long Thaw. Princeton, NJ. Princeton University Press. 2009. Pp. 156–157.

3. Taillant, Jorge Daniel. Glaciers: The Politics of Ice. Oxford University Press. 2015. Pp. 219–220.

4. See: https://www.youtube.com/watch?v=aCacMeSOxAg; see also: https://www.cnn.com/style/article/ice-stupa-sonam-wangchuk/index.html.

5. Tveiten, Ingvar Norstegard. Glacier Growing – A Local Response to Water Scarcity in Baltistan and Gilgit, Pakistan. Norwegian University of Life Sciences. Master's Thesis. 2007. See: http://www.umb.no/statisk/noragric/publications/master/2007_ingvar_tveiten.pdf
    See also: https://www.akdn.org/project/glacier-growing

6. See: Taillant, Jorge Daniel. Glaciers: The Politics of Ice. Oxford University Press. 2015. P. 219.

7. See: https://www.motherjones.com/environment/2019/09/arctic-ice-is-melting-faster-than-expected-these-scientists-have-a-radical-idea-to-save-it/.

8. For a wonderful TED Talk by M Jackson on people and ice, see: https://www.youtube.com/watch?v=QFNzKPVirt4.

9. See: https://www.npr.org/2020/11/13/934257652/kotchakorn-voraakhom-how-can-we-better-design-cities-to-fight-floods.

10. See: Taillant, Jorge Daniel. Glaciers: The Politics of Ice. Oxford University Press. 2015.

11. For the unofficial English translation, see: http://center-hre.org/wp-content/uploads/Argentine-National-Glacier-Act-Traducción-de-CEDHA-no-oficial.pdf; for the Spanish original, see: http://center-hre.org/wp-content/uploads/Ley-de-Glaciares-Argentina.pdf.

12. For the text of the Paris Agreement see: https://unfccc.int/files/essential_background/convention/application/pdf/english_paris_agreement.pdf.

13. For the text of the UN Framework Convention on Climate Change (signed in 1992) see: https://unfccc.int/files/essential_background/background_publications_htmlpdf/application/pdf/conveng.pdf.

14. See: http://www.igsd.org/documents/PrimeronShort-LivedClimatePollutantsNovember Electronicversion.pdf.

15. This content of this section is adapted from the IGSD Primer on Short Lived Climate Pollutants.

16. See: https://www.epa.gov/ghgemissions/overview-greenhouse-gases.

17. See: https://www.scientificamerican.com/article/how-bad-of-a-greenhouse-gas-is-methane/.

18. See: https://ww2.arb.ca.gov/our-work/programs/aliso-canyon-natural-gas-leak.

19. See: https://en.wikipedia.org/wiki/Aliso_Canyon_gas_leak.

20. See: https://www.youtube.com/watch?v=sVQ-WZ8f0C8.

Chapter 10

1. Wu X., Nethery R. C., Sabath M. B., Braun D., and Dominici, F. Air Pollution and COVID-19 Mortality in the United States: Strengths and Limitations of an Ecological Regression Analysis. Science Advances. Vol. 6. No. 45. 2020. Article Number p.eabd4049.

2. Lenton T. M., et al. Climate Tipping Points—Too Risky to Bet Against. Nature. Comment. Vol. 575. 2019. Pp. 592–595, 592 ("Models suggest that the Greenland ice sheet could be doomed at 1.5 °C of warming, which could happen as soon as 2030. ... The world's remaining emissions budget for a 50:50 chance of staying within 1.5 °C of warming is only about 500 gigatonnes (Gt) of CO_2. Permafrost emissions could take an estimated 20% (100 Gt CO_2) off this budget, and that's without including methane from deep permafrost or undersea hydrates. If forests are close to tipping points, Amazon dieback could release another 90 Gt CO_2 and boreal forests a further 110 Gt CO_2. With global total CO_2 emissions still at more than 40 Gt per year, the remaining budget could be all but erased already. ... We argue that the intervention time left to prevent tipping could already have shrunk toward zero, whereas the reaction time to achieve net zero emissions is 30 years at best. Hence we might already have lost control of whether tipping happens. A saving grace is that the rate at which damage accumulates from tipping—and hence the risk posed—could still be under our control to some extent.").

3. Allen, M., et al. (2018) Summary for Policymakers, in IPCC (2018) Global Warming of 1.5 °C, 6. ("Human activities are estimated to have caused approximately 1.0°C of global

warming above pre-industrial levels, with a likely range of 0.8°C to 1.2°C. Global warming is likely to reach 1.5°C between 2030 and 2052 if it continues to increase at the current rate (high confidence).") In addition to cutting CO_2 emissions and emissions of the super climate pollutants, the IPCC 1.5°C Report also calculates the need for significant CO_2 removal. Ibid., 17. ("C.3. All pathways that limit global warming to 1.5°C with limited or no overshoot project the use of carbon dioxide removal (CDR) on the order of 100–1000 Gt CO_2 over the 21st century.").

4. Feyrer, James, and Sacerdote, Bruce. Did the Stimulus Stimulate? Real Time Estimates of the Effects of the American Recovery and Reinvestment Act. National Bureau of Economic Research. 2011.

See: https://www.nber.org/papers/w16759.

Bibliography

Aber, James. Glacial Isostasy and Eustasy. ES 331/767. Lecture 9. See: http://www.geospectra. net/academic/glacial/lec09/lec9.htm

Ackerman, Daniel. Ancient Life Awakens Amid Thawing Ice Caps and Permafrost. Washington Post. July 7, 2019. See: https://www.washingtonpost.com/science/ancient-life-awakens-amid-thawing-ice-caps-and-permafrost/2019/07/05/335281f8-7108-11e9-9f06-5fc2ee80027a_story.html

Aedo, Maria Paz, and Teresa Montesinos. Glaciares Andinos: Recursos Hídricos y Cambio Climático: Desafíos para la Justicia Climática en el Cono Sur. Chile Sustentable. 2011.

Ahumada, Ana Lia. Periglacial Phenomena in the High Mountains of Northwestern Argentina. South African Journal of Science 98, March/April 2002. Pp.166–170.

Ahumada, Ana Lia, S. V. Paez, and G. Ibañez Palacios. Los Glaciares de Escombros en la Alta Cuenca del Río Andalgalá, SE de la Sierra de Aconquija, Catamarca. VIII Congreso Geológico Argentino. May 2011.

Ahumada, Ana Lia, et al. El Permafrost Andino, Reducto de la Criósfera en el Borde Oriental de la Puna, NO de Argentina. In Asociación Argentina de Geofísicos y Geodestas. Ciencias d ela Tierra. Pp. 249–255. 2009.

Allen, M., et al. Summary for Policymakers. In IPCC (2018) Global Warming of 1.5 °C. 2018. UNEP, Geneva Switzerland. P. 6. 2018.

Anderson, Don, and Carl Benson. The Densification and Diagenesis of Snow. In Ice and Snow. Ed. W. D. Kingery. Cambridge, Mass. MIT Press. Pp. 391–411. 1963. See: https://core.ac.uk/download/pdf/33109832.pdf

Archer, David. The Long Thaw: How Humans are Changing the Next 100,000 Years of Earth's Climate. Princeton, NJ. Princeton University Press. 2009.

Arenson, Lukas, Silvio Pastore, and Dario Trombotto Liaudat. Characteristics of Two Rock Glaciers in the Dry Argentinean Andes Based on Initial Surface Investigations. GEO, 2010, Calgary Alberta. Pp.1501–1508. See: http://pubs.aina.ucalgary.ca/cpc/CPC6-1501.pdf

Armitage, Thomas W. K., et al. Enhanced Eddy Activity in the Beaufort Gyre in Response to Sea Ice Loss. Nature Communications 11, Article Number 761. 2020. See: https://www.nature. com/articles/s41467-020-14449-z.

Astrum. Narrated by Alex McColgan. What did NASA's New Horizons Discover around Pluto. January 28, 2018. See: https://www.youtube.com/watch?v=6l4kr36TzQ4&feature=youtu.be.

Azócar, Guillermo, and Alexander Brenning. Intervenciones en Glaciares Rocosos en Minera Los Pelambres, Región de Coquimbo, Chile. University of Waterloo, Ontario, Canada. 2008. See: https://www.researchgate.net/publication/242295733_IN_THE_LOS_PELAMBRES_MINE_COQUIMBO_REGION_CHILE

Bahr, D. B., and V. Radic. Significant Total Mass Contained in Small Glaciers. In the Cryosphere Discussions 6, 737–758. 2012. See: https://tc.copernicus.org/preprints/6/737/2012/tcd-6-737-2012.pdf

Bandt, Richard E., et al. Surface Albedo of the Antarctic Sea Ice Zone. Journal of Climate. Vol. 18, No. 17. September 1, 2005. Pp. 3606–3622 See: https://journals.ametsoc.org/view/journals/clim/18/17/jcli3489.1.xml

Barnard, P. L. et al. Dynamic Flood Modeling Essential to Assess the Coastal Impacts of Climate Change. Scientific Reports 9, 4309. 2019. Pp. 1–13. See: https://www.nature.com/articles/s41598-019-40742-z.

Barsch, Dietrich. Rock-Glaciers: Indicators for the Present and Former Geoecology in High Mountain Environments. Springer. Berlin, 1996.

BBC. Exploding Methane Gas Bubbles. See: https://www.science.org.au/curious/video/exploding-methane-gas-bubbles.

Befus, K. M., et al. Increasing Threat of Coastal Groundwater Hazards from Sea-Level Rise in California. Nature Climate Change 10. Pp. 946–952. October 2020. See: https://doi.org/10.1038/s41558-020-0874-1.

Benn, Douglas I., and D. Evans. Glaciers and Glaciation. Arnold, Hodder Headline Group. London, 1998.

Berwyn, Bob. Polar Vortex: How the Jet Stream and Climate Change Bring on Cold Snaps. Inside Climate News. February 2008. See: https://insideclimatenews.org/news/02022018/cold-weather-polar-vortex-jet-stream-explained-global-warming-arctic-ice-climate-change/

BGC Engineering. Pascua Lama Permafrost Characterization Study. BGC Engineering. 2009.

Bianchini, Flaviano. Impactos de los Emprendimientos Veladero y Pascua Lama sobre los Recursos Hídricos de la Provincia de San Juan. CEDHA, Cordoba, Argentina. 2011.

Bidle, Kay D., et al. Fossil Genes and Microbes in the Oldest Ice on Earth. Proceedings of the National Academy of Sciences of the United States of America 104, 33. 2007. Pp. 13455–13460. See: https://www.pnas.org/content/104/33/13455

Blum, Michael D et.al. Ups and Downs of the Mississippi Delta. In Geology. Vol. 36, No. 9. 2008. Pp. 675–678. See: https://pubs.geoscienceworld.org/gsa/geology/article-abstract/36/9/675/29795/Ups-and-downs-of-the-Mississippi-Delta?redirectedFrom=fulltext

Blumenfeld, Jared. King Tides. Podship Earth. Episode 70. 2019. See: https://www.podshipearth.com.

Blumenfeld, Jared. Rock Glacier. Podship Earth. Episode 67. 2019. See: https://www.podshipearth.com.

Bodin, Xavier, et al. Two Decades of Responses (1986–2006) to Climate by the Laurichard Rock Glacier, French Alps. Permafrost and Periglacial Processes 20. Pp. 331–344. 2009. See: https://onlinelibrary.wiley.com/doi/abs/10.1002/ppp.665

Bórquez, Roxana., S. Larraín, R. Polanco, and J. C. Urquidi. Glaciares Chilenos: Reservas Estratégicas de Agua Dulce para la sociedad, los ecosistemas y la economía. Chile Sustentable. 2006.

Bosson, J. B., M. Huss, and E. Osipova. Disappearing World Heritage Glaciers as a Keystone of Nature Conservation in a Changing Climate. Earth's Future. DOI. 10.1029/2018EF001139. 2018. See: https://agupubs.onlinelibrary.wiley.com/doi/epdf/10.1029/2018EF001139

Brand, Uwe, et al. Methane Hydrate: Killer Cause of Earth's Greatest Mass Extinction. Paleoworld. Vol. 25, No. 4. Pp. 496–507. December 2016. https://doi.org/10.1016/j.palwor.2016.06.002.

Brenning, Alexander, and Guillermo Azócar. Minería y glaciares rocosos: Impactos ambientales, antecedentes políticos y legales, y perspectivas futuras. Revista de Geografía Norte Grande, No. 47. Pp. 143–158. 2010. See: https://scielo.conicyt.cl/scielo.php?script=sci_arttext&pid=S0718-34022010000300008

Broccoli, A. J., and S. Manabe. The Influence of Continental Ice, Atmospheric CO_2, and Land Albedo on the Climate of the Last Glacial Maximum. Climate Dynamics Vol. 1, No. 2. February 1987. Pp. 87–99. See: https://www.researchgate.net/publication/225903571_The_influence_of_continental_ice_atmospheric_CO2_and_land_albedo_on_the_climate_of_the_last_glacial_maximum

Cabrera, Gabriel, and Juan Carlos Leiva. Monitoreo de Glaciares del Paso Conconta, Iglesia. San Juan Argentina. Conicet. 2008.

Caine, Nel. Recent Hydrological Change in a Colorado Alpine Basin: An Indicator of Permafrost Thaw? Annals of Glaciology 51, 56. 2010. Pp. 130–134. See: https://www.

cambridge.org/core/journals/annals-of-glaciology/article/recent-hydrologic-change-in-a-colorado-alpine-basin-an-indicator-of-permafrost-thaw/AA500D47A22B42E02755C6BB
B77E8B36

Carey, Mark. In the Shadow of Melting Glaciers: Climate Change and Andean Society. Oxford University Press. New York, 2010.

Center for Human Rights and Environment (CHRE). Derechos Humanos y Ambiente en la República Argentina: Propuestas para una Agenda Nacional. Advocatus, Cordoba, Argentina. 2005.

Champion, Marc. Melting Ice Redraws the World Map and Starts a Power Struggle. Bloomberg. October 11, 2019. See: https://apple.news/AQW7P1XhzSZmZnm3JswtY-g.

Chen, Jenny. Melting Glaciers Are Wreaking Havoc on Earth's Crust. Smithsonian Magazine. September 1, 2016. See: https://www.smithsonianmag.com/science-nature/melting-glaciers-are-wreaking-havoc-earths-crust-180960226/

Cheng, Lijing, et al. Continues Record Global Ocean Warming. Advances in Atmospheric Sciences Vol. 36, No. 3. Pp. 249–252. 2018. March 2019. See: https://link.springer.com/article/10.1007%2Fs00376-019-8276-x.

Ciraci, E., et al. Continuity of the Mass Loss of the World's Glaciers and Ice Caps from the GRACE and GRACE Follow-on Missions. Geophysical Research Letters. 10.1029/2019GL086926. 2019. See: https://agupubs.onlinelibrary.wiley.com/doi/abs/10.1029/2019GL086926

Clark Howard, Brian. West Antarctica Glaciers Collapsing, Adding to Sea Level Rise. National Geographic. May 2014. See: https://news.nationalgeographic.com/news/2014/05/140512-thwaites-glacier-melting-collapse-west-antarctica-ice-warming/.

Climate Central. Flooded Future: Global Vulnerability to Sea Level Rise Worse than Previously Understood. October 29, 2019.

Climate One Come Together. Can Greenland's Melting Affect the Weather? 2013. See: https://www.youtube.com/watch?v=d9Urq1mZ5KA.

Cortázar, Julio. Rayuela. Editorial Sudamericana. Buenos Aires, Argentina. 1963.

Corte, Arturo E. Geocriología: El Frío en la Tierra. Ediciones Culturales de Mendoza. Mendoza, Argentina, 1983.

Criss, Doug. A Piece of Antarctica Twice the Size of New York May Soon Break Off. CNN. February 26, 2019. See: https://www.cnn.com/2019/02/25/world/iceberg-antarctica-wxc-trnd

Cruikshank, Julio. Glaciers and Climate Change: Perspectives from Oral Tradition. Arctic 54, 4. Pp. 377–393. December 2001. See: https://journalhosting.ucalgary.ca/index.php/arctic/article/view/63852

Davis, Neil. Permafrost: A Guide to Frozen Ground in Transition. Fairbanks. University of Alaska Press. 2001.

Dean, Cornelia. As Alaska's Glaciers Melt, It's Land that Is Rising. New York Times. May 17, 2009. See: https://www.nytimes.com/2009/05/18/science/earth/18juneau.html

DeJong, Benjamin, Paul Bierman, Wayne Newell, Tammy Rittenour, Shannon Mahan, Greg Balco, and Dylan Rood. Pleistocene Relative Sea Levels in the Chesapeake Bay Region and Their Implications for the Next Century. GSA Today Vol. 25, No. 8. August 2015. Pp. 4–10. See: https://www.geosociety.org/gsatoday/archive/25/8/article/i1052-5173-25-8-4.htm

de Magalhaes Neto, Newton, Heitor Evangelista, Thomas Condom, Antoine Rabatel, and Patrick Ginot. Amazonian Biomass Burning Enhances Tropical Andes Glaciers Melting. Scientific Reports 9. Article Number 16914. November 28, 2019. See: https://www.nature.com/articles/s41598-019-53284-1.

Dempster, Allison. Wildfire Soot Darkening Glaciers Could Speed up Melt Rate, Scientists Fear. CBS News. January 6, 2019. See: https://www.cbc.ca/news/canada/calgary/darkening-glaciers-wildfires-melt-rate-1.4963754.

Department of Natural Resources State of Alaska. Volcanoes and Glaciers. See: https://dggs. alaska.gov/popular-geology/volcanoes-glaciers.html.

Dickerman, Kenneth, and James Whitlow Delano. How Climate Change is Affecting Life in the Italian Alps. Washington Post. August 23, 2019. See: https://www.washingtonpost.com/photography/2019/08/23/how-climate-change-is-affecting-quality-life-italian-alps/.

Doucleff, Michaeleen. Are There Zombie Viruses in the Thawing Permafrost. NPR. January 24, 2018. See: https://www.npr.org/sections/goatsandsoda/2018/01/24/575974220/are-there-zombie-viruses-in-the-thawing-permafrost.

Dubey, Saket. Glacier Lake Outburst Flood Hazard, Downstream Impact, and Risk Over the Indian Himalayas. Water Resources Research Vol. 56, No. 4. April 2020. See: https://agupubs.onlinelibrary.wiley.com/doi/abs/10.1029/2019WR026533.

Dufresne, A., et al. The 2016 Lamplugh Rock Avalanche, Alaska: Deposit Structures and Emplacement Dynamics. Landslides 16. Pp. 2301–2319. July 2019. See: https://doi.org/10.1007/s10346-019-01225-4.

Duguay, Maxime, et al. Quantifying the Significance of the Hydrological Contribution of a Rock Glacier: A Review. For GeoQuebec. Quebec, Canada, 2015.

Dumont, M., J. Gardelle, P. Sirguey, A. Guillot, D. Six, A. Rabatel, and Y. Arnaud. Linking Glacier Annual Mass Balance and Glacier Albedo Retrieved from MODIS Data. The Cryosphere, Copernicus 2012, 6. Pp. 1527–1539. 2012. See: https://tc.copernicus.org/articles/6/1527/2012/

Ekman, Martin. The Changing Level of the Baltic Sea during 300 Years: A Clue to Understanding the Earth. Aland Islands. Summer Institute for Historical Geophysics. 2009. See: https://www.baltex-research.eu/publications/Books%20and%20articles/The%20Changing%20Level%20of%20the%20Baltic%20Sea.pdf

Englander, John. High Tide on Main Street. Boca Raton. The Science Book Shelf. 2012.

Espizua, Lydia et.al. Ambiente y Procesos Glaciares y Periglaciales en Lama-Veladero. Conicet, San Juan Argentina. 2006. See: https://www.conicet.gov.ar/new_scp/detalle.php?keywords=Buscar&id=30950&convenios=yes&detalles=yes&conv_id=374752

Espizua, Lydia, and Pierre Pitte. The Little Ice Age Glacier Advance in the Central Andes (35° S), Argentina. Palaeoclimatology, Palaeoecology 281. Pp. 345–350. 2009. See: https://www.researchgate.net/publication/248290057_The_Little_Ice_Age_glacier_advance_in_the_Central_Andes_35S_Argentina

Fauqué, L., and D. Azcurra. Condiciones periglaciales en la vertiente occidental de los Nevados del Aconquija, Catamarca. Argentina. XII Congreso Geológico Chileno. 2009.

Feyrer, James, and Bruce Sacerdote. Did the Stimulus Stimulate? Real Time Estimates of the Effects of the American Recovery and Reinvestment Act. For National Bureau of Economic Research. Working Paper Series, No. 16759. Cambridge Massachusetts. February, 2011. See: https://www.nber.org/system/files/working_papers/w16759/w16759.pdf

Flavia, Croce, and Juan Pablo Milana. Desarrollo de Sistemas Geocriogénicos en la Zona del Paso Agua Negra y su Importancia en Geología Aplicada. Actas del XV Congreso Geológico Argentino. El Calafate. 2002. See: http://docplayer.es/46521964-Desarrollo-de-sistemas-geocriogenicos-en-la-zona-del-paso-agua-negra-y-su-importancia-en-geologia-aplicada-flavia-a-croce-y-juan-p.html

Fox-Skelly, Jasmin. BBC. There are Diseases Hidden in Ice, and They Are Waking Up. May 4, 2017. See: http://www.bbc.com/earth/story/20170504-there-are-diseases-hidden-in-ice-and-they-are-waking-up.

Fountain, Henry. When the Glaciers Disappear, Those Species Will Go Extinct. New York Times. April 17, 2019. See: https://www.nytimes.com/interactive/2019/04/16/climate/glaciers-melting-alaska-washington.html.

Fountain, Henry, and Ben Solomon. Where Glaciers Melt Away Switzerland Sees Opportunities. NY Times, Feb 13, 2019. See: https://www.nytimes.com/interactive/2019/04/17/climate/switzerland-glaciers-climate-change.html

Francou, Bernarad. Montaña y Glaciares. In Montaña y Glaciares. América Natural. Eds. Antonio Vizcaino and Ximena de la Macorra. Mexico. Pp. 32–37. 2011. See: https://center-hre.org/wp-content/uploads/2012/03/BFRANCOU-AMERICA-NATURAL-8mai2011.pdf

French, Hugh M. The Periglacial Environment. Third Edition. Wiley, Sussex, England. 2007.

Ganopolski, Andrey, et al. Critical Insolation-CO_2 Relation for Diagnosing Past and Future Glacial Inception. Nature 534, S–19–S20 (2016). See: https://www.nature.com/articles/nature18452.

Gascoin, S., et al. Glacier Contribution to Streamflow in Two Headwaters of the Huasco River, Dry Andes of Chile. The Cryosphere 5. Pp. 1099–1113. 2011. See: https://tc.copernicus.org/articles/5/1099/2011/

Gleike, Peter H. (Ed.) Water in Crisis: A Guide to the World's Fresh Water Resources. Oxford University Press. 1993.

Glick, Daniel. The Big Thaw. National Geographic. See: https://www.nationalgeographic.com/environment/global-warming/big-thaw/.

Gosnell, Mariana. Ice: The Nature, the History, and the Uses of an Astonishing Substance. New York. Alfred A. Knopf. 2005.

Green, Clare, et al. Simulating the Impact Of Freshwater Inputs and Deep-Draft Icebergs Formed during a MIS 6 Barents Ice Sheet Collapse. Paleoceanography and Paleoclimatology. Vol. 26, No.2. May 2011. Pp.1–16. See: https://agupubs.onlinelibrary.wiley.com/doi/full/10.1029/2010PA002088

Gruber, S. Derivation and Analysis of High-Resolution Estimate of Global Permafrost Zonation. The Cryosphere 6. Pp. 221–233. 2012. See: https://tc.copernicus.org/articles/6/221/2012/tc-6-221-2012.pdf

Guardian. Antarctica Logs Hottest Temperature on Record with a Reading of 18.3C. Guardian. February 7, 2020. See: https://www.theguardian.com/world/2020/feb/07/antarctica-logs-hottest-temperature-on-record-with-a-reading-of-183c.

Guardian. Pizzly or Grolar bear: Grizzly-Polar Hybrid is a New Result of Climate Change. May 18, 2016. See: https://www.theguardian.com/environment/2016/may/18/pizzly-grolar-bear-grizzly-polar-hybrid-climate-change

Guyton, Bill. Glaciers of California: Modern Glaciers, Ice Age Glaciers, the Origin of Yosemite Valley, and a Glacier Tour in the Sierra Nevada. University of California Press, Berkeley. 1998.

Haemmig, Christopher, et al. Hazard Assessment of Glacier Lake Outburst Floods from Kyagar Glacier, Karakoram Mountains, China. Annals of Glaciology Vol. 55, No. 66. 2014. Pp. 34–44. See: https://www.cambridge.org/core/journals/annals-of-glaciology/article/hazard-assessment-of-glacial-lake-outburst-floods-from-kyagar-glacier-karakoram-mountains-china/82534375EF47FF7DC53E999783CD973C

Hallegatte, Stephane, et al. Future Flood Losses in Major Coastal Cities. Nature Climate Change 3. Pp. 802–806. 2013. See: https://www.nature.com/articles/nclimate1979.

Hambrey, Michael, and Jurg Alean. Glaciers. Cambridge University Press, Cambridge. 1992.

Hansen, James, and Larissa Nazarenko. Soot Climate Forcing via Snow and Ice Albedos. Proceedings of the National Academy of Sciences of the United States of America 101, 2. Pp. 423–428. January 13, 2004. See: https://www.pnas.org/content/101/2/423

Hays, J. D. et al. Variations in the Earth's Orbit: Pacemaker of the Ice Ages. Science Vol. 194, No. 4270. Pp. 1121–1132. December 10, 1976. See: https://science.sciencemag.org/content/194/4270/1121.

HIMAP. The Hindu Kush Himalaya Assessment—Mountains, Climate Change, Sustainability and People. Sprinter Nature. Switzerland AG. Cham. See: https://doi.org/10.1007/978-3-319-92288-1.

Hoffman, Paul F., and Daniel P. Schrag. The Snowball Earth. Harvard. (Paper published at Harvard, no citation available). 1999. See: https://imedea.uib-csic.es/master/cambioglobal/Modulo_V_cod101619/HoffmanSchragSnowballEarth.pdf

Hoganboom,Melissa.InSiberiaThereisaHugeCraterandIt'sGettingBigger.BBC.February24,2017. See: http://www.bbc.com/earth/story/20170223-in-siberia-there-is-a-huge-crater-and-it-is-getting-bigger.

Horstmann, Britta. Glacial Lake Outburst Floods in Nepal and Switzerland: New Threats Due to Climate Change. Germanwatch. Bonn Germany. 2004. See: https://germanwatch.org/sites/default/files/publication/3647.pdf

Hughes, Holly. "A Sense of Real Fear": Climate Change Photog Katie Orlinsky on Documenting Arctic Melt. PDNPULSE. August 30, 2018. See: https://pdnpulse.pdnonline.com/2018/08/a-sense-of-real-fear-climate-change-photog-katie-orlinsky-on-documenting-arctic-melt.html.

Humlum, Ole. The Climatic and Palaeoclimatic Significance of Rock Glaciers. Permafrost and Periglacial Processes 9(4). Pp. 375–395. 1999. See: https://onlinelibrary.wiley.com/doi/abs/10.1002/%28SICI%291099-1530%28199810/12%299%3A4%3C375%3A%3AAID-PPP301%3E3.0.CO%3B2-0

Humlum, Ole. The Climatic and Palaeoclimatic Significance of Rock Glaciers. Project description document, funded by University Courses on Svalbard (UNIS) 2000–2005, Department of Geology, Svalbard, Norway. 12/22/2010 See: https://center-hre.org/wp-content/uploads/2011/10/climate-significance-of-rock-glaciers-Homlum.pdfHunziker, Robert. The Permafrost Nightmare Turns More Real. Counter Punch. June 10, 2019. See: https://www.counterpunch.org/2019/06/10/the-permafrost-nightmare-turns-more-real/

Huss, Matthias, and Regine Hock. Global-Scale Hydrological Response to Future Glacier Mass Loss. Nature Climate Change 8. Pp. 135–140. 2018. See: https://www.nature.com/articles/s41558-017-0049-x

ICIMOD. Glacier Lakes and Glacier Lake Outburst Floods in Nepal. ICIMOD. 2011. See: https://lib.icimod.org/record/27755

ICIMOD. Inventory of Glacier Lakes and Identification of Potentially Dangerous Glacial Lakes in Koshi, Gandaki, and Karnali River Basins of Nepal, the Tibet Autonomous Region of China, and India. 2020. See: https://www.np.undp.org/content/dam/nepal/docs/reports/environment%20and%20energy/Inventory_Glacial_Lakes_2020_Full2.pdf.

Immerzeel, W. W., et al. Importance and Vulnerability of the World's Water Towers. Nature 577. January 16, 2020. Pp. 364–369. See: https://www.nature.com/articles/s41586-019-1822-y?proof=t

Institute for Governance and Sustainable Development (IGSD). Primer on Polar Warming and Implications for Global Climate Change. May 2019. See: http://www.igsd.org/wp-content/uploads/2019/05/Primer-on-Polar-Warming.pdf.

IPCC. Fifth Assessment Report. Glaciers and Glaciation. March 2014.

IPCC. Summary for Policymakers. In IPCC Special Report on the Ocean and Cryosphere in a Changing Climate. Eds. H.-O. Pörtner, D. C. Roberts, V. Masson-Delmotte, P. Zhai, M. Tignor, E. Poloczanska, K. Mintenbeck, M. Nicolai, A. Okem, J. Petzold, B. Rama, N. Weyer. UNEP, Geneva, Switzerland. 2019.

Iza, Alejandro, and Marta Brunilda Rovere. Apectos Jurídicos de la Conservación de los Glaciares. IUCN, Gland Switzerland. 2006.

Jackson, M. The Secret Lives of Glaciers. Green Writers Press. Brattleboro, Vermont. 2019.

Jamail, Dahr. The End of Ice: Bearing Witness and Finding Meaning in the Path of Climate Disruption. The New Press, New York. 2019.

Johansson, Emma. The Melting Himalayas: Examples of Water Harvesting Techniques. Bachelor's Thesis. Lund University. 2012. See: https://lup.lub.lu.se/luur/download?func=downloadFile&recordOId=2760100&fileOId=2760103

Johnson, Scott. Alaskan Warming Permafrost Experiment Produces Surprising Results. Ars Technica. July 3, 2019. See: https://arstechnica.com/science/2019/07/permafrost-experiment-shows-surprising-amount-of-co2-release/

Jones, Darren B., et al. Rock Glaciers and Mountain Hydrology: A Review. Earth-Sciences Reviews 193. Pp. 66–90. 2019. See: https://www.sciencedirect.com/science/article/pii/S0012825218305609

Kasischke, Eric S., et al. The Arctic-Boreal Vulnerability Experiment: A Concise Plan for a NASA-Sponsored Field Campaign. A Nasa Field Report. October 2010. See: https://cce.nasa.gov/terrestrial_ecology/pdfs/ABoVE%20Final%20Report.pdf

Kellerer, Andreas, Heinz Slupetzky, and Michael Avian. Ice-Avalanche Impact Landforms. The Event in 2003 at the Glacier Nördliches Bockkarkees, Hohe Tauern Range, Austria. Goegrafiska Annaler. Series A, Physical Geography 94, 1. 2012. Pp. 97–115. See: https://www.tandfonline.com/doi/abs/10.1111/j.1468-0459.2011.00446.x

Knorr, Paul Octavius. The Case for High-Order Pleistocene Sea-Level Fluctuations in Southwest Florida. Master's Thesis. 2006. University of South Florida. See: https://scholarcommons.usf.edu/cgi/viewcontent.cgi?article=3587&context=etd.

Komori, Jiro, et al. Glacier Lake Outburst Events in the Bhutan Himalayas. Global Environmental Research 16. Pp.59–70. 2012. See: https://www.researchgate.net/publication/265385010_Outline_of_research_project_on_glacial_lake_outburst_floods_in_the_Bhutan_Himalayas

Kronenberg, Jakub. Linking Ecological Economics and Political Ecology to Study Mining, Glaciers and Global Warming. Environmental Policy and Governance 23. Pp. 75–90. 2013. See: https://onlinelibrary.wiley.com/doi/abs/10.1002/eet.1605

Kurter, Ajun. Glaciers of Turkey: Glaciers of the Middle East and Africa. US Geological Survey Professional Paper 1386-G-1. 1988.

Lem, Pola. Batagaika Crater Expands. NASA Earth Observatory. 2016. See: https://earthobservatory.nasa.gov/images/90104/batagaika-crater-expands.

Lenton T. M., et al. Climate Tipping Points—Too Risky to Bet Against. Nature, Comment 575. Pp. 592–595. 2019. See: https://www.nature.com/articles/d41586-019-03595-0

Lliboutry, Luis. Nieves y Glaciares de Chile: Fundamentos de Glaciología. Ediciones de la Universidad de Chile. 1956.

Lovett, Richard A. Melting Glaciers Mean Double Trouble for Water Supplies. National Geographic. December 21, 2011. See: https://www.nationalgeographic.com/science/article/1112-melting-glaciers-mean-double-trouble-for-water-supplies

Macdougall, Doug. Frozen Earth: The Once and Future Story of Ice Ages. University of California Press. 1994.

Mann, Michael, et al. Influence of Anthropogenic Climate Change on Planetary Wave Resonance and Extreme Weather Events. Scientific Reports 7. Article number 45242. 2017. See: https://www.nature.com/articles/srep45242

Mark, Brian G., et al. Glaciers as Water Resources. A chapter of the book The High-Mountain Cryosphere Environmental Changes and Human Rights. Eds. Huggel C. et al., First Edition. Cambridge University Press, Cambridge. 2014.

Maslin, Mark. Ice Ages Have Been Linked to the Earth's Wobbly Orbit—but When is the Next One? Phys.org. December 12, 2016. See: https://phys.org/news/2016-12-ice-ages-linked-earth-orbitbut.html.

Maza, Jorge. Crecida de Diseño Generada por la Rotura del Endicamiento Glaciar Grande del Nevado del Plomo en el Río Plomo. Instituto Nacional del Agua. IT 175-CRA. 2016. See: https://www.researchgate.net/publication/326579273_CRECIDA_DE_DISENO_ GENERADA_POR_LA_ROTURA_DEL_ENDICAMIENTO_GLACIAR_GRANDE_ DEL_NEVADO_DEL_PLOMO_EN_EL_RIO_PLOMO

McGowan, Jake. Arctic Ice Effects Global Temperatures, Jet Streams. Daily Targum. March 31, 2019. See: https://dailytargum.com/article/2019/04/arctic-ice-effects-global-temperatures-jet-streams

McGuire, Bill. Will Global Warming Triggera a New Ice Age? In YaleGlobal Online. November 13, 2003. See: https://yaleglobal.yale.edu/content/will- global-warming-trigger-new-ice-age

McNamara, et.al Climate Adaptation and Policy Induced Inflation of Coastal Property Values. PLOS One. Vol. 10, No. 3. P. e0121278. doi:10.1371/journal.pone.0121278. See: https://journals.plos.org/plosone/article?id=10.1371/journal.pone.0121278

Millar, Constance, and Robert Westfall. Geographic, Hydrological, and Climate Significance of Rock Glaciers in the Great Basin, USA. Arctic, Antarctic, and Alpine Research: An Interdisciplinary Journal. 51:1. 2019. Pp. 232–249. See: https://www.tandfonline.com/doi/full/10.1080/15230430.2019.1618666

Miller, Alan; Zaelke, Durwood; Andersen, Stephen. Resetting Our Future: Cut Super Pollutants Now! Changemakers Books. Winchester, UK. 2021.

Miller, Jeremy. The Dying Glaciers of California. Earth Island Journal 28, 2. Pp. 48–53. Summer 2013. See: https://www.earthisland.org/journal/index.php/magazine/entry/the_dying_glaciers_of_california/

Milner, Alexander M., et al. Glacier Shrinkage Driving Global Changes in Downstream Systems. Proceedings of the National Academy of Sciences of the United States of America (PNAS). 114 (37). September 5, 2017. Pp. 9770–9778. See: https://www.pnas.org/content/114/37/9770

MIM Argentina Exploraciones. Informe de Impacto Ambiental: Etapa de Exploración Proyecto Filo Colorado, para Xstrata Copper. June 2005.

Mook, Pradeep, et al. Glacial Lakes and Glacial Lake Outburst Floods in Nepal. ICIMOD, Kathmandu. 2011.

Mooney, Chris. Unprecedented Data Confirms that Antarctica's Most Dangerous Glacier Is Melting from Below. Washington Post. January 30, 2020. See: https://www.washingtonpost.com/climate-environment/2020/01/30/unprecedented-data-confirm-that-antarcticas-most-dangerous-glacier-is-melting-below/

Mortensen, Anette, Matthias Bigler, Karl Grönvold, Jørgen Steffensen, and Sigfús Johnsen. Volcanic Ash Layers from the Last Glacier Termination in the NGRIP Ice Core. Journal of Quaternary Science Vol. 20, No. 3. March 2005. Pp. 209–219. See: https://onlinelibrary.wiley.com/doi/abs/10.1002/jqs.908

Muhs, Daniel, et al. Sea-level History of the Past Two Interglacial Periods: New Evidence from U-Series Dating of Reef Corals from South Florida. Quaternary Science Reviews 30. Pp. 570–590. 2011. See: https://pubs.er.usgs.gov/publication/70036701

Muhs, Daniel R., et al. Quaternary Sea-Level History of the United States. Developments in Quaternary Science 1. December 2003. DOI: 10.1016/S1571-0866(03)01008-X. See: https://www.researchgate.net/publication/239568104_Quaternary_sea-level_history_of_the_United_States

Muir, John. My First Summer in the Sierra. Penguin Books, New York. 1911.

Muschitiello, Francesco, Francesco Pausata, James Lea, Douglass Mair, and Barbara Wohlfarth. Enhanced Ice Sheet Melting Driven by Volcanic Eruptions during the Last Deglaciation. Nature Communications 8. Article Number 1020. 2017. See: https://www.nature.com/articles/s41467-017-01273-1..

NASA and US Geological Survey. Earth's Albedo and Global Warming. See: http://d3tt741pwxqwm0.cloudfront.net/WGBH/conv18/ipy07_int_albedo/ipy07_int_albedo.html.

National Snow and Ice Data Center (NSIDC). Arctic Sea Ice News and Analysis: Rapid Ice Loss in Early April Leads to New Record Low. NSIDC. May 2, 2019 See: http://nsidc.org/arcticseaicenews/2019/05/rapid-ice-loss-in-early-april-leads-to-new-record-low/.

Owen, Lewis, and John England. Observations on Rock Glaciers in the Himalayas and Karakoram Mountains of Northern Pakistan and India. Geomorphology 26. Pp. 199–213. 1998 See: http://webcentral.uc.edu/eProf/media/attachment/eprofmediafile_437.pdf

Painter, Thomas H., Mark G. Flanner, Georg Kaser, Ben Marzeion, Richard A. VanCuren, and Waleed, Abdalati. End of Little Ice Age in the Alps Forced by Industrial Black Carbon. Proceedings of the National Academy of Sciences of the United States of America. Vol. 110, No. 38. Pp. 15216–15221. September 17, 2013. See: https://www.pnas.org/content/110/38/15216.

Palmer, Tim. California Glaciers. Rocklin California. Sierra College Press. 2012.

Parkinson, Claire L. A 40-Y Record Reveals Gradual Antarctic Sea Ice Increases Followed by Decreases at Rates Far Exceeding the Rates Seen in the Arctic. Proceedings of the National Academy of Sciences of the United States of America 116, 29. Pp. 14414–14423. July 16, 2019. See: https://www.pnas.org/content/116/29/14414

Paul, Frank, Horst Machguth, and Andreas Kääb. On the Impacts of Glacier Albedo Under Conditions of Extreme Glacier Melt: The Summer of 2003 in the Alps. EARSeL eProceedings 4, 2. 2005. Pp. 139–149. See: http://eproceedings.uni-oldenburg.de/website/vol04%5F2/04_2_paul1.pdf

Pikuta, Elena V., et al. *Canobacterium pleistocenium* sp. nov., a Novel Psychrotolerant, Facultative Anaerobe Isolated from Permafrost of the Fox Tunnel in Alaska. International Journal of Systematic and Evolutionary Microbiology Vol. 55, No. 1. 2005. Pp. 473–478. See: https://www.microbiologyresearch.org/docserver/fulltext/ijsem/55/1/473.pdf?expires=1619715827&id=id&accname=guest&checksum=6B7FF793A6A475AF871853579EDF1098

Pirazzini, Roberta. Surface Albedo Measurements over Antarctica Sites in Summer. Journal of Geophysical Research 109, D20018. 2004. DOI: 10.1029/2004JD004617. Pp 1-15. See: https://agupubs.onlinelibrary.wiley.com/doi/epdf/10.1029/2004JD004617

Pistone, et. al. Observational Determination of Albedo Decrease Caused by Vanishing Arctic Sea Ice. PNAS. Vol. 111. No. 9. March 4, 2014. Pp. 3322–3326. See: https://www.pnas.org/content/pnas/111/9/3322.full.pdf

Pokrovsky, Oleg. Editor. Permafrost: Distribution, Composition and Impacts on Infrastructure and Ecosystems. Nova Publishers, New York. 2014.

Post, Austin, and Edward Lachapelle. Glacier Ice. Revised Edition. Seattle: University of Washington Press. 1971.

Qian, Y., M. G. Flanner, L. R. Leung, and W. Wang. Sensitivity Studies on the Impacts of Tibetan Plateau Snowpack Pollution on the Asian Hydrological Cycle and Monsoon Climate. Atmospheric Chemistry and Physics 11. Pp. 1929–1948. 2011. See: https://acp.copernicus.org/articles/11/1929/2011/

Radford, Tim. Arctic Sea Ice Loss Affects the Jet Stream. Climate News Network. 2019. See: https://climatenewsnetwork.net/arctic-sea-ice-loss-affects-the-jet-stream/

Rasul, G., et al. Glaciers and Glacial Lakes under Changing Climate in Pakistan. Pakistan Journal of Meteorology. Vol. 8, No., 15. 2012. Pp. 1-8. See: http://www.pmd.gov.pk/rnd/rnd_files/vol8_issue15/1_Glaciers%20and%20Glacial%20Lakes%20under%20Changing%20Climate%20in%20Pakistan.pdf

Raub, William, Suzanne Brown, and Austin Post. Inventory of Glaciers in the Sierra Nevada, California. United States Geological Survey, Reston, Virginia. 2006. See: https://pubs.er.usgs.gov/publication/ofr20061239

Riffle, Adam. Internal Composition, Structure and Hydrological Significance of Rock Glaciers in the Eastern Cascades, Washington. Master's Thesis. Central Washington University. 2018. See: https://digitalcommons.cwu.edu/etd/1010/

Robinson, Charles, and Peter Dea. Quaternary Glacial and Slope Failure Deposits of the Crested Butte Area, Gunnison County Colorado. New Mexico Geological Society Guidebook, 32nd Conference, Western Slope Colorado. 1981. Pp. 155–164. See: https://nmgs.nmt.edu/publications/guidebooks/downloads/32/32_p0155_p0163.pdf

Romanowsky, Erik, et al. The Role of Stratospheric Ozone for Arctic-Midlatitude Linkages. Scientific Reports 9. Article Number: 7962. 2019. See: https://www.nature.com/articles/s41598-019-43823-1

Romanovsky, Vladimir, et al. Frozen Ground. Global Outlook for Ice and Snow 7. Pp. 181–200. See: https://wedocs.unep.org/bitstream/handle/20.500.11822/14478/GEO_C7_LowRes.pdf?sequence=1&isAllowed=y

Rosier, Sebastian et.al. The tipping points and early warning indicators for Pine Island Glacier, West Antarctica. In The Cryosphere. Vol. 15. Pp. 1501–1516, 2021. See: https://doi.org/10.5194/tc-15-1501-2021

Ross, Rachel. What Are the Different Types of Ice Formations Found on Earth? Life Sciences, January 8, 2019. See: https://www.livescience.com/64444-ice-formations.html

Sasgen, Ingo, et al. Return to Rapid Ice Loss in Greenland and Record Loss in 2019 Detected by the GRACE-FO Satellites. Communications Earth and Environment 1. Article number: 8. August 2020. Pp. 1–8 See: https://www.nature.com/articles/s43247-020-0010-1.

Scanu, Marcelo. Leyendas de los Andes Argentinos. Cruzpampa Editores, Buenos Aires. 2012.

Schriber, Michael. "Snowball Earth" Might Have Been Slushy. National Aeronautics and Space Administration. Research Features. August 2015. See: https://www.giss.nasa.gov/research/features/201508_slushball/

Schoen, Erik R., et al. Future of Pacific Salmon in the Face of Environmental Change. Lessons from One of the World's Remaining Productive Salmon Regions. Fisheries 42. 2017. See: https://www.tandfonline.com/doi/full/10.1080/03632415.2017.1374251.

Scientific American. As the Earth Warms, the Diseases that May Lie within Permafrost Become a Bigger Worry. November 1, 2016. See: https://www.scientificamerican.com/article/as-earth-warms-the-diseases-that-may-lie-within-permafrost-become-a-bigger-worry/

Shakhova, Natalia, et al. Current Rates And Mechanisms of Subsea Permafrost Degradation in the East Siberian Arctic Shelf. Nature Communications, 8. Article Number 15872. 2017. See: https://www.nature.com/articles/ncomms15872.

Shakhova, Natalia, et al. The East Siberian Arctic Shelf: Towards Further Assessment of Permafrost-Related Methane Fluxes and Role of Sea Ice. The Royal Society 373. October 13, 2015. See: https://doi.org/10.1098/rsta.2014.0451.

Simon, Matt. Permafrost is Thawing so Fast, It's Gouging Holes in the Arctic. Wired. February 3, 2020. See: https://www.wired.com/story/abrupt-permafrost-thaw/.

Snay, Richard, Jeffrey Freymueller, Michael Craymer, Chris Pearson, and Jarir Saleh. Modeling 3-D Crustal Velocities in the United States and Canada. JGR Solid Earth Vol. 121, No. 7. July 2016. Pp. 5365–5388. See: https://agupubs.onlinelibrary.wiley.com/doi/full/10.1002/2016JB012884

Steffen, Will, et al. Trajectories of the Earth's System in the Anthropocene. Proceedings of the National Academy of Sciences of the United States of America 115, 33. Pp. 8252–8259. August 14, 2018. See: https://doi.org/10.1073/pnas.1810141115.

Stokes, C. R., et al. Recent Glacier Retreat in the Caucasus Mountains, Russia, and Associated Increase in Supraglacial Debris Cover and Supra-/Proglacial Lake Development. Annals of Geology 46 (1). 2007. Pp. 195–203. See: https://www.cambridge.org/core/journals/annals-of-glaciology/article/recent-glacier-retreat-in-the-caucasus-mountains-russia-and-associated-increase-in-supraglacial-debris-cover-and-supraproglacial-lake-development/37F104B58C7468FB673BCA44D252D99B

Swift, Darrel A., Simon Cook, Tobias Heckmann, Jeffrey Moore, Isabelle Gärtner-Roer, and Oliver Korup. Ice and Snow as Landforming Agents. Chapter in book Snow and Ice-Related Hazards, Risks and Disasters. Eds Shroder, John, Wilfried Haeberli and Colin Whiteman. Elsevier Book, Chapter R2. 2015.

Taillant, Jorge Daniel. Barrick's Glaciers: Technical Report on the Impacts by Barrick Gold on Glaciers and Periglacial Environment at Pascua Lama and Veladero. CEDHA, Cordoba, Argentina. 2013.

Taillant, Jorge Daniel. La Democratización de los Glaciares. Hydria, 41. Pp. 17–19. 2012.

Taillant, Jorge Daniel. Glaciers and Mining in the Province of La Rioja, Argentina. CEDHA, Cordoba, Argentina. 2012.

Taillant, Jorge Daniel. Glaciers and Periglacial Environments in Diaguita-Huascoaltino Indigenous Territory, Chile. CEDHA, Cordoba, Argentina. 2012.

Taillant, Jorge Daniel. Glaciers: The Politics of Ice. Oxford University Press, New York. 2015.

Taillant, Jorge Daniel. The Human Right ... to Glaciers? Journal of Environmental Law and Litigation 28, 1. Pp. 59–78. 2013. See: https://scholarsbank.uoregon.edu/xmlui/bitstream/handle/1794/13587/Taillant.pdf?sequence=1

Taillant, Jorge Daniel. Impactos en Glaciares de Roca y en Ambiente Periglacial de los Proyectos Mineros Filo Colorado (Xstrata) y Agua Rica (Yamana Gold). CEDHA, Cordoba, Argentina. 2011.

Taillant, Jorge Daniel. Impacts to Rock Glaciers and Periglacial Environments by El Pachón (Xstrata). CEDHA Cordoba, Argentina. 2011.

Taillant, Jorge Daniel, and Peter Collins. Cryoactivism. LASA Forum Vol. XLVII, No. 4. Pp. 34–38. Fall 2016. See: https://forum.lasaweb.org/files/vol47-issue4/Debates7.pdf

Talke, S. A., et al. Relative Sea Level, Tides, and Extreme Water Levels in Boston Harbor from 1825 to 2018. Journal of Geophysical Research: Oceans. 2018. DOI: 10.1029/2017JC013645. See: https://agupubs.onlinelibrary.wiley.com/doi/full/10.1029/2017JC013645

Tapscott, Don, and Anthony Williams. Wikinomics: How Mass Collaboration Changes Everything. Portfolio, New York. 2007.

Teckle, Nescha, and Krishna Vatsa. GLOF Risk Reduction through Community-Based Approaches. UNDP, New Delhi, India. 2015. 38 pp. See: https://www.undp.org/content/dam/undp/library/crisis%20prevention/disaster/asia_pacific/Booklet%20GLOF%20CBDP%20Final.pdf

Tedesco, Marco. The Hidden Life of Ice: Dispatches from a Disappearing World. Translated by Denise Muir. Il Saggiatore-The Experiment, New York. 2020.

Thomas, J. L., C. M. Polashenski, A. J. Soja, L. Marelle, K. A. Casey, H. D. Choi, et al. Quantifying Black Carbon Deposition over the Greenland Ice Sheet from Forest Fires in Canada. Geophysical Research Letters Vol. 44, No. 15. August 16, 2017. Pp. 7965–7974. See: https://agupubs.onlinelibrary.wiley.com/doi/full/10.1002/2017GL073701

Troianovsky, Anton, and Chris Mooney. Radical Warming in Siberia Leaves Millions on Unstable Ground. Washington Post. October 3, 2019. See: https://www.washingtonpost.com/graphics/2019/national/climate-environment/climate-change-siberia/.

Trombotto, Dario. Mapping of Permafrost and Periglacial Environments, Cordón del Plata, Argentina. IANIGLA, Mendoza Argentina. 2003.

Trombotto, Dario. Survey of Cryogenic Processes, Periglacial Forms and Permafrost Conditions in South America. Revista do Instituto Geológico Sao Paulo 21, 1/2. Pp. 33–55. 2000. See: https://core.ac.uk/download/pdf/159285286.pdf

Trombotto, Dario and E. Borzotta. Indicators of Present Global Warming through Changes in Active Layer-Thickness, Estimation of Thermal Diffusivity and Geomorphological Observations in the Morenas Coloradas Rockglacier, Central Andes of Mendoza, Argentina. Cold Regions Science and Technology 55. Pp. 321–330. 2009. See: https://www.researchgate.net/publication/223562106_Indicators_of_present_global_warming_through_changes_in_active_layer-thickness_estimation_of_thermal_diffusivity_and_geomorphological_observations_in_the_Morenas_Coloradas_rockglacier_Central_Andes_of

Tveiten, Ingvar Norstegard. Glacier Growing – A Local Response to Water Scarcity in Baltistan and Gilgit, Pakistan. Norwegian University of Life Sciences. Master's Thesis. 2007. See: http://www.umb.no/statisk/noragric/publications/master/2007_ingvar_tveiten.pdf

UCSB Science Line. On the Heat Absorption of Color. March 23, 2018. See: https://scienceline.ucsb.edu/getkey.php?key=3873).

Union of Concerned Scientists. Underwater: Rising Seas, Chronic Floods, and the Implications for US Coastal Real Estate. 2018. See: https://www.ucsusa.org/sites/default/files/attach/2018/06/underwater-analysis-full-report.pdf.

United Nations Environment Program. Recent Trends in Melting Glaciers, Tropospheric Temperatures over the Himalayas and Summer Monsoon Rainfall over India. 2009. See: https://na.unep.net/siouxfalls/publications/Himalayas.pdf.

United National Environmental Program and Climate and Clean Air Coalition. Global Methane Assessment; Benefits and Costs of Mitigating Methane Emissions. United Nations Environmental Program. Nairobi. 2021. https://www.ccacoalition.org/en/file/7941/download?token=q_bCnfYV

United States Geological Survey. Satellite Image Atlas of Glaciers of the World. South America. Professional Paper 1386-I. 1988. See: https://pubs.er.usgs.gov/publication/pp1386

Vergara, Walter, et al. Assessment of the Impacts of Climate Change on Mountain Hydrology: Development of a Methodology Through A Case Study in the Andes of Peru. The World Bank, Washington, DC. 2011. See: https://openknowledge.worldbank.org/handle/10986/2278

Vick. Steven G. Morphology and the Role of Landsliding in the Formation of Some Rock Glaciers in the Mosquito Range, Colorado. Geological Society of America Bulletin, Part 1 92. Pp.75–84. 1981.

Watanbe, Teiji, and Daniel Rothacher. Mountain Chronicles: The 1994 Lugge Tsho Glacial Lake Outburst Flood, Bhutan Himalaya. Mountain Research and Development 16, 1. Pp. 77–81. 1996.

Whalley, W. Brian, and Fethi Azizi. Rock Glaciers and Protalus Land Forms: Analogous Forms and Ice Sources on Earth and Mars. Journal of Geophysical Research 108, E4. P. 8032. 2003. See: https://agupubs.onlinelibrary.wiley.com/doi/full/10.1029/2002JE001864

White, Christopher. The Melting World: A Journey across America's Vanishing Glaciers. St. Martin's Press. New York. 2013.

Wilkinson, Jerry. Keys Geology. Online information about the Florida Keys by General Keys History. See: http://www.keyshistory.org/keysgeology.html.

Wu X., R. C. Nethery, M. B. Sabath, D. Braun, and F. Dominici. Air Pollution and COVID-19 Mortality in the United States: Strengths and Limitations of an Ecological Regression Analysis. Science Advances Vol. 6. No. 45. p.eabd4049. 2020. See: https://advances.sciencemag.org/content/6/45/eabd4049

Xstrata Copper. Proyecto El Pachón: Reporte de Sustentabilidad 2010. 2010.

Xu, Baiqing, Junji Cao, James Hansen, Tandong Yao, Daniel R. Joswia, Ninglian Wang, Guangjing Wu, Mo Wang, Huabiao Zhao, Wei Yang, Xianqin Liu, and Jianqiao He. Black Soot and the Survival of Tibetan Glaciers. Proceedings of the National Academy of Sciences of the United States of America 106, 52. 22114:22118. December 29, 2009. See: https://www.pnas.org/content/106/52/22114

Yafeng, Shi, et al. Glaciers of Asia: Glaciers of China. U.S. Geological Survey Professional Paper. Eds. Richard Williams Jr. and Jane Ferrigno. 1386-F-2. 2005. See: https://pubs.usgs.gov/pp/p1386f/pdf/Asia_front_pgs.pdf

Yeung, Jessie. CNN. Melting Glaciers in the Russian Arctic Reveal Five New Islands. October 23, 2019. See: https://apple.news/Ag3qL6eVdRGy2lgbJ4GuSxA.

Young, James, and Stefan Hastenrath. Glaciers of Africa: Glaciers of the Middle East and Africa. US Geological Survey Professional Paper 1386-G-3. 1987. https://pubs.usgs.gov/pp/p1386g/africa.pdf

Zhou, Chunxia, et al. The Characteristics of Surface Albedo Change Trends over the Antarctic Sea Ice Region during Recent Decades. Remote Sensing 11, 821. DOI: 10.3390/rs11070821. April 2019. See: https://www.mdpi.com/2072-4292/11/7/821

Zumbühl, H. J., and S. U. Nussbaumer. Little Ice Age Glacier History of the Central and Western Alps from Pictorial Documents. Geographical Research Letters 44, 1. Pp. 115–136. 2018.

About the Author

Jorge Daniel Taillant, or Daniel as most of his friends and colleagues call him, was born in Buenos Aires, Argentina, in 1968, but learned to walk in San Francisco following a family migration to California. After a frustrated but very rewarding attempt to become a professional soccer player (for which he moved briefly back to Buenos Aires in 1986) he visited his first glacier, the Perito Moreno in Glacier National Park in Patagonia. While he was awed by its grandeur and beauty he didn't realize at the time that glaciers would become an integral part of his identity. Out of money and out of soccer, he hopped on a plane back to California, where he dove into a career in political science, at the University of California, at Berkeley. A portion of that degree was completed in Europe in Lyon, France, where he studied at the Institute d'Etudes Politiques. Still not sure where to settle, he moved east to Washington, DC, to Georgetown, where he obtained a masters in Latin American Studies with a focus on political economy and governance (with another interlude at the Universidad Católica of Chile). Glaciers would come later.

After graduating, Daniel worked for numerous government and international agencies, including the United States National Oceanic Atmospheric Administration (NOAA), the World Bank, the United Nations, and the Organization of American States (OAS) as well as a policy advisor to governments. He would move three more times, to Peru, Haiti, and Cambodia, before he and his wife co-founded the globally recognized Center for Human Rights and Environment (CHRE or CEDHA in Spanish), in Cordoba, Argentina, with a satellite office in Patagonia. CHRE won the Sierra Club's most prestigious international advocacy prize, the Earth Care Award, in 2007, sharing it that year with Al Gore and Thomas Friedman of the New York Times, a recognition for their unique contributions to sustainable development and environmental protection.

It was during the period 2006–2008 that Daniel discovered his passion for protecting glaciers and educating society about their spectacular beauty and the critical role they play in the Earth's planetary ecosystem; his focus was on policies and actions to address their extreme vulnerability to climate change and to anthropogenic industrial activity. He fell into all things glaciers when he learned that a mining company was dynamiting glaciers in the Central Andes to get at gold.

Daniel teamed up once more with his wife, Romina Picolotti, by then secretary of the environment in Argentina, to devise a way to curtail mining operations in sensitive glacier ecosystems (or as Daniel refers to them with a term he has coined, *glaciosystems*). This effort evolved into an international movement to protect glacier resources. Along with dozens of environmental advocacy organizations, congressional representatives, and a broad social movement supporting the cause, he helped achieve the passage of the world's first glacier protection law, after an earlier version of the law had been vetoed by the president in order to defend an international mining company—*the one dynamiting glaciers*. The law has been firmly in place since 2010. It was around then that, together with his good friend and glaciologist Bernard Francou, he coined the term *cryoactivism* (environmental activism to protect the world's cryosphere—its frozen environment). Daniel has been working ever since to advance cryoactivism in Argentina and around the world.

After Argentina's glacier law was enacted and firmly in place, concerned that the government was not doing enough to comply with the law, including to get Argentina's tens of thousands of glaciers inventoried and to identify and stop industrial impacts to them, Daniel single-handedly inventoried *thousands* of glaciers and rock glaciers in mining impact areas, and produced numerous reports on the impacts of mining on glaciers, permafrost, and glaciosystems throughout Argentina and Chile. He has developed tech-friendly ways to see, understand, identify, inventory, and appreciate glaciers through social media and modern technology. He has also developed education materials for children and regularly visits communities (including indigenous communities) to teach about glacier relevance and vulnerability. His latest glacier excursion was to visit California's Sierra Nevada Mountains where he is trying to get Californians to recognize and learn about the extensive but invisible subsurface rock glaciers that provide critical water supplies to Californians and to agriculture.

In 2015 he published *Glaciers: The Politics of Ice*, which recounts the rollercoaster trials and tribulations he experienced in Argentina when working to get the world's first glacier law enacted. In that book, he describes in easy to read prose the nature and characteristics of glaciers and the periglacial environment (permafrost). That same year, after threats on his family's life and well-being, and facing political persecution in retaliation to his and his wife's environmental activism, Daniel and his family moved to Florida and relocated CHRE to the United States. He now works in collaboration with the Institute for Governance and Sustainable Development to promote climate policy at the global and local levels, developing global and local strategies to phase out the most polluting climate contaminants while at the same time addressing the historical social, economic, and racial inequities of systemic environmental injustice.

Meltdown is the latest educational endeavor of his very personal cryoactivism, aimed at raising society's awareness of climate change, the tragedy of rapidly melting glaciers, and the subsequent impacts to society and to ecosystems because of their melt.

Contact: jdtaillant@gmail.com

Index

For the benefit of digital users, indexed terms that span two pages (e.g., 52–53) may, on occasion, appear on only one of those pages.

Figures are indicated by *f* following the page number